Optimization and Energy Maximizing Control Systems for Wave Energy Converters II

Optimization and Energy Maximizing Control Systems for Wave Energy Converters II

Editors

Giuseppe Giorgi
Mauro Bonfanti

MDPI

Basel • Beijing • Wuhan • Barcelona • Belgrade • Novi Sad • Cluj • Manchester

Editors
Giuseppe Giorgi
Politecnico di Torino
Turin
Italy

Mauro Bonfanti
Politecnico di Torino
Turin
Italy

Editorial Office
MDPI AG
Grosspeteranlage 5
4052 Basel, Switzerland

This is a reprint of articles from the Special Issue published online in the open access journal *Journal of Marine Science and Engineering* (ISSN 2077-1312) (available at: https://www.mdpi.com/journal/jmse/special_issues/cl_wave_energy_converters_II).

For citation purposes, cite each article independently as indicated on the article page online and as indicated below:

Lastname, A.A.; Lastname, B.B. Article Title. *Journal Name* **Year**, *Volume Number*, Page Range.

ISBN 978-3-7258-1899-0 (Hbk)
ISBN 978-3-7258-1900-3 (PDF)
doi.org/10.3390/books978-3-7258-1900-3

Cover image courtesy of Giuseppe Giorgi

Contents

About the Editors

Giuseppe Giorgi

Giuseppe Giorgi, Ph.D., assistant professor, is currently working at the Marine Offshore Renewable Energy Lab (MOREnergy Lab) of Politecnico di Torino, Italy. He holds a Ph.D. in Engineering, obtained from the Centre for Ocean Energy Research (COER) of Maynooth University, Ireland, with major contributions in the nonlinear hydrodynamic modelling of wave energy conversion under controlled conditions. He is a Marie Curie individual fellow alumnus, working on computationally efficient nonlinear Froude–Krylov force calculation methods, applied to control and parametric resonance detection. His current research interests involve floating offshore wind and wave energy, earth observation measurements and metocean in-situ surveying, holistic techno-economic modeling and optimization, and energy harvesting leveraging nonlinear phenomena of bistability and parametric resonance.

Mauro Bonfanti

Mauro Bonfanti was born in Crema (Italy) in 1992. He received a B.Sc. degree in Mechanical Engineering from the Politecnico di Milano (Italy), an M.Sc. degree in Mechanical Engineering from the Politecnico di Torino (Italy), and a Ph.D. degree in Mechanical Engineering from the Politecnico di Torino, in 2014, 2016, and 2021, respectively. He is currently an academic researcher (RTDA) at the Politecnico di Torino, with a focus on wave energy conversion systems, hydrofoil-equipped boats, autonomous navigation applied to the waterborne mobility, and zero-emission hybrid propulsion systems for marine vessels.

Preface

The reprint "Optimization and Energy Maximizing Control Systems for Wave Energy Converters II" brings together cutting-edge research in the field of wave energy conversion, focusing on optimizing control systems to maximize energy harvesting and minimize costs. This compilation spans a broad spectrum of topics, from hydraulic power take-off (PTO) systems and spectral-domain modeling to advanced control strategies such as model predictive control and reinforcement learning. The goal is to provide a comprehensive overview of the latest advancements and methodologies, highlighting the crucial role of control systems in enhancing the efficiency and commercial viability of wave energy converters (WECs).

The motivation behind this reprint is to address the pressing need for sustainable and renewable energy sources, emphasizing the significant potential of wave energy. By collating these pioneering studies, we aim to foster a deeper understanding and inspire further innovation in the field. This reprint is intended for researchers, technology developers, and industry stakeholders who are engaged in the development and optimization of marine energy technologies.

As Guest Editors, we acknowledge the invaluable support of the Marine Offshore Renewable Energy Lab (MOREnergy Lab) of Politecnico di Torino. The MOREnergy Lab is at the forefront of marine energy research, with extensive expertise in design, numerical modeling, control systems development, and testing in both tanks and open sea environments. The MOREnergy Lab's collaborative efforts with industrial stakeholders and national and international partners have significantly advanced the field of marine energy, and their work is prominently featured in this reprint.

We extend our gratitude to all the authors involved in this Special Issue for their insightful research and dedication. Their collective efforts have culminated in a comprehensive resource that we believe will drive the future of wave energy technology.

Giuseppe Giorgi and Mauro Bonfanti
Editors

Journal of
Marine Science and Engineering

MDPI

Editorial

Optimization and Energy Maximizing Control Systems for Wave Energy Converters II

Giuseppe Giorgi * and Mauro Bonfanti *

Marine Offshore Renewable Energy Lab (MOREnergy Lab), Department of Mechanical and Aerospace Engineering, Politecnico di Torino, 10129 Turin, Italy
* Correspondence: giuseppe.giorgi@polito.it (G.G.); mauro.bonfanti@polito.it (M.B.)

1. Introduction

Over the past decade, the renewable energy sector has witnessed remarkable growth, which has been attributed to government support [1], financial incentives [2], and technological advancements [3], leading to significant cost reduction opportunities [4] and experience in quantifying cost drivers [5]. In the context of the design of a wave energy converter (WEC), the process involves intricate optimization stages covering each component of the conversion chain, such as the floater, internal mechanics, and energy conversion [6].

The availability of effective control systems for maximal energy harvesting is a critical milestone in the pathway towards the commercialization of WECs, as they have significant potential to reduce the associated levelized cost of energy. Various control methods have been proposed in the literature, encompassing both optimization-based approaches [7] and non-optimization-based strategies [8]. Focusing on the latter, simple controllers based on the impedance-matching (IM) principle uses different architectures [9], encompassing both feedforward structures (which require knowledge of the so-called wave excitation force [10]) and feedback structures, some of which incorporate constraint-handling mechanisms. However, research on energy-maximizing controllers has been mostly restricted to linear time-invariant (LTI) models, lacking systematic and efficient control synthesis for nonlinear systems, which is possible via spectral modelling [11] or appropriate system identification [12]. This is particularly important because nonlinearities are inherently common in most wave energy converters [13], but also because nonlinear mechanisms have the potential to enhance the overall performance of WECs [14].

Moreover, concerning the design and optimization of a WEC, it can be asserted that there exists a lack of comprehensive and systematic research addressing the idea of robustness [15] and stochastic modelling [16]. Precisely, the neighbouring reliability-based design optimization (RBDO) field of research directed its attention to WECs, concerning only structural and maintenance cost uncertainties [17] and the relationship between reliability and hull geometry [18]. Research efforts were directed towards the pervasive shape optimization of single WECs [19] or array layouts [20]. Metaheuristic algorithms became popular options in the field, thanks to their ability to handle the typically high dimensionality of the optimization space [21]. Finally, at the cross-roads between optimization and control, much attention has recently been paid to control co-design [22], based on the awareness that these two aspects (control and design optimization) are intimately intertwined [23].

The book *Optimization and Energy Maximizing Control Systems for Wave Energy Converters II* aims to tackle a wide range of challenges in wave energy control and optimization, while introducing cutting-edge methods and techniques. It serves as a valuable resource for researchers and technology developers alike, offering a blend of theoretical insights and practical solutions aimed at overcoming current obstacles in the field of wave energy technology development.

Citation: Giorgi, G.; Bonfanti, M. Optimization and Energy Maximizing Control Systems for Wave Energy Converters II. *J. Mar. Sci. Eng.* **2024**, *12*, 1297. https://doi.org/10.3390/jmse12081297

Received: 26 June 2024
Accepted: 29 July 2024
Published: 1 August 2024

2. An Overview of Published Articles

In "Maximum Power Control Algorithm for Power Take-Off System Based on Hydraulic System for Floating Wave Energy Converters" (contribution 1), a hydraulic-based power take-off (PTO) system for floating wave energy converters (FWECs) was investigated, with a focus on achieving maximum power control. The research involved modelling a hydraulic system generator power converter to analyze and optimize the performance of the PTO system under various load control algorithms. Unlike previous studies that primarily focused on input power performance, this study emphasized electrical load control strategies for maximizing power output. The study applied two main algorithms for load control: the perturb and observe (P&O) algorithm for speed control and an optimal torque control algorithm. These algorithms were evaluated to determine their effectiveness in maximizing the power generation performance of the hydraulic-based PTO system. By analyzing the system's characteristics and power generation efficiency under different control variables, the study proposed an optimal torque control algorithm specifically suited for maximizing power output. The key findings included the following: the hydraulic system type is advantageous for FWECs due to its capability to operate efficiently at low speeds and generate high torque; the proposed optimal torque control algorithm demonstrated effectiveness in achieving maximum power control for the hydraulic-based PTO system; and the simulation results were validated against actual sea test data, confirming the feasibility and effectiveness of the proposed load control algorithms. In conclusion, the study contributed to advancing the understanding and application of hydraulic-based PTO systems in wave energy conversion. The findings serve as a basis for optimizing PTO system performance in real-world applications, offering insights for future developments in wave energy technology.

Using MATLAB/Simulink with the Simscape fluids toolbox, a comprehensive model of the wave absorber device with both conventional and improved hydraulic PTO (HPTO) units was developed in "An Improved Hydraulic Power Take-Off Unit Based on Dual Fluid Energy Storage for Reducing the Power Fluctuation Problem in the Wave Energy Conversion System" (contribution 2). The study focused on evaluating the performance of the improved HPTO unit under various irregular wave conditions and investigating the impact of high-pressure accumulator (HPA) pressure constraints on its performance. The key findings from the simulation analysis were as follows: the optimization of control strategy parameters using a genetic algorithm significantly enhanced the performance of the improved HPTO unit, reducing unnecessary fluctuations in generated electrical power; the integration of dual HPA and fluid energy control (FEC) modules improved stability, allowing the HPTO unit to achieve a stable electrical power output up to 87.3% of the time in irregular sea states; compared to conventional HPTO units, the improved model demonstrated superior performance across different sea states, maintaining rated capacity operation during significant portions of wave conditions; and variations in HPA pressure constraints had notable effects on power generation and fluctuation occurrences, with adjustments influencing power output and stability. This study contributed valuable insights for WEC design and optimization, offering a promising solution to mitigate power fluctuation issues through advanced PTO technology.

In "Downsizing the Linear PM Generator in Wave Energy Conversion for Improved Economic Feasibility" (contribution 3), the authors integrated hydrodynamic and generator models into the WEC optimization process to comprehensively assess WEC power production. Three sizing methods for the linear generator were compared, emphasizing their effects on techno-economic metrics. The study incorporated wave resource data from three European sea sites to highlight the influence of PTO sizing on WEC performance under varying environmental conditions. The key findings indicated that PTO sizing significantly affected the power conversion efficiency of linear generators in point-absorber WECs. The study underscored that accurately sizing the generator is critical for optimizing the levelized cost of energy (LCOE). Neglecting to adjust for varying generator efficiencies can lead to substantial errors in estimating annual energy production (AEP) and optimal PTO capacity, with errors reaching up to 10% and 29%, respectively. A comparative analysis revealed that optimizing main machine parameters alongside scaling laws provided a more realistic reflection of WEC techno-economic potential during PTO sizing. Moreover, the study highlighted the pronounced impact of wave resource variability on generator efficiency and LCOE, stressing the need for tailored PTO designs across different sea sites.

The work "Improving Computational Efficiency in WEC Design: Spectral-Domain Modelling in Techno-Economic Optimization" (contribution 4) focused on improving efficiency through spectral-domain modelling (SDM) for the ISWEC (inertial sea wave energy converter), a gyroscopic-type WEC with a hydraulic power take-off (HPTO). Unlike traditional time-domain modelling (TDM), SDM offers both speed and precision, which is crucial for early-stage design in high-energy sea sites. The research evaluated ISWEC performance with HPTO in locations like the Balder site in the North Sea, highlighting the shift to hydraulic transmissions for efficiency. SDM optimizes multi-objective functions: maximizing annual energy production (AEP) and minimizing the cost of energy (CoE). The results revealed trade-offs, where maximizing AEP favours larger, costlier setups, while CoE minimization prefers smaller, economically viable configurations. This study emphasized integrated optimization tools that consider technical feasibility and economic viability, which are essential for accelerating WEC development. Future work includes refining SDM for HPTO dynamics, comparing PTO configurations, and incorporating uncertainty analysis to enhance the design robustness and economic feasibility of ISWEC systems.

The slope-pendulum wave energy conversion (S-PWEC) device was designed and studied in "Design and Research of Slope-Pendulum Wave Energy Conversion Device" (contribution 5). This study focused on the Zhejiang sea area, employing numerical simulations and experimental array testing under 66 different sea conditions to evaluate the S-PWEC's performance. The results demonstrated that incorporating a slope structure at the device's bottom significantly enhances its motion response capability and resilience against extreme sea conditions. In regular and irregular wave tests, the electrical power output efficiency improved by 13.24% and 10.06%, respectively. Array experiments revealed that optimal spacing between floating plates is critical, influencing both diffraction and radiation effects to maximize power output. The key findings included the following: the slope structure effectively enhanced wave energy capture and overall efficiency at a reduced cost; the device demonstrated robustness in handling large wave heights and maintained operational stability with a "self-locking" mechanism; optimal spacing between arrayed floating plates was crucial, with 25 metres identified as maximizing electrical output. Future research will delve into optimizing the bottom oblique plate angle under varying sea conditions and further refining array configurations to enhance overall wave energy absorption capacity through improved design and active control strategies.

The study "Comparison of Offline, Real-Time Models and Hardware-in-the-Loop Test Results of a Power Take-Off for Wave Energy Applications" (contribution 6) focused on comparing different modelling approaches for a power take-off (PTO) system in wave energy converters (WECs), crucial for optimizing energy extraction and reducing costs. Two modelling methods were discussed: an offline detailed model and a simplified real-time (RT) model, both of which were applied to a 250 kW modular electromechanical PTO connected to an oscillating wave surge converter (OWSC). The offline model, implemented in Matlab Simulink (v2018a) with WEC-Sim, accurately captured system dynamics, including wave–flap interaction and PTO mechanics. It served as a benchmark against the RT model, which ran up to 10 times faster and included simplified control algorithms like maximum torque per ampere (MTPA) and field weakening (FW). The study also incorporated hardware-in-the-loop (HIL) tests to validate these models against real-world conditions. The results showed that, while the RT model effectively replicated the dynamics of the offline model, discrepancies existed in force and torque outputs due to control loop discretization and FW algorithm application. The HIL tests demonstrated closer alignment with real-world performance, highlighting the importance of considering real-time dynamics and hardware integration in validating PTO concepts. Understanding these modelling approaches is crucial for WEC design and validation, ensuring accurate performance assessment and technology readiness assessment with appropriate simplifications.

The study "Model Predictive Energy-Maximising Tracking Control for a Wavestar-Prototype Wave Energy Converter" (contribution 7) addressed the challenge of achieving energy-maximizing control to reduce the levelized cost of energy (LCOE) for wave energy converters (WECs). It presented a model predictive velocity tracking control method for a Wavestar-like device in the WEC-SIM benchmark. The control system has a hierarchical structure: the first part estimates the wave excitation moment (WEM) using a Kalman filter (KF) and an extended Kalman filter (EKF) to obtain the amplitude and angular frequency of the WEM. This information is used to compute the reference velocity. The second part was a model predictive control (MPC) method that ensured the WEC tracked the optimal reference velocity for maximum energy extraction from irregular waves. Two Gaussian Process models predicted future WEM and reference velocity, which are crucial for MPC design. The proposed strategy demonstrated excellent tracking performance with minimal errors, pushing the WEC into near-resonance conditions for optimal power extraction. The MPC system also effectively handled input constraints, reducing negative power and protecting the power take-off (PTO) system from large bidirectional energy flows. Future work will focus on enhancing the robustness of the KF and MPC to improve performance under various conditions.

The work entitled "On the Effect of Wave Direction on Control and Performance of a Moored Pitching Wave Energy Conversion System" (contribution 8) examined how wave directionality and mooring dynamics affected the control and performance of a moored pitching wave energy converter (WEC), specifically using the pendulum wave energy converter (PeWEC) as a case study. The research highlighted the importance of incorporating mooring dynamics and wave directionality into the control synthesis and performance evaluation of WECs. It was found that neglecting wave directionality could lead to an overestimation of device performance by up to 50%, even if a predominant wave direction existed at the site. The study used a tailored data-based model to evaluate the PeWEC's performance under different wave directions and site conditions. The results emphasized that, for accurate performance assessment and control synthesis, it is crucial to consider wave directionality, particularly during the initial design stages of wave energy systems. This approach is necessary to advance the commercial viability of wave energy converters.

In "Comparison of Advanced Control Strategies Applied to a Multiple-Degrees-of-Freedom Wave Energy Converter: Nonlinear Model Predictive Controller versus Reinforcement Learning" (contribution 9), the authors compared nonlinear model predictive controller (NMPC) and reinforcement learning (RL) techniques for controlling a multiple-degrees-of-freedom wave energy converter (WEC). Both approaches were tested on a nonlinear WEC–power take-off (PTO) system using the WEC-Sim™ toolbox under various conditions. The results indicated that RL, with optimal agent selection and training, outperformed NMPC, especially under stringent conditions where NMPC struggled. RL also showed better computational efficiency, reducing the task execution time (TET). The case study on Dehlsen Associates' CENTIPOD WEC highlights that RL-DDPG achieves superior power extraction, robust performance, and design flexibility compared to NMPC. This demonstrates RL's potential in enhancing WEC efficiency and operational robustness.

In the work "Measuring the Robustness of Optimal Design Solutions for Wave Energy Converters via a Stochastic Approach" (contribution 10), the author addressed the challenge of ensuring that wave energy converters (WECs) are not only efficient in energy extraction but also robust in the face of uncertain and harsh marine conditions. Unlike traditional approaches, this research employed a stochastic method using Monte Carlo simulations and Gaussian process regression (GPR) to account for uncertainties. The results showed significant deviations in robustness between the nominally optimal designs and those optimized for robustness. For example, applying robustness metrics led to a 20% reduction in performance but improved reliability. A case study comparing two WEC designs (D1 and D2) revealed that D1, although more costly, had better performance and reduced sensitivity to uncertainties. This emphasizes the importance of incorporating robust optimization in WEC design, balancing performance and robustness, and suggests future research should focus on the impact of control logic and environmental uncertainties to optimize techno-economic indices like the levelized cost of energy (LCoE).

Conflicts of Interest: The authors declare no conflicts of interest.

List of Contributions:

1. Roh, C. Maximum Power Control Algorithm for Power Take-Off System Based on Hydraulic System for Floating Wave Energy Converters. *J. Mar. Sci. Eng.* **2022**, 10, 603. https://doi.org/10.3390/jmse10050603.
2. Jusoh, M.A.; Yusop, Z.M.; Albani, A.; Daud, M.Z.; Ibrahim, M.Z. An Improved Hydraulic Power Take-Off Unit Based on Dual Fluid Energy Storage for Reducing the Power Fluctuation Problem in the Wave Energy Conversion System. *J. Mar. Sci. Eng.* **2022**, 10, 1160. https://doi.org/10.3390/jmse10081160.
3. Tan, J.; Wang, X.; Polinder, H.; Laguna, A.J.; Miedema, S.A. Downsizing the Linear PM Generator in Wave Energy Conversion for Improved Economic Feasibility. *J. Mar. Sci. Eng.* **2022**, 10, 1316. https://doi.org/10.3390/jmse10091316.
4. Bonfanti, M.; Giorgi, G. Improving Computational Efficiency in WEC Design: Spectral-Domain Modelling in Techno-Economic Optimization. *J. Mar. Sci. Eng.* **2022**, 10, 1468. https://doi.org/10.3390/jmse10101468.
5. Wan, Z.; Li, Z.; Zhang, D.; Zheng, H. Design and Research of Slope-Pendulum Wave Energy Conversion Device. *J. Mar. Sci. Eng.* **2022**, 10, 1572. https://doi.org/10.3390/jmse10111572.
6. Castellini, L.; Gallorini, F.; Alessandri, G.; Alves, E.F.; Montoya, D.; Mudigonda, B.; Tedeschi, E. Comparison of Offline, Real-Time Models and Hardware-in-the-Loop Test Results of a Power Take-Off for Wave Energy Applications. *J. Mar. Sci. Eng.* **2022**, 10, 1744. https://doi.org/10.3390/jmse10111744.
7. Li, D.; Patton, R. Model Predictive Energy-Maximising Tracking Control for a Wavestar Prototype Wave Energy Converter. *J. Mar. Sci. Eng.* **2023**, 11, 1289. https://doi.org/10.3390/jmse11071289.
8. Paduano, B.; Faedo, N.; Mattiazzo, G. On the Effect of Wave Direction on Control and Performance of a Moored Pitching Wave Energy Conversion System. *J. Mar. Sci. Eng.* **2023**, 11, 2001. https://doi.org/10.3390/jmse11102001.
9. Haider, A.S.; Bubbar, K.; McCall, A. Comparison of Advanced Control Strategies Applied to a Multiple-Degrees-of-Freedom Wave Energy Converter: Nonlinear Model Predictive Controller versus Reinforcement Learning. *J. Mar. Sci. Eng.* **2023**, 11, 2120. https://doi.org/10.3390/jmse11112120.
10. Giorcelli, F.; Sirigu, S.A.; Giorgi, G.; Faedo, N.; Bonfanti, M.; Ramello, J.; Giorcelli, E.; Mattiazzo, G. Measuring the Robustness of Optimal Design Solutions for Wave Energy Converters via a Stochastic Approach. *J. Mar. Sci. Eng.* **2024**, 12, 482. https://doi.org/10.3390/jmse12030482.

References

1. O'hagan, A.; Huertas, C.; O'callaghan, J.; Greaves, D. Wave energy in Europe: Views on experiences and progress to date. *Int. J. Mar. Energy.* **2016**, 14, 180–197. [CrossRef]
2. Gonzalez, A.T.; Dunning, P.; Howard, I.; McKee, K.; Wiercigroch, M. Is wave energy untapped potential? *Int. J. Mech. Sci.* **2021**, 205, 106544. [CrossRef]
3. Paduano, B.; Carapellese, F.; Pasta, E.; Bonfanti, M.; Sirigu, S.A.; Basile, D.; Pizzirusso, D.; Faedo, N.; Mattiazzo, G. Experimental and Numerical Investigation on the Performance of a Moored Pitching Wave Energy Conversion System. *IEEE J. Ocean. Eng.* **2024**, 49, 802–820. [CrossRef]
4. Ruiz-Minguela, P.; Noble, D.R.; Nava, V.; Pennock, S.; Blanco, J.M.; Jeffrey, H. Estimating future costs of emerging wave energy technologies. *Sustainability* **2022**, 15, 215. [CrossRef]
5. Giglio, E.; Petracca, E.; Paduano, B.; Moscoloni, C.; Giorgi, G.; Sirigu, S.A. Estimating the cost of wave energy converters at an early design stage: A bottom-up approach. *Sustainability* **2023**, 15, 6756. [CrossRef]
6. Bonfanti, M.; Faedo, N.; Mattiazzo, G. Towards efficient control synthesis for nonlinear wave energy conversion systems: Impedance-matching meets the spectral-domain. *Nonlinear Dyn.* **2024**, 112, 11085–11109. [CrossRef]

7. Faedo, N.; Olaya, S.; Ringwood, J.V. Optimal control, MPC and MPC-like algorithms for wave energy systems: An overview. *IFAC J. Syst. Control* **2017**, *1*, 37–56. [CrossRef]
8. Coe, R.G.; Bacelli, G.; Forbush, D. A practical approach to wave energy modelling and control. *Renew. Sustain. Energy Rev.* **2021**, *142*, 110791. [CrossRef]
9. Carapellese, F.; Pasta, E.; Paduano, B.; Faedo, N.; Mattiazzo, G. Intuitive LTI energy-maximising control for multi-degree of freedom wave energy converters: The PeWEC case. *Ocean Eng.* **2022**, *256*, 111444. [CrossRef]
10. Bonfanti, M.; Carapellese, F.; Sirigu, S.A.; Bracco, G.; Mattiazzo, G. Excitation forces estimation for non-linear wave energy converters: A neural network approach. *IFAC-Pap.* **2020**, *53*, 12334–12339. [CrossRef]
11. Bonfanti, M.; Sirigu, S.A. Spectral-domain modelling of a non-linear wave energy converter: Analytical derivation and computational experiments. *Mech. Syst. Signal Process.* **2023**, *198*, 110398. [CrossRef]
12. Faedo, N.; Giorgi, G.; Ringwood, J.V.; Mattiazzo, G. Optimal control of wave energy systems considering nonlinear Froude–Krylov effects: Control-oriented modelling and moment-based control. *Nonlinear Dyn.* **2022**, *109*, 1777–1804. [CrossRef]
13. Davidson, J.; Costello, R. Efficient nonlinear hydrodynamic models for wave energy con-verter design—A scoping study. *J. Mar. Sci. Eng.* **2020**, *8*, 35. [CrossRef]
14. Daqaq, M.F.; Giorgi, G.; Khasawneh, M.A. Deliberate introduction of a nonlinear restoring element to point wave energy absorbers: A review and assessment. *Nonlinear Dyn.* **2024**, *112*, 13695–13731. [CrossRef]
15. Farajvand, M.; Grazioso, V.; García-Violini, D.; Ringwood, J.V. Uncertainty estimation in wave energy systems with applications in robust energy maximising control. *Renew. Energy* **2023**, *203*, 194–204. [CrossRef]
16. Ambühl, S.; Kofoed, J.P.; Sørensen, J.D. Stochastic modeling of long-term and extreme value estimation of wind and sea conditions for probabilistic reliability assessments of wave energy devices. *Ocean Eng.* **2014**, *89*, 243–255. [CrossRef]
17. Ambühl, S.; Kramer, M.; Sørensen, J.D. Reliability-based structural optimization of wave energy converters. *Energies* **2014**, *7*, 8178–8200. [CrossRef]
18. Clark, C.E.; DuPont, B. Reliability-based design optimization in offshore renewable energy systems. *Renew. Sustain. Energy Rev.* **2018**, *97*, 390–400. [CrossRef]
19. Garcia-Teruel, A.; Forehand, D.I. A review of geometry optimisation of wave energy converters. *Renew. Sustain. Energy Rev.* **2021**, *139*, 110593. [CrossRef]
20. Teixeira-Duarte, F.; Clemente, D.; Giannini, G.; Rosa-Santos, P.; Taveira-Pinto, F. Review on layout optimization strategies of offshore parks for wave energy converters. *Renew. Sustain. Energy Rev.* **2022**, *163*, 112513. [CrossRef]
21. Shadmani, A.; Nikoo, M.R.; Gandomi, A.H.; Chen, M.; Nazari, R. Advancements in optimizing wave energy converter geometry utilizing metaheuristic algorithms. *Renew. Sustain. Energy Rev.* **2024**, *197*, 114398. [CrossRef]
22. Grasberger, J.; Yang, L.; Bacelli, G.; Zuo, L. Control co-design and optimization of oscillating-surge wave energy converter. *Renew. Energy* **2024**, *225*, 120234. [CrossRef]
23. Abdulkadir, H.; Abdelkhalik, O. Control co-design optimization of nonlinear wave energy converters. *Ocean Eng.* **2024**, *304*, 117827. [CrossRef]

Journal of
Marine Science and Engineering

MDPI

Article

Maximum Power Control Algorithm for Power Take-Off System Based on Hydraulic System for Floating Wave Energy Converters

Chan Roh

Department of Energy Engineering, Inje University, 197 Inje-ro, Gimhae-si 50834, Korea; rohchan@inje.ac.kr

Abstract: In this study, a hydraulic system generator power converter was modeled to verify the performance of a hydraulic-based power take-off (PTO) system. Moreover, the characteristics and output performance of the PTO system were analyzed with various load control algorithms applied for maximum power control. The simulation performance was verified through a comparison with actual sea test results. Unlike previous studies on hydraulic-based PTO system control for input power performance, the performance of a hydraulic-based PTO system was analyzed through electrical load control in this study. The electrical load control was analyzed by applying a speed control algorithm based on the perturb and observe algorithm and an optimal torque control algorithm. A load control algorithm suitable for maximum power control of the PTO system was proposed by analyzing the characteristics and power generation performance of the system according to the control variables of each algorithm. The proposed optimal torque control algorithm proved to be suitable for maximum power control of the considered PTO system.

Keywords: floating wave energy converter; power take-off system; hydraulic system; PTO force; maximum power control; electrical load control; P&O algorithm; speed control; optimal torque control; efficiency; power performance

Citation: Roh, C. Maximum Power Control Algorithm for Power Take-Off System Based on Hydraulic System for Floating Wave Energy Converters. *J. Mar. Sci. Eng.* **2022**, *10*, 603. https://doi.org/10.3390/jmse10050603

Academic Editors: Giuseppe Giorgi and Mauro Bonfanti

Received: 22 March 2022
Accepted: 27 April 2022
Published: 29 April 2022

1. Introduction

Wave Energy Converters (WECs) are an emerging technology. Therefore, technological growth for device development and testing is continuously required [1,2]. Several models have already been implemented at sea for the commercialization of wave power plants [3,4]. However, in order to enter the commercialization stage, it is necessary to study a control strategy that can extract maximum power by adjusting the power take-off in response to irregular marine environment conditions. Therefore, PTO studies for various WEC devices have recently been conducted.

The floating wave energy converter (FWEC) concept is based on extracting wave energy using floating bodies in the sea [5,6]. In an FWEC, the floating body moves by wave energy and a power take-off (PTO) system generates power using this movement. Because the wave period in an actual sea is generally long, FWEC motion tends to be slow. For instance, Wavestar, a representative FWEC, requires a bidirectional load torque of 1 MNm to extract an average of 27 kW of power [7,8]. Therefore, the average and peak wave energy converter (WEC) power levels may differ by a factor of 10 or more [9]. However, it cannot be simply discarded because the peak power significantly contributes to the overall production. Consequently, energy storage is required to store the energy peak and maintain the wave energy variability. Regarding PTO system implementation for FWEC, a study was conducted on a linear generator directly driven using the movement of a floating body. However, owing to the slow motion of the floating body, the existing permanent-magnet linear generator was very large [8]. Wavestar requires a load of 400 kN, but the general linear generator void shear stress level is 20–25 kN/m^2 [10]. In other words, even if the necessary support structure is ignored, the weight becomes excessive, making it impossible to implement a linear generator [11]. Direct drive linear generators have been proposed many times because of their advantages due to their ability to directly convert linear input motion into electrical energy. In this way, the energy conversion chain is greatly simplified using a limited number of components [9]. Instead, we used another system that was more prevalent in real-world applications. Other systems first store sea wave energy, pressurize the fluid through a piston pump, and then convert this energy into electricity using an alternator driven by a hydraulic turbine or volumetric expander [12]. In other words, linear generators are currently the subject of ongoing research [13,14], as direct drive solutions have sufficient advantages.

A method of using mechanical transmission (gearbox, belts, etc.) was also studied, but the size was large owing to the high gearing ratio and bidirectional movement characteristics, making it difficult to implement.

Hydraulic systems are suitable for PTO systems for FWECs. A hydraulic system can control high forces at low speeds and can easily control movement in both directions with a small actuator.

Figure 1a shows the initial hydraulic system of a PTO system for an FWEC. The cylinders act as a pump to create bidirectional flow that drives the hydraulic motor. The generator rotates in one direction through the hydraulic motor drive. In [15], a PTO system was optimized from wave energy to grid power using this model in the case of Wavestar. At the optimum point, the overall PTO power conversion efficiency was 65%, but in small waves, it quickly dropped to 45%. However, the system shown in Figure 1a does not allow energy storage or smoothing. Therefore, as shown in Figure 1b, a PTO system with a cylinder operated by a pump at a constant pressure was used in [16,17], where energy smoothing could be performed using an accumulator. This approach enables hydraulic motors and generators to operate at fairly constant loads, resulting in a PTO efficiency of up to 80%. However, the cylinder is limited to providing a Coulomb damping force, which reduces the amount of energy absorbed [18]. Therefore, in [19], a hydraulic transformer capable of controlling the force of the cylinder was proposed to overcome the force load. However, the load efficiency of a hydraulic transformer was poor because there were two variable displacement pumps or motors in series. In [20], discrete control of hydraulic cylinders was applied. A mobile hydraulic system was investigated, showing a promising increase in energy efficiency compared to a conventional load system. Meanwhile, [21] discussed a PTO system using two asymmetric cylinders. The efficiency of cylinder force control by pressure movement was between 88% and 94%, excluding the friction of the cylinder. In addition, [22] described a hydraulic transmission with an accumulator for energy smoothing that operated at 70% efficiency, and [23] proposed a system similar to that in [17]. Furthermore, [24] presented a scaled version capable of applying a 16 kNm torque to the system of [23]. The overall efficiency was estimated to be 69–80%.

In this study, the hydraulic transmission system shown in [16,17] was used. The maximum output control was analyzed through electrical load control of the power converter, rather than mechanical control using the discrete displacement cylinder system shown in [18,25]. The load control method described in [16,17] is difficult to directly apply to an actual system because it provides theoretical load control. Therefore, in this study, the force of the PTO system and amount of power generated were calculated by analyzing the PTO system characteristics based on the hydraulic system according to the load control of the power converter. There is a lack of research on hydraulic-based PTO system analysis with the application of a maximum power control algorithm using the existing power conversion method. Various maximum power control algorithms exist in wind power systems that have already been studied [26–30]. Among them, the most frequently applied algorithms can be classified into two categories. The perturb and observe (P&O) algorithm shown in [29] does not rely on knowledge of the characteristic curves of wind turbines. This method observes turbine speed fluctuations and checks for changes in the output power accordingly. That is, this approach can simply control the maximum output. However, it reduces the convergence rate at the maximum power point (MPP) depending on the turbine inertia and control period, and the MPP may not be reached.

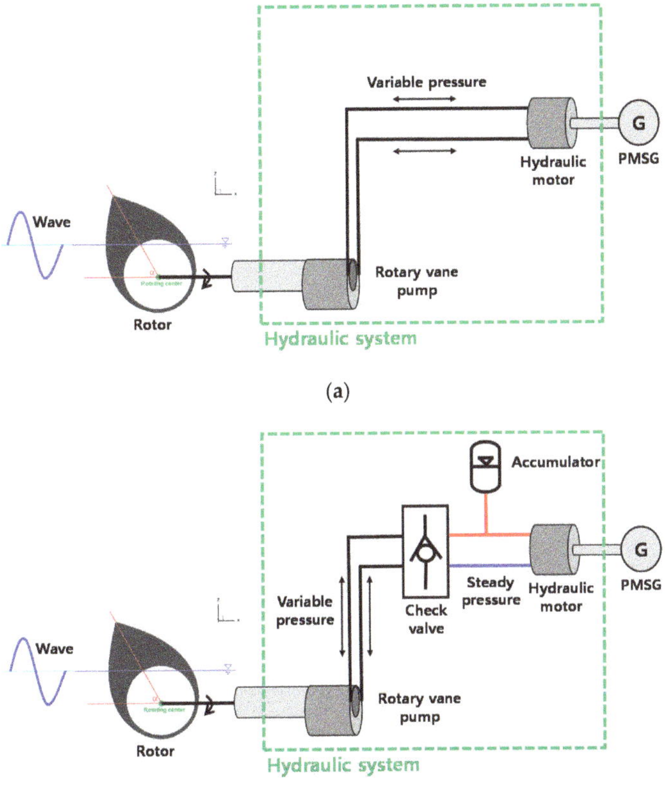

Figure 1. Basic hydraulic transmission system model: (**a**) direct hydraulic motor system and (**b**) rectifying valve accumulator motor.

Recently, a study was conducted to ensure that the generated power reaches the maximum power based on the parameters of the hydraulic PTO because it is difficult to adapt to various sea conditions when the parameters of the hydraulic PTO are fixed [30]. However, it is difficult to follow the electrical maximum power point tracking(MPPT) control response. The MPPT control strategy is a well-known adaptive control strategy with great success in other renewable energy industries (mainly solar and wind). In order to apply this in wave power generation, there have been some studies on linear generators for MPPT. However, there are few studies using MPPT technology for hydraulic PTO so far, and it is mainly used for research on linear generators [31–35].

Therefore, this method is mainly used for solar power generation with little change or small-capacity wind power generation, requiring a simple controller configuration. The power signal feedback algorithm shown in [29] can converge to the MPP and produce a fast convergence speed. This method is controlled based on the turbine curve for maximum power, depending on the turbine speed. However, to converge to the MPP, a characteristic curve for a wind turbine must be obtained, which must be done experimentally. This method was applied to improve the efficiency and convergence of large wind power generation. Maximum FWEC power control can be derived based on the algorithm that is most often applied to wind power generation systems. Therefore, a load control algorithm suitable for a hydraulic-based PTO system was developed in this study by comparing various load algorithms and control variables for maximum power control. The characteristic analysis of the PTO system was conducted using a simulation that included a hydraulic system, generator, and power converter. The developed simulator was verified using the power generation data obtained from an actual sea test, and it was proven that the proposed algorithm can be directly applied to actual sea tests.

2. PTO System Configuration and Modeling

2.1. FWEC Power Conversion Chain

An FWEC converts wave energy into relative linear or rotational motion. The power conversion chain converts mechanical power into electrical power. Figure 2 illustrates the power conversion chain used in this study. According to the Office of Energy Efficiency and Renewable Energy WEC database, there are 34 power conversion chain structures with technology readiness levels (TRL) of 5 or higher. Among them, the most commonly used PTO system is a hydraulic system [26,27].

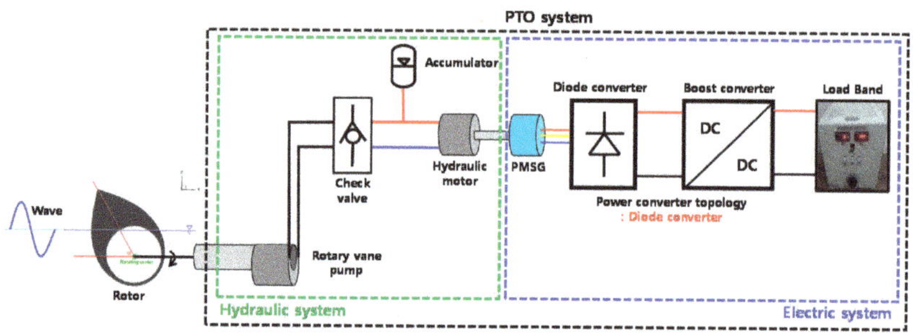

Figure 2. FWEC power conversion chain.

The input energy variability of an FWEC is large, so the development of an efficient and reliable PTO system for power conversion is important. In other words, using a hydraulic system as the PTO system of an FWEC can provide stable power generation. Therefore, a hydraulic system was used as the PTO system in this study. The floating body used Salter's duck [36] to transform blue energy into pitch motion. The pitch motion of the floating body was converted into bidirectional rotational motion using a rotary vane pump. The bidirectional flow generated by the rotary vane pump was converted into unidirectional flow through a check valve. The flow of the check valve was directed towards the accumulator and hydraulic motor, and the flow rate to the hydraulic motor was determined by the electrical torque. The number of revolutions per minute (rpm) of the generator was determined by the flow rate of the hydraulic motor, and the power converter current was controlled to perform appropriate load control according to the rpm. In other words, the amount of power generated by the FWEC could be adjusted through appropriate load current control according to the input energy. The next section describes the load control algorithm in detail.

2.2. Hydraulic System Modeling

A hydraulic system was used as a suitable PTO system for FWEC because it can provide low speed and high torque. In particular, it can generate stable power using an accumulator even with irregular wave energy [4]. The flow rate generated by the rotary vane pump using the pitch angle of the floating body θ_{pitch} can be expressed as follows:

$$\omega_{pitch} = \frac{d}{dt}\theta_{pitch} \tag{1}$$

$$Q_{pump} = D_{pump} * \omega_{pitch} \tag{2}$$

The pitch angle of the floating body moves according to the input energy but is also affected by the PTO force generated by the hydraulic system, which is given by

$$F_{PTO} = D_{pump} * (P_H - P_L) * sign\left(\omega_{pitch}\right) \tag{3}$$

The flow generated by the rotary vane pump is bidirectional and can be converted into unidirectional flow through a check valve, as follows:

$$Q_{check} = \left|Q_{pump}\right| \tag{4}$$

The flow rate of the check valve is the sum of the flow rates of the accumulator and hydraulic motor:

$$Q_{check} = Q_{accu} + Q_{motor} \tag{5}$$

The hydraulic system circuit pressure can be calculated based on the accumulator flow rate. The accumulator volume changes according to the accumulator flow rate, and the pressure increases accordingly. Because the accumulator pressure is the same as the hydraulic motor input, the pressure is the same. The change in pressure according to the accumulator flow rate can be calculated as follows:

$$V_{accu} = \int Q_{accu} dt \tag{6}$$

$$P_H = \frac{P_{H\_pre}}{\left(V_{accu}/V_{accu\_pre}\right)^{\gamma}} \tag{7}$$

In this study, several accumulators were used to reduce the rate of change in pressure. To check the pressure change with multiple accumulators, the pressure change with two accumulators can be expressed as follows:

$$V_{accu1} = V_{accu1\_pre} * \left(P_{H\_pre}/P_H\right)^{\frac{1}{\gamma}} \tag{8}$$

$$Q_{accu1} = \frac{d}{dt} V_{accu1} \tag{9}$$

$$Q_{accu2} = Q_{accu} - Q_{accu1} \tag{10}$$

$$V_{accu2} = \int Q_{accu2} dt \tag{11}$$

$$P_H = \frac{P_{H\_pre}}{\left(V_{accu2}/V_{accu2\_pre}\right)^{\gamma}} \tag{12}$$

Because accumulators 1 and 2 are the same node, the pressure is the same, and the accumulator pressure is the same as the pressure in the circuit. Using this information, the mechanical torque of the hydraulic motor can be calculated as follows:

$$T_m = D_{motor} * (P_H - P_L) \tag{13}$$

The dynamic state of the hydraulic motor assumes that there is no loss due to friction, and modeling is performed as follows, using the difference between the mechanical and electrical torques of the generator:

$$\frac{d}{dt}\omega_{motor} = \frac{1}{J}(T_m - T_e). \tag{14}$$

The rpm of the hydraulic motor is changed by the electrical torque, and the hydraulic motor flow rate can be calculated as follows:

$$Q_{motor} = D_{motor} * \omega_{motor} \tag{15}$$

That is, the rpm and flow rate of the hydraulic motor change according to the electrical load, and the flow rate of the accumulator changes according to Equation (5), so the pressure in the circuit changes. The pressure in the circuit affects the motion of the floating body according to Equation (3), which affects the overall power generation.

2.3. Permanent Magnet Synchronous Generator (PMSG) Modeling

Because the PMSG is directly coupled to the hydraulic motor, the mechanical dynamic state of the PMSG can be expressed using Equation (14). The PMSG-based model can be expressed by applying the Kirchhoff voltage law. The generator model based on PMSG is displayed using the dq-axis coordinate system, which is a rotating coordinate system synchronized with the speed of the generator's rotor:

$$V_{sd} = L_{sq}\omega_e i_{sq} - L_{sd}\frac{di_{sd}}{dt} - R_s i_{sd} \tag{16}$$

$$V_{sq} = \omega_e \Psi_{pm} - L_{sd}\omega_e i_{sd} - L_{sq}\frac{di_q}{dt} - R_s i_{sq} \tag{17}$$

L_{sd} and L_{sq} represent the stator dq-axis inductances, R_s is the stator resistance, ω_e is the electrical angular frequency, Ψ_{pm} is the flux linkage of the permanent magnet, and V_{sd} and V_{sq} are the stator dq-axis terminal voltages.

The instantaneous power generation, P_e of PMSG, can be expressed as follows

$$P_e = \frac{3}{2}\left(V_{sd}i_{sd} + V_{sq}i_{sq}\right) \tag{18}$$

In other words, if the loss due to resistance is very small, it can be expressed in terms of P_e using Equations (11) and (12) as follows:

$$P_e = \frac{3}{2}\left[\omega_e \Psi_{pm} i_{sq} - \omega_e\left(L_{sd} - L_{sq}\right)i_{sd}i_{sq}\right] \tag{19}$$

In the PMSG considered in this study, L_{sd} and L_{sq} are the same because the rotor uses a motorized structure. Thus, Equation (14) can be written as follows:

$$P_e = \frac{3}{2}\omega_e \Psi_{pm} i_{sq} \tag{20}$$

Electrical torque T_e can be calculated as follows, using the relationship between the instantaneous power generation and the angular velocity of the turbine:

$$T_e = \frac{P_e}{\omega_m} = \frac{3}{2}\frac{\omega_e \Psi_{pm} i_{sq}}{\omega_{motor}} = \frac{3}{2}N_p \Psi_{pm} i_{sq} \tag{21}$$

The ratio of the electrical angular frequency ω_e to the angular velocity ω_m of the turbine is equal to the number of dipoles N_p of the rotor. From Equation (16), it can be seen that the electrical torque changes according to the q-axis current of the generator. Figure 3 shows the configuration diagram of the applied hydraulic-based PTO system in this paper.

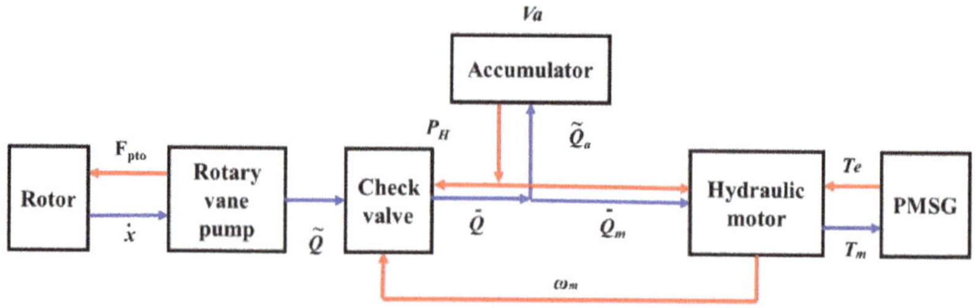

Figure 3. Hydraulic-based PTO system.

2.4. Power Converter Configuration and Modeling

The FWEC power converter supplies output power of variable voltage and variable frequency to the grid in the form of constant voltage and frequency. In general, a back-to-back converter is used to perform this function, and the back-to-back converter consists of an AC/DC converter and a DC/AC inverter. The AC/DC converter mainly applies electrical torque control for generator speed control, and the DC/AC inverter controls the system output power factor for grid connection. In this study, the generator-side converter was analyzed because the power generation performance and the maximum power control algorithm were compared through the generator-side load control. The structure of the generator-side converter of the wave power generator uses a diode DC/DC converter structure. This structure is used in small-capacity renewable energy systems. Because the structure and control are simple and the price is low, this structure is mainly used for small capacity systems with little power generation, even if the efficiency is rather low. The variable voltage and variable frequency power of the generator is converted into constant DC power using a diode rectifier, and the DC/DC converter boosts the DC output. Because this structure uses a diode rectifier, the input power factor is low because it does not directly control the AC output, but rather passively converts it into a DC output. This approach can reduce the power generation; however, because the PTO system used in this study employs a hydraulic transmission system, the variability of the generator rpm was not as large as that of the input energy. Therefore, even if the structure of a diode DC/DC converter is used, the electrical torque control performance is not significantly reduced. As shown in Figure 4, the power converter calculates the reference voltage using the reference and inductor currents, and the reference voltage creates a duty ratio through comparison with the carrier.

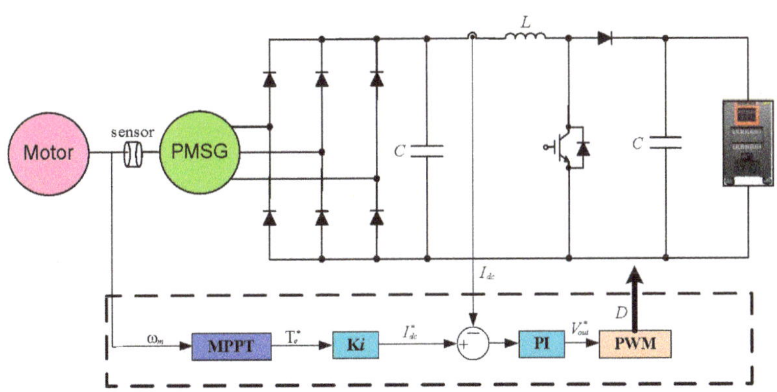

Figure 4. DC/DC boost converter control block diagram.

The output for converter operation can be expressed using the duty ratio as follows:

$$(V_{in})DT_{samping} = (V_{out} - V_{in})(1 - D)T_{samping} \tag{22}$$

$$V_{out} = \frac{1}{1 - D}V_{in} \tag{23}$$

3. Load Control Algorithm for Maximum Power Control of FWEC

Owing to the FWEC power generation characteristics, a large amount of input energy must be absorbed to increase the amount of power generation. The power absorbed by the FWEC can be calculated using the floating body movement and PTO force. The floating body movement changes according to the PTO force, and part of it becomes the MPP. The PTO force of the hydraulic transmission system is proportional to the pressure in the circuit, and the PTO force can be calculated using Equation (3).

This section presents the circuit pressure and PTO force of the hydraulic system according to various load control algorithms for maximum power control. The load control algorithm for maximum FWEC power control was applied to the P&O algorithm-based speed control algorithm and the optimal torque control algorithm, and performance analysis of each algorithm was conducted to develop a maximum FWEC power control algorithm.

3.1. Speed Control Algorithm-Based P&O Algorithm

Through electrical load control, the speed of the hydraulic motor and generator of the PTO system can be controlled. The pressure in the circuit of the hydraulic transmission system fluctuates according to the changes in generator speed. That is, as the generator speed increases with the same input energy, the pressure and PTO force decrease. Through speed control, it is possible to change the circuit pressure of the hydraulic transmission system; thus, the PTO force can be adjusted. Similar to existing renewable energy systems such as solar and wind power generation systems, FWEC can yield the maximum power by using the power generation characteristic curve according to the PTO force, as shown in Figure 5.

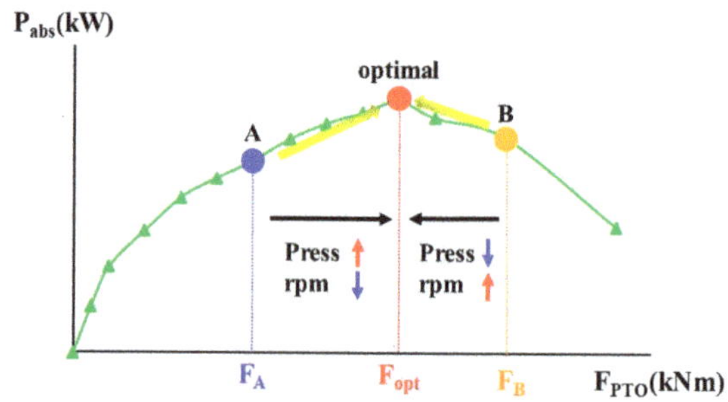

Figure 5. P&O algorithm-based power versus PTO force.

The specific point of the PTO force can be changed according to the period and wave height of the input wave, but the controller can be designed by applying the P&O algorithm using the characteristic curve shown in Figure 5. As the speed increases or decreases, the pressure in the circuit of the hydraulic system changes, and the PTO force that affects the movement of the floating body changes. In other words, the reference speed increases or decreases according to the increase or decrease in the amount of power generated, and accordingly, the PTO force that can yield the maximum power is sought. Owing to the large wave energy variability, there may be limitations in applying the P&O algorithm, but because the PTO system used in this study is a hydraulic transmission system, the speed variability can be significantly reduced. Figure 6 shows a flow chart of the P&O algorithm used in this study.

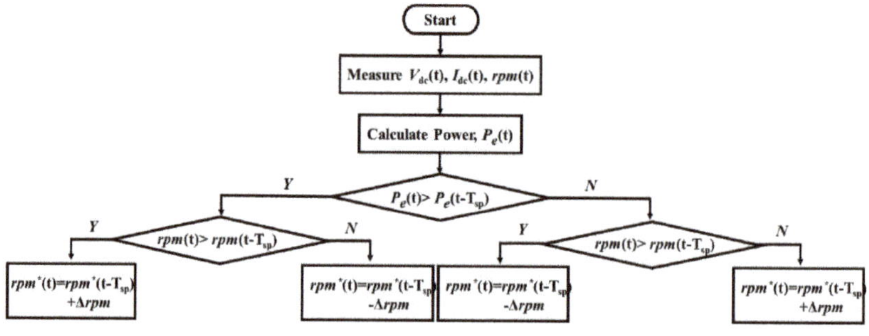

Figure 6. Flow chart of speed control algorithm based on the P&O algorithm for FWEC.

3.2. Optimal Torque Control Algorithm

Optimal torque control can yield the maximum power by using the PTO force to satisfy the FWEC resonance condition [28–31]. To obtain the maximum power, the wave must have the resonance frequency, in which case the wave frequency can be expressed as:

$$\omega_{wave} = \omega_{reson} = \sqrt{A\rho g/(M + \mu(\omega))} \tag{24}$$

When the wave frequency satisfies Equation (24), the power absorbed by the floating body can be summarized as:

$$P_{abs} = \frac{1}{2} B_{pto} |F_{exc}|^2 / \{B_{rad}(\omega) + B_{vis} + B_{PTO}\}^2 \tag{25}$$

For the absorbed power to be the maximum power, the following equation must be satisfied:

$$\frac{d}{dt} P_{abs} = \frac{1}{2} |F_{exc}|^2 \frac{B_{rad}(\omega) + B_{vis} - B_{PTO}}{\{B_{rad}(\omega) + B_{vis} + B_{PTO}\}^3} \tag{26}$$

$$\frac{d}{dt} P_{abs} = 0 \tag{27}$$

Using Equations (26) and (27), the PTO damping factor for following the maximum power can be summarized as:

$$B_{PTO} = B_{rad}(\omega) + B_{vis} \tag{28}$$

The optimal PTO damping coefficient in the resonance period can be expressed as Equation (28), but the PTO damping coefficient changes every period. Figure 7a shows the optimal PTO damping coefficient for each period. Figure 7b presents the output power generation performance obtained using the optimal PTO damping coefficient for each period and the PTO damping coefficient of the resonance period. When the optimal PTO damping coefficient for each period is applied, the generation power performance is high, but it is almost identical to that obtained using the other method near the resonance period. In particular, because the input wave conditions are irregular, it is difficult to change the PTO damping coefficient according to the period. Therefore, the optimal torque control adjusts the load by using the optimal PTO damping coefficient in the resonance period.

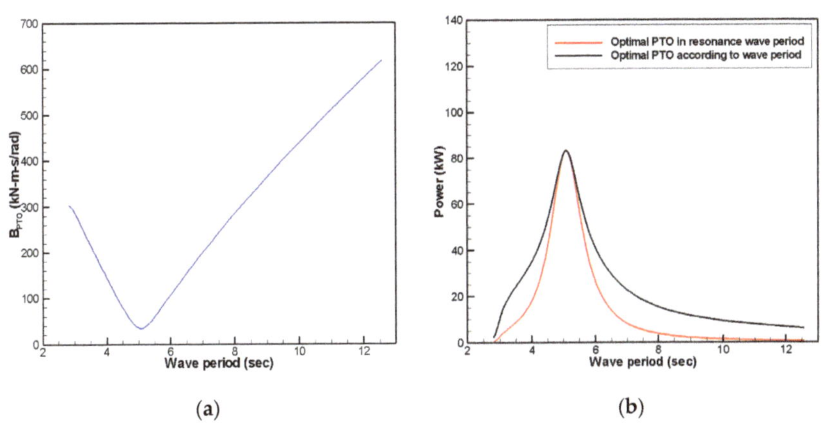

Figure 7. (**a**) Optimal PTO damping coefficient for each period. (**b**) Output power generation performance according to the optimal PTO damping coefficient for each period and resonance period PTO damping coefficient.

To obtain the maximum power by using the optimal PTO damping coefficient, it is necessary to control the electrical torque of the generator accordingly. If the loss of the PTO system is not considered, the mechanical power can be expressed as the movement of the floating body and the PTO force:

$$P_m = F_{PTO} * \omega_{pitch} \tag{29}$$

The PTO force can be written as follows if expressed as a simplified model using the PTO damping coefficient:

$$F_{PTO} = B_{PTO} * \omega_{pitch} \tag{30}$$

As shown in Equation (28), if the mechanical power is expressed using the optimal PTO damping coefficient calculated to obtain the maximum power, it can be written as:

$$P_{m\_opt} = B_{PTO\_opt} * \left(\omega_{pitch}\right)^2 \tag{31}$$

If the above equation is expressed using the rotary vane pump volume, hydraulic motor volume, and rpm of the hydraulic motor, it can be written as:

$$P_{m\_opt} = B_{PTO\_opt} * \left(\frac{D_{motor}}{D_{pump}}\right)^2 \omega_{motor}{}^2 \tag{32}$$

Using the relationship between torque and power, the optimum mechanical torque can be calculated as:

$$T_{m\_opt} = \frac{T_{m\_opt}}{\omega_{motor}} = B_{PTO\_opt} * \left(\frac{D_{motor}}{D_{pump}}\right)^2 \omega_{motor} \tag{33}$$

$$k_{opt} = B_{PTO\_opt} * \left(\frac{D_{motor}}{D_{pump}}\right)^2 \tag{34}$$

The electrical torque reference value for obtaining the maximum output can be calculated as:

$$T_e^* = T_{m\_opt} = k_{opt} * \omega_{motor} \tag{35}$$

However, under irregular wave input conditions, the optimal PTO damping coefficient may include components of various periods rather than the resonance period. Therefore, in this study, the output performance was compared and verified according to the PTO damping coefficient by utilizing various parameters.

4. Results

Through hydraulic-based PTO system modeling, the performance of the maximum power control algorithm with speed control based on the P&O algorithm and optimal torque control was verified. The floating body movement changed by reflecting the PTO system characteristics according to the MPP tracking (MPPT) algorithm under the same wave conditions and the overall power generation performance changed accordingly. Through computational fluid dynamics (CFD) analysis, the motion of the sub-fluid under electrical load control was applied, as shown in Figure 8a. CFD analysis was performed through a commercial program called STAR CCM+, and the simulation was performed considering the fluid viscosity of the floating body. The model was verified by performing model tests using the Froude scale [37].

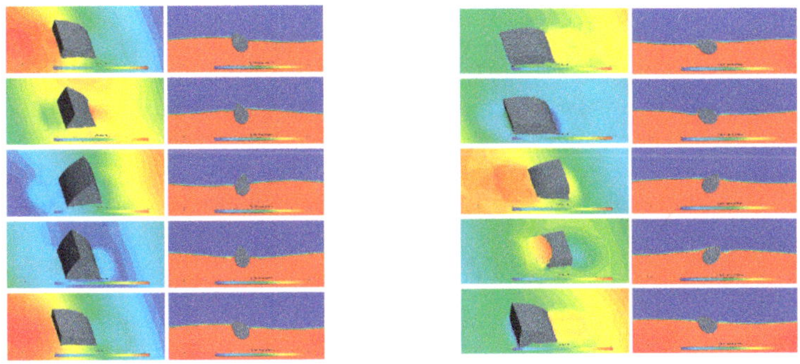

(**a**)

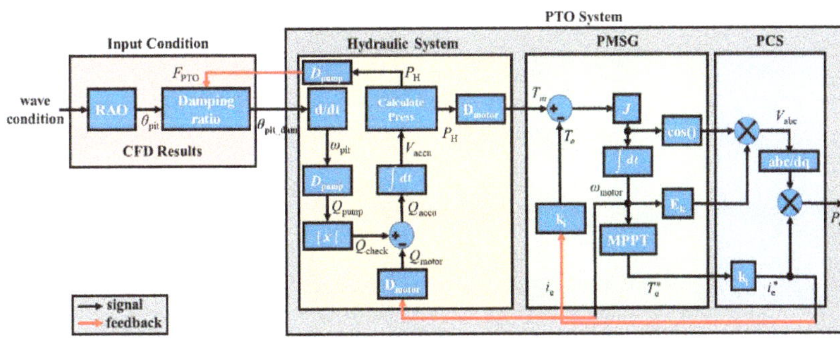

(**b**)

Figure 8. (**a**) CFD analysis plot for input conditions [37]. (**b**) Simulation block diagram for comparison of load control algorithm performance.

The simulation was conducted under regular and irregular wave conditions, and the power generation performance according to the variable change of the MPPT algorithm was compared under each set of wave conditions. Figure 8b provides a block diagram of this simulation. To ensure the accuracy of the simulation, the model was verified through lab tests.

4.1. Simulation Results Obtained under Regular Wave Conditions

The input conditions in the regular wave simulation were a resonance period of 0.75 m wave height and 4.75 s wave period. Under the input conditions, the pitch angle of the floating body according to the load and the absorbed power were calculated using CFD, and Figure 9 presents the results.

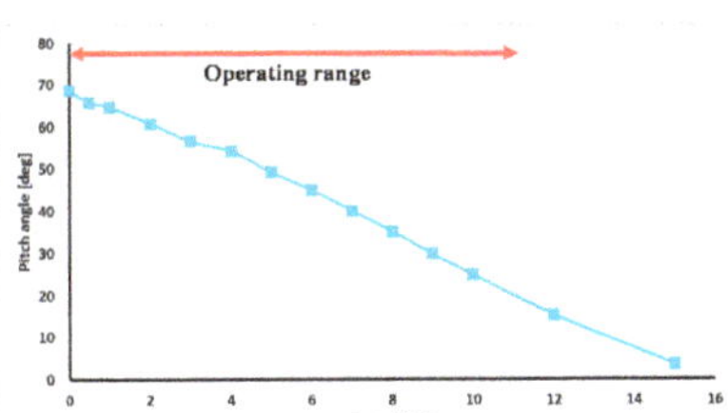

(**a**)

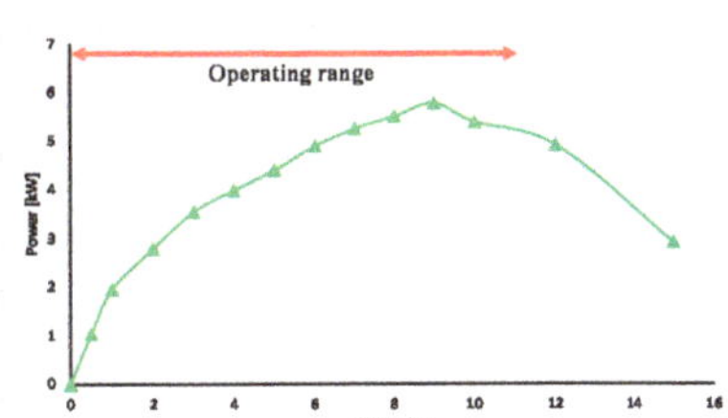

(**b**)

Figure 9. (**a**) Heave motion and (**b**) absorbed power versus PTO force under regular wave conditions.

Figure 9 shows the floating body movement and absorbed power generation according to the PTO force acting on the floating body. As the PTO force increases, the floating body movement decreases. In addition, there is a load that yields an optimal amount of power generation as the PTO force increases and floating body movement decreases. Because the PTO force can be calculated according to the pressure of the hydraulic system, the operating range is determined by the maximum pressure in the circuit, as shown in Figure 9. The power generation performance was compared by applying the P&O algorithm-based speed control algorithm and optimal torque control algorithm for maximum power control under the following input conditions, and a maximum power control algorithm was developed that is suitable for hydraulic-based PTO systems.

Figure 10 shows the motion characteristics and PTO force of the floating body according to the initial reference rpm of the speed control algorithm based on the P&O algorithm under regular wave conditions. The load control and pressure in the circuit change according to the initial reference rpm, and the PTO force changes accordingly. As shown in Figure 10a,b, the pressures all differ according to the initial reference rpm, and as the PTO force increases, the floating body movement decreases. As shown in Figure 10b, the PTO force was operated in a one-way manner to increase the floating body movement. In other words, the part where the pitch angle of the floating body is affected is where the PTO force operates. As shown in Figure 10c, the floating body velocity changes according to its movement, which affects the flow rate generated in the circuit, as shown in Figure 10d. In other words, when the floating body movement decreases, the velocity and generated flow rate also decrease.

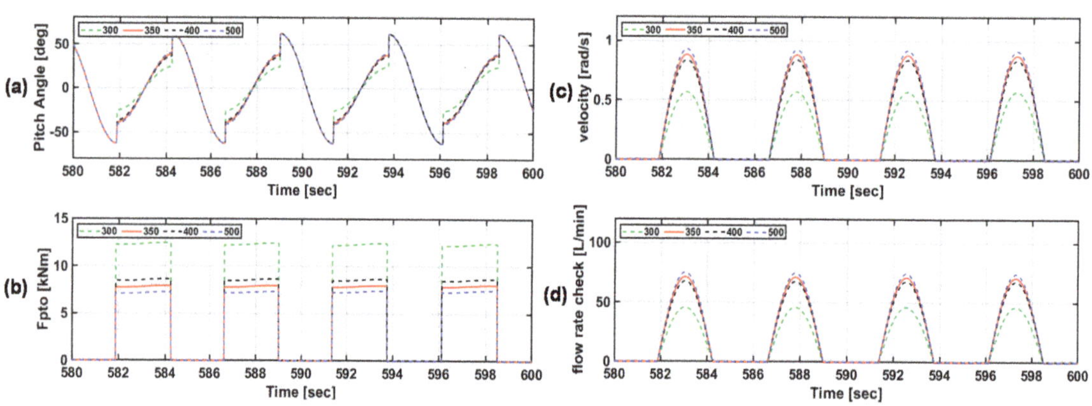

Figure 10. (**a**) Floating body motion, (**b**) PTO force, (**c**) floating body speed, and (**d**) generated flow rate according to the initial reference rpm of the speed control algorithm under regular wave conditions (legend indicates initial rpm).

Figure 11 shows the characteristics of the hydraulic and electrical systems according to the initial reference rpm of the P&O algorithm-based speed control algorithm under regular wave conditions. The pressure in the circuit and rpm of the hydraulic motor change according to the input energy and load. Because the speed control algorithm based on the P&O algorithm changes the reference rpm according to the amount of power generation, the rpm continuously changes, as shown in Figure 11b. However, as shown in Figure 11c,d, the lower the initial reference rpm, the higher the load at the same input energy; thus, the pressure increases. As the pressure increases, the PTO force increases, and as shown in Figure 11b, the generation amount can be increased until a constant PTO force is reached. In other words, the pressure in the circuit can be adjusted by applying a speed control algorithm based on the P&O algorithm, and the maximum power generation can be obtained by adjusting the PTO force.

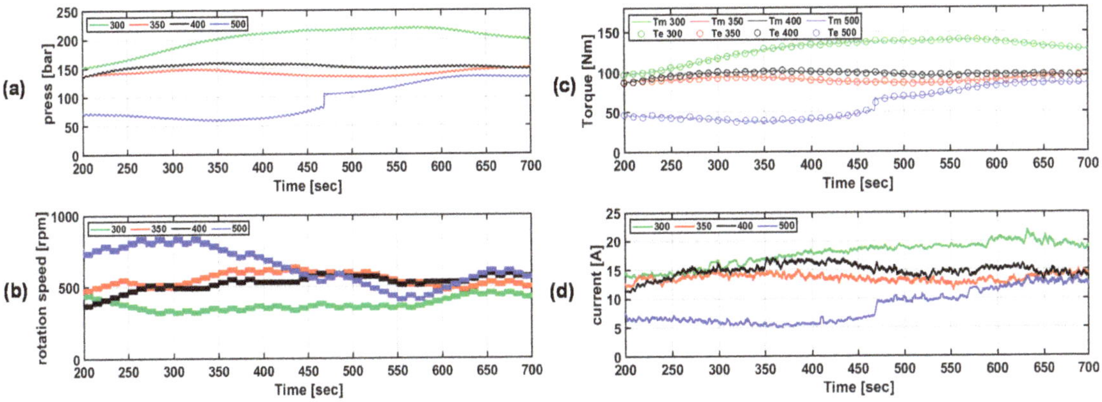

Figure 11. (**a**) Pressure in the circuit, (**b**) hydraulic motor rpm, (**c**) mechanical and electrical torques, and (**d**) load current according to the initial reference rpm of the speed control algorithm under regular wave conditions (legend indicates initial rpm).

In conclusion, the speed control algorithm based on the P&O algorithm can perform maximum power control only by adjusting the initial reference rpm setting and rpm change rate according to the system and environmental characteristics.

Figure 12 shows the input power generation and mechanical and electrical power generation according to the initial reference rpm of the speed control algorithm based on the P&O algorithm under regular wave conditions. The input power generated can be obtained according to the pitch angular velocity and PTO force of the floating body, the mechanical power generated can be obtained from the flow rate and pressure of the hydraulic motor, and the electrical power generated can be calculated from the output power of the power converter. The power generation can be increased by controlling the PTO force, which can absorb as much energy as possible through load control. As shown in Figure 12, the pressure changes according to the initial reference rpm, and the closer to the PTO force that can absorb the maximum power, the more power that can be obtained. As shown in Figure 9, in the case of an initial reference of 400 rpm, the PTO force was 9 kNm (160 bar), which yielded the greatest power absorption, and the highest power generation of 5.35 kWh was obtained. Table 1 lists the input power, output power, and efficiency of the PTO system for each case.

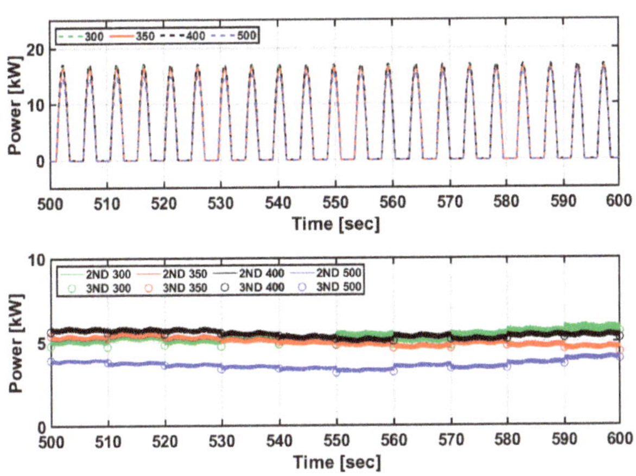

Figure 12. Comparison of power generation results according to the initial reference rpm of the speed control algorithm under regular wave conditions (legend indicates initial rpm).

Table 1. Input and output power of the PTO system and efficiency of the PTO system according to the initial reference rpm of the speed control algorithm based on the P&O algorithm.

Case (rpm)	Absorbed Power (kWh)	Output Power (kWh)	Efficiency (%)
300	5.45	5.09	92.99
350	5.33	5.22	97.94
400	5.45	5.35	98.12
500	4.11	3.75	91.22

Figure 13 shows the motion characteristics and PTO force of the floating body according to the torque coefficient of the optimal torque control algorithm under regular wave conditions. The load and pressure in the circuit change according to the torque coefficient, and the PTO force changes accordingly. As shown in Figure 13a,b, the pressures are all different according to the torque coefficient, and the movement of the floating body decreases as the PTO force increases. As shown in Figure 13b, the PTO force operates one way using the same method as the speed control algorithm. As shown in Figure 13c,d, the movement of the floating body affects the float velocity and the flow rate generated in the circuit. In other words, the optimal torque control algorithm can produce the maximum amount of power generation by changing the torque coefficient.

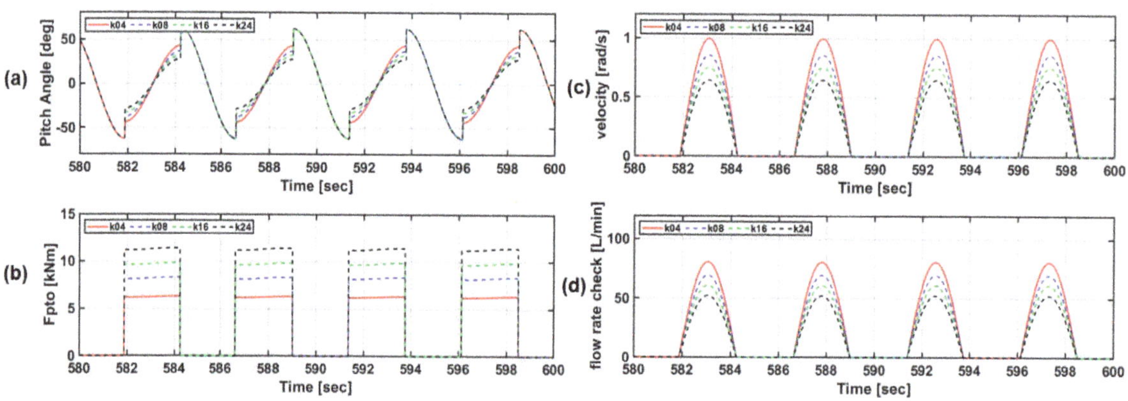

Figure 13. (**a**) Float pitch angle, (**b**) F_{pto}, (**c**) float velocity, and (**d**) generated flow rate according to the torque coefficient of the optimal torque control algorithm under regular wave conditions (legends indicate torque coefficients).

Figure 14 shows the characteristics of the hydraulic and electrical systems according to the torque coefficient of the optimal torque control algorithm under regular wave conditions. The pressure in the circuit and rpm of the hydraulic motor change according to the input energy and load control. Because the load control changes according to the torque coefficient, the pressure and speed are different under the same input conditions. As shown in Figure 14c,d, as the torque coefficient increases, the pressure increases. As the pressure increases, the PTO force increases, and as shown in Figure 9, it is possible to control the load that can absorb the maximum power generation. In other words, the pressure in the circuit can be adjusted according to the torque coefficient of the optimal torque control, and the maximum power generation can be obtained by adjusting the PTO force.

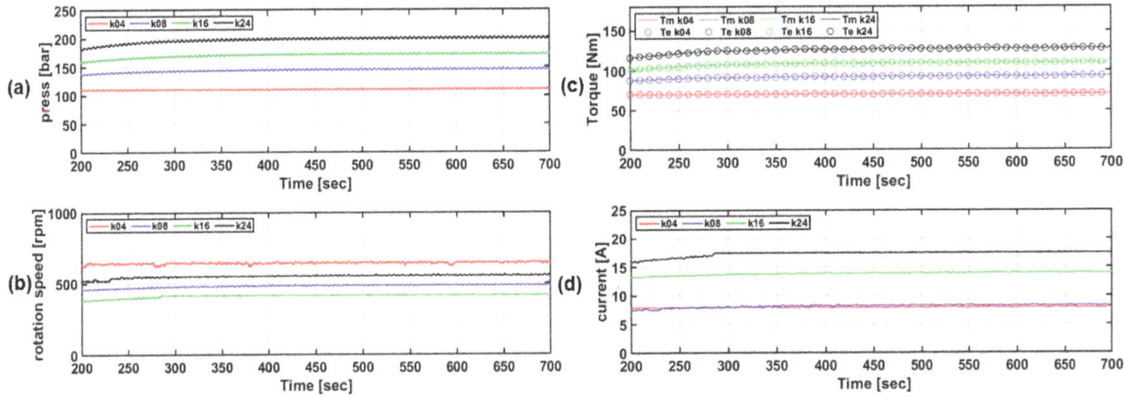

Figure 14. (**a**) Circuit pressure, (**b**) hydraulic motor rpm, (**c**) mechanical and electrical torques, and (**d**) load current according to the torque coefficient of the optimal torque control algorithm under regular wave conditions (legends indicate torque coefficients).

In conclusion, the maximum power generation can be obtained by adjusting the torque coefficient according to the system and environmental characteristics of the optimal torque control algorithm.

Figure 15 shows the input, mechanical, and electrical power generation according to the torque coefficient of the optimal torque control algorithm under regular wave conditions. The overall efficiency of the PTO system is quite high, so absorbing as much energy as possible through load control can increase the amount of power generated. As shown in Figure 14, as in the speed control algorithm, the pressure changes according to the torque coefficient, and the maximum power can be obtained as the PTO force approaches the force that can absorb the most power. As shown in Figure 9, the torque coefficient k close to 9 kNm (160 bar), which is the PTO force capable of absorbing the most power, was 1.6, and the greatest power generation was obtained at 5.53 kWh. Table 2 shows the input power, output power, and efficiency of the PTO system for each case.

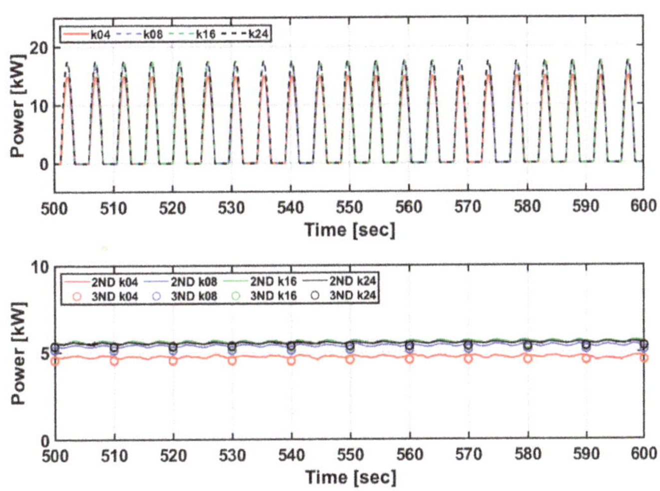

Figure 15. Power generation comparison according to torque coefficient of optimal torque control algorithm under regular wave conditions (legends indicate torque coefficients).

Table 2. Input and output power according to the torque coefficient of the optimal torque control algorithm and the efficiency of the PTO system.

Case (k)	Absorbed Power [kWh]	Output Power (kWh)	Efficiency (%)
0.4	4.74	4.53	95.5
0.8	5.38	5.14	95.5
1.6	5.59	5.35	95.7
2.4	5.51	5.34	96.9

4.2. Simulation Results Obtained under Irregular Wave Conditions

The input conditions for the irregular wave simulation were $H_s = 0.75$ m and $T_p = 4.75$ s. The pitch RAO for each period of the floating body and damping ratio of the floating body motion according to the PTO force in the irregular wave input conditions were calculated using CFD and are shown in Figure 16.

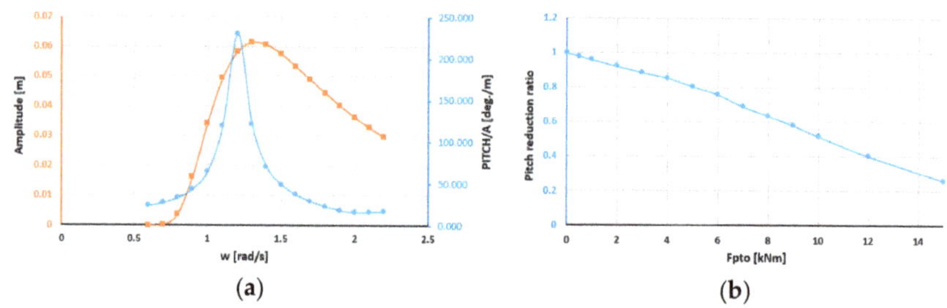

Figure 16. (**a**) Wave spectrum and pitch RAO and (**b**) floating body pitch motion damping ratio according to the PTO force under irregular wave input conditions ($H_s = 0.75$ m, $T_p = 4.75$ s).

Figure 16a presents the input wave spectrum and pitch RAO of the float obtained under the irregular wave input conditions ($H_s = 0.75$ m, $T_p = 4.75$ s). Figure 16b shows the damping ratio of the pitch motion of the floating body according to the load. According to the irregular wave input conditions, the pitch angle of the floating body can be calculated as the product of the wave spectrum and pitch PAO of the floating body. The damping of the pitch angle of the floating body due to the increase in F_{pto} as the pressure of the hydraulic system increases can be calculated using the damping ratio shown in Figure 16b. In the simulation for the maximum power control algorithm under irregular wave conditions, the speed control algorithm based on the P&O algorithm and the optimal torque control algorithm were applied. The power generation performance and characteristics of the PTO system according to each algorithm were analyzed under irregular input conditions, and a suitable load control algorithm for a PTO system applying a hydraulic system was developed. Because actual sea wave energy is an irregular input condition, the performance of the proposed maximum power control algorithm for a hydraulic-based PTO system can be verified by analyzing the control algorithm under irregular wave input conditions.

Figure 17 shows the motion characteristics and PTO force of the floating body according to the initial reference rpm of the speed control algorithm based on the P&O algorithm under irregular wave conditions ($H_s = 0.75$ m, $T_p = 4.75$ s). Unlike the regular wave conditions, because the input energy changes rapidly, a large difference occurs in the load control value and the pressure change according to the initial reference rpm; accordingly, a large difference occurs in the PTO force. As shown in Figure 17c,d, because the pitch angle of the floating body decreases with increasing PTO force, the velocity of the floating body and the flow rate generated in the circuit also decrease. As the PTO force increases and the floating body motion decreases, there is a PTO force that can yield the maximum power, similar to the regular wave simulation. That is, because the actual input wave cannot be predicted, the speed control algorithm based on the P&O algorithm can control the maximum output power by appropriately adjusting the initial reference rpm and the rpm change rate according to the input wave energy.

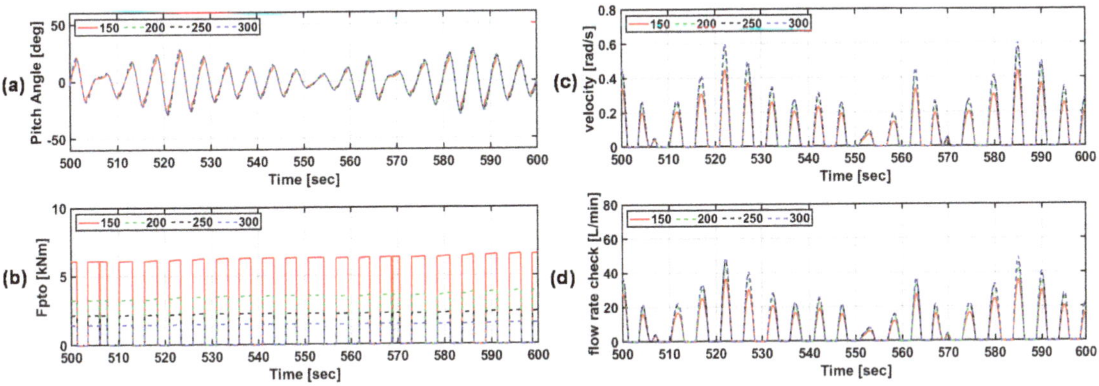

Figure 17. (**a**) Float pitch angle, (**b**) *Fpto*, (**c**) float velocity, and (**d**) generated flow rate according to the initial rpm of the speed control algorithm under irregular wave conditions (H_s = 0.75 m, T_p = 4.75 s) (legend indicates initial rpm).

Figure 18 shows the characteristics of the hydraulic and electrical systems according to the initial reference rpm of the speed control algorithm based on the P&O algorithm under irregular wave conditions (H_s = 0.75 m, T_p = 4.75 s). If the initial reference rpm is high, the pressure cannot increase because a significant flow rate is used to drive the hydraulic motor. Therefore, owing to the low pressure, the PTO force required to absorb a large amount of input power cannot be achieved. However, when the initial reference rpm is low, the pressure in the circuit increases, and accordingly, the necessary PTO force can be obtained, which can further increase the absorbed power. As the pressure in the circuit increases, the electrical load can be increased, as shown in Figure 18c,d, and, accordingly, the output power can be increased.

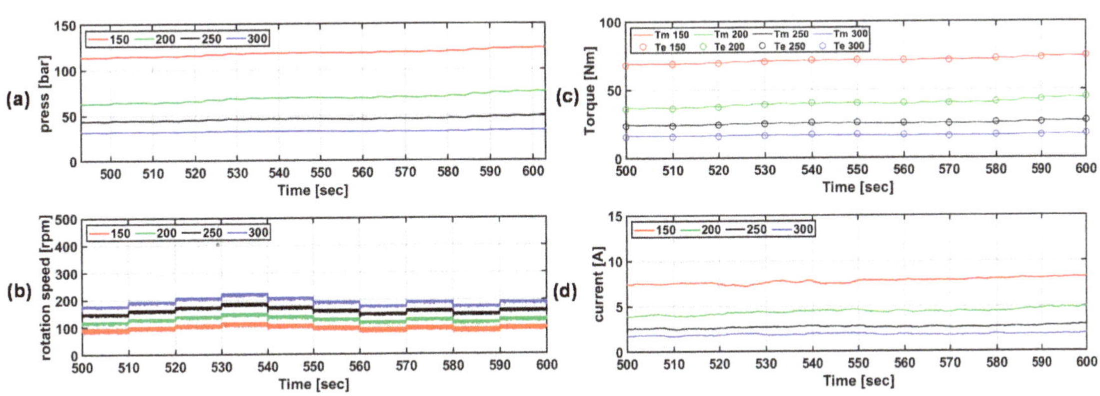

Figure 18. (**a**) Pressure in the circuit, (**b**) hydraulic motor rpm, (**c**) mechanical and electrical torque, and (**d**) load current according to the initial rpm of the speed control algorithm under irregular wave conditions (H_s = 0.75 m, T_p = 4.75 s) (legend indicates initial rpm).

In conclusion, the speed control algorithm based on the P&O algorithm can achieve the maximum power generation by selecting the appropriate initial rpm according to the input energy and environmental characteristics.

Figure 19 shows the input, mechanical, and electrical power generation according to the initial reference rpm of the P&O algorithm-based speed control algorithm under irregular wave conditions. By controlling the pressure and PTO force to absorb as much energy as possible through load control, the output power can be increased. Under the input conditions of $H_s = 0.75$ m, $T_p = 4.75$ s, using 150 rpm initially corresponded to 0.602 kWh, which yielded the highest power generation. At ≤150 rpm initially, the pressure in the circuit increased further, and more power could be obtained. However, at ≤80 rpm, the power supply of the power converter did not operate; therefore, it was not considered as a control reference value.

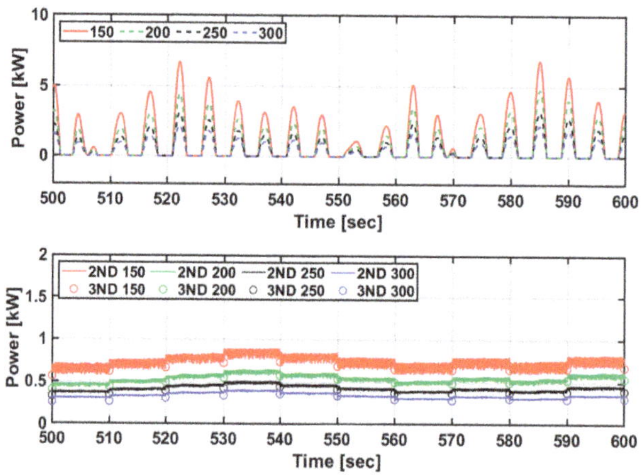

Figure 19. Comparison of the amount of power generated according to the initial rpm of the speed control algorithm under irregular wave conditions (H_s = 0.75 m, T_p = 4.75 s) (legend indicates initial rpm).

Figure 20 shows the motion characteristics and PTO forces of the floating body according to the torque coefficient of the optimal torque control algorithm under irregular wave conditions ($H_s = 0.75$ m, $T_p = 4.75$ s). Similar to the speed control algorithm based on the P&O algorithm, because the input energy changes rapidly, there are large differences in the load control value and pressure variability with the torque coefficient; accordingly, the PTO force also varies considerably. As shown in Figure 20c,d, because the movement of the floating body decreases according to the PTO force, the speed of the floating body and the flow rate generated in the circuit also decrease. These tendencies are the same as those observed using the speed control algorithm. The higher the torque coefficient, the more the load is applied at the same rpm; thus, the pressure increases further.

In other words, as in the speed control algorithm, adjusting the torque coefficient appropriately according to the input wave energy is a way to obtain the maximum power.

Figure 21 shows the characteristics of the hydraulic and electrical system according to the torque coefficient of the optimal torque control algorithm under irregular wave conditions (H_s = 0.75 m, T_p = 4.75 s). If the torque coefficient is low, the initial rpm is high, and accordingly, the flow rate for driving the hydraulic motor is high. That is, the pressure in the circuit is low, making the PTO force far from the optimal value for absorbing the input energy. However, as the torque coefficient increases, the pressure in the circuit increases, and accordingly, the PTO force that can further increase the absorbed power can be obtained. As the pressure in the circuit increases, the electrical load can be increased, as shown in Figure 21c,d, and accordingly, the amount of output power can be increased. However, as shown in Figure 21, when the torque coefficient increases to a certain value ($k = 0.34$), the output power generation decreases because the rpm decreases as the load increases. That is, as the torque coefficient increases and the pressure in the circuit increases, the movement of the floating body decreases. Therefore, the maximum power generation can only be obtained by selecting an appropriate torque coefficient according to the input energy and environmental characteristics.

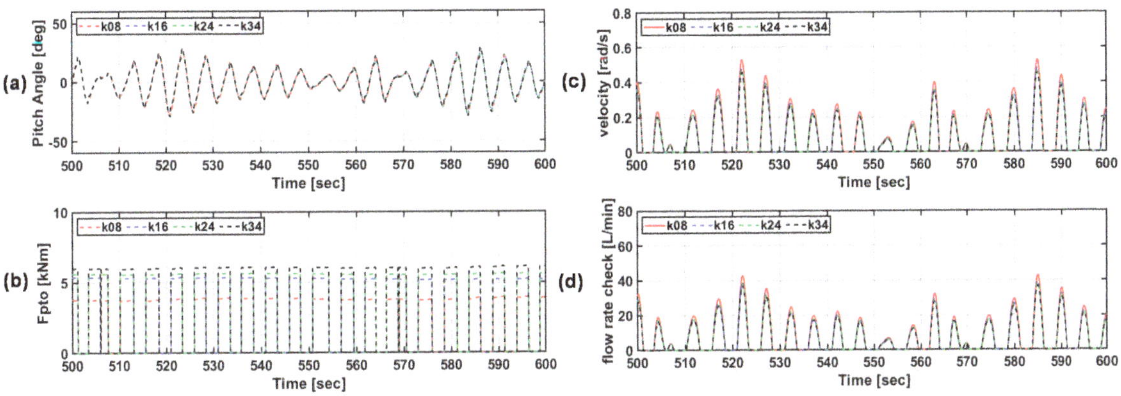

Figure 20. (**a**) Floating body pitch angle, (**b**) PTO force, (**c**) floating body velocity, and (**d**) occurrence flow rate according to torque coefficient of optimal torque control algorithm flux under irregular wave conditions (H_s = 0.75 m, T_p = 4.75 s) (legends indicate torque coefficients).

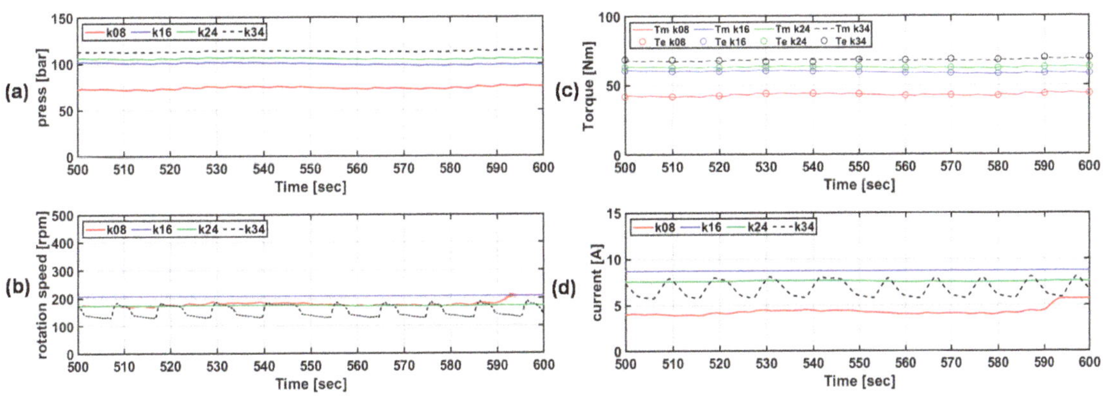

Figure 21. (**a**) Pressure in the circuit, (**b**) hydraulic motor rpm, (**c**) mechanical and electrical torque, and (**d**) electrical load current according to the torque coefficient of the optimal torque control algorithm under irregular wave conditions (H_s = 0.75 m, T_p = 4.75 s) (legends indicate torque coefficients).

Figure 22 shows the input power generation and mechanical and electrical power generation according to the torque coefficient of the optimal torque control algorithm under irregular wave conditions. Similar to the speed control algorithm, the amount of power generated can be increased by controlling the pressure and the PTO force to absorb as much energy as possible through load control. However, if the torque coefficient increases above a certain value, the PTO force increases and the input power increases, but the rpm decreases owing to an increase in pressure, and the output power generated may decrease. In conclusion, under input conditions of H_s = 0.75 m, T_p = 4.75 s, the torque coefficient for obtaining the maximum power generation of the optimal torque control algorithm was 0.24, and 0.911 kWh of power could be generated.

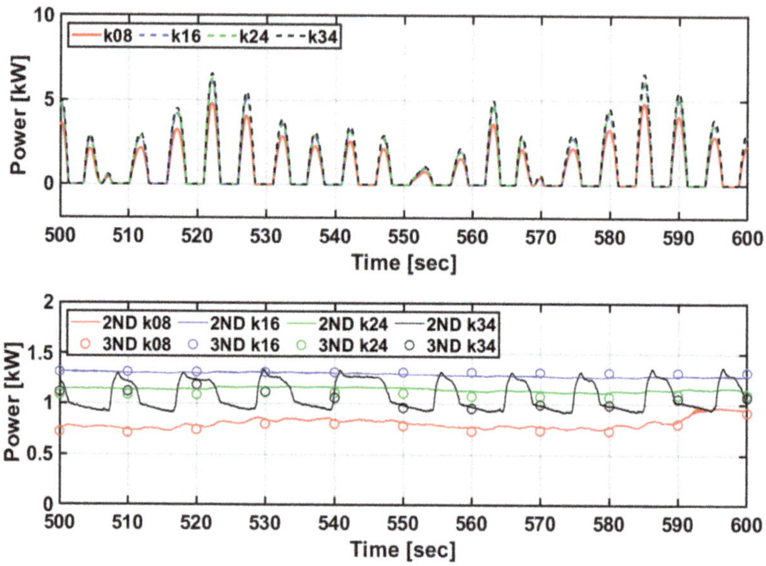

Figure 22. The result of comparing the generation amounts according to the torque coefficient of the optimal torque control algorithm under irregular wave conditions (H_s = 0.75 m, T_p = 4.75 s) (legends indicate torque coefficients).

Figure 23 shows the power generation performance according to the torque coefficient of the optimal torque control algorithm under irregular wave conditions. In case A (k = 0.08) with the lowest torque coefficient, the pressure was also low because the load was low, so the input power was low; thus, the output power generation was low. In case B (k = 0.16) with an increased torque coefficient, the pressure was increased compared to that in case A because the load value was increased, and both the input generation amount and the output generation amount were increased. That is, when the torque coefficient increases, the pressure in the circuit increases, which increases the input power, and the load value increases, thereby increasing the output power generation.

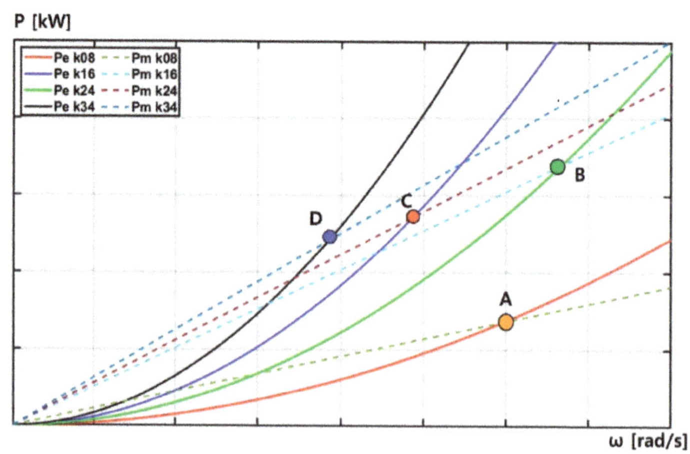

Figure 23. Comparison of output power generation performance according to torque coefficient of optimal torque control algorithm (legends indicate torque coefficients).

In addition, in cases C and D ($k = 0.24$ and $k = 0.34$, respectively), the torque coefficient was increased, but the output power generation was decreased compared to that in case B. As the torque coefficient increases, the pressure increases, but the rpm decreases, and the output power generation decreases. However, as the circuit pressure increases, the amount of input power generated can increase according to the optimal PTO force of the rotor. As shown in Figure 23, under irregular wave conditions, the maximum output can be obtained using the torque coefficient.

Figure 24 shows the power generation performance and efficiency for each algorithm under the same irregular wave input conditions. In the speed control algorithm based on the P&O algorithm, the pressure changes according to the initial reference rpm, and the input absorbed power appears to differ accordingly. The lower the initial reference rpm, the higher the pressure that can be obtained, and the higher the input absorbed power. However, the speed control algorithm based on the P&O algorithm has a lower output power than the optimal torque control algorithm because it is difficult to change the reference speed appropriately according to the input energy, and load control is performed to maintain the reference speed for a certain period of time. The optimal torque control algorithm yielded greater power generation through appropriate load control according to the input energy than the speed control algorithm based on the P&O algorithm. In addition, high efficiency and output energy are increased through appropriate circuit pressure control according to the input energy. However, as the torque coefficient increases with the input energy, the circuit pressure increases; thus, the input energy may increase, but the efficiency and power generation of the PTO system are reduced.

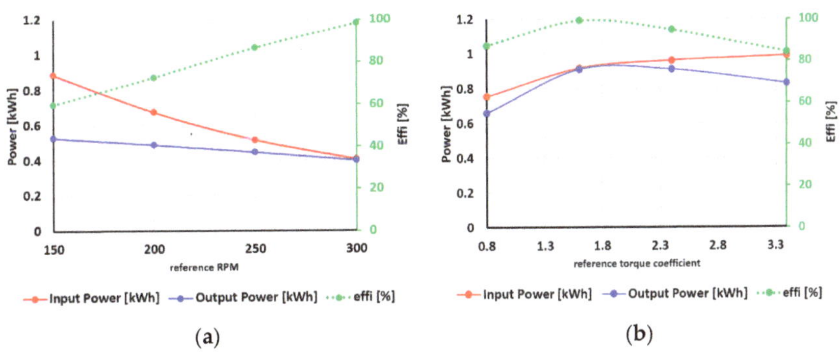

Figure 24. Input and output power generation and PTO system efficiency according to (**a**) speed control algorithm and (**b**) optimal torque control algorithm under the same irregular wave input conditions ($H_s = 0.75$ m, $T_p = 4.75$ s).

Figure 25 compares the power generation of the algorithms under various input energy conditions to verify the performance of each algorithm. Input energy was performed under irregular wave conditions, and results were derived based on 30 min data. It was confirmed that the optimal torque control algorithm obtained high power generation under all input conditions.

In conclusion, under the irregular wave conditions, the optimum torque control could produce high power generation, which enabled the appropriate load change according to the change in input energy. In addition, it is possible to increase the input absorbed power by increasing the circuit pressure and increasing the load; however, if a large amount of energy is used to increase the pressure, the efficiency and power generation of the PTO system may be reduced.

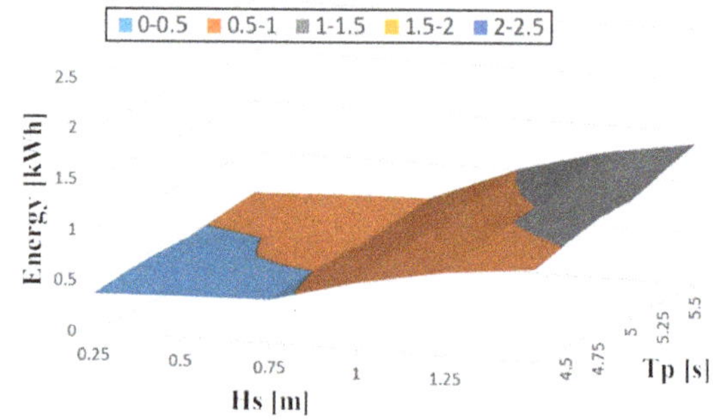

(a)

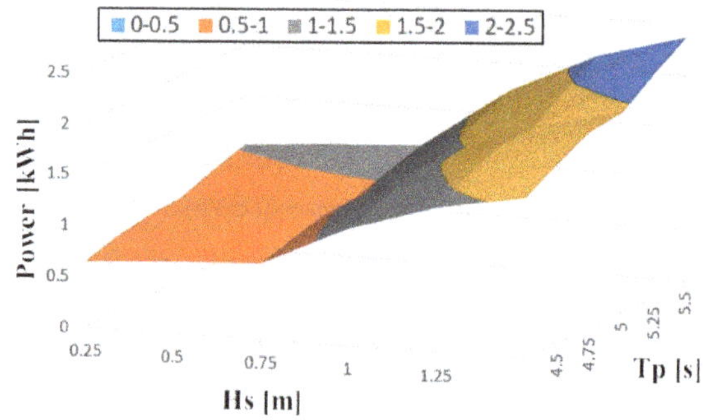

(b)

Figure 25. Performance of (**a**) P&O algorithm-based speed control algorithm and (**b**) optimal torque control algorithm according to input condition change.

5. Comparison of Real Sea Test and Simulation

The simulator developed in this study was verified based on the output data of an actual FWEC sea test conducted in the western sea of Jeju Island, Korea. The PTO system of the simulator was modeled based on the specifications used in the actual sea test.

Figure 26 shows the FWEC rotor, PTO systems, and measurement and control systems used in the actual sea test. The data used for the simulator verification were verified using approximately 400 s of actual sea test power generation data. Data measurement was performed at 20 Hz and measurement was performed using NI DAQ equipment. For the verification method, the characteristics of the simulator PTO system were compared under the same floating body pitch motion input conditions as in the actual sea test. In addition, the validity of applying the load control algorithm was verified by comparing the results of the simulator and applying the load control algorithm for maximum power control with actual sea data. Because an actual sea test is limited in terms of the variety of data that can be obtained, a simulation with various input conditions could be conducted and used as basic data for the actual sea test later.

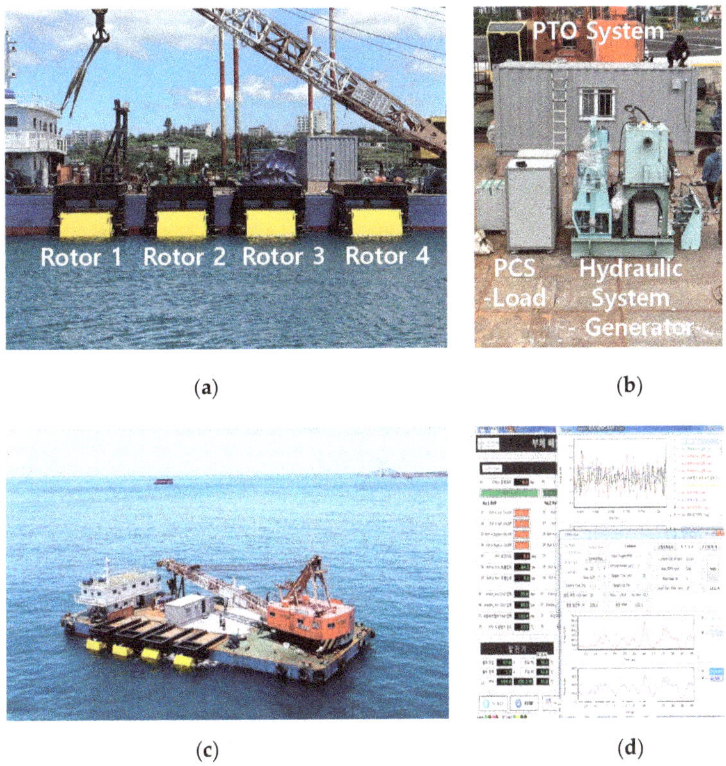

Figure 26. Actual FWEC sea test photos: (**a**) rotors, (**b**) PTO system (including the hydraulic system, generator, and PCS), (**c**) actual sea experiment, and (**d**) measurement and control system.

5.1. Comparison of P&O Algorithm-Based Speed Control Algorithm

As shown in Figure 27, the simulator used the same input conditions as the actual sea test to analyze the characteristics of the PTO system. To verify the simulator performance based on actual sea data, the load control algorithm was analyzed by applying the speed control algorithm based on the P&O algorithm and optimal torque control algorithm described in the previous section. Firstly, the speed control algorithm based on the P&O algorithm selects an initial reference speed based on the speed data of the actual sea test and manages control while changing the reference speed according to the P&O algorithm. The hydraulic system, electrical load, and power generation characteristics were compared between the P&O algorithm-based speed control algorithm results and those of a PTO system with actual sea data.

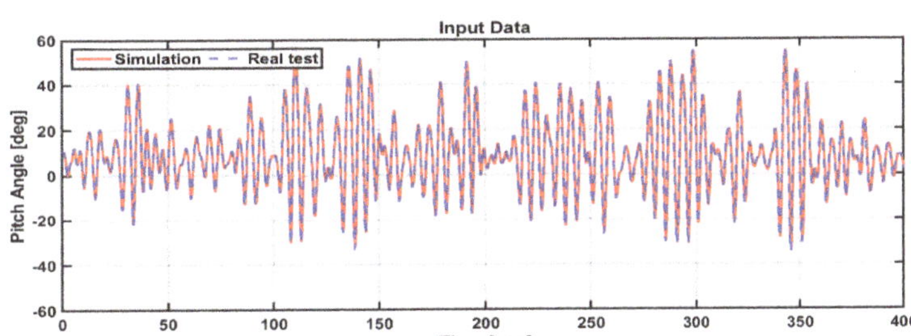

Figure 27. Simulator input data results obtained using pitch angle data of a floating body in an actual sea test.

Figure 28 shows the actual sea test data and the results of the speed control algorithm based on the simulator P&O algorithm. By applying the speed control algorithm, it was possible to obtain results similar to the pressure data of the actual sea test. It was possible to verify the PTO system results of the simulator through load control under the same input conditions based on the data obtained in the actual sea test. In addition, as shown in Figure 28b–d, the mechanical pressure (Figure 28b) and flow rate (Figure 28c) characteristics of the PTO system according to the speed control and electrical load current (Figure 28d) were also compared with the actual sea test results. Some numerical differences may exist due to physical losses and the limited ability to simulate the load control in the same way as in the actual test, but the trends of the data are similar.

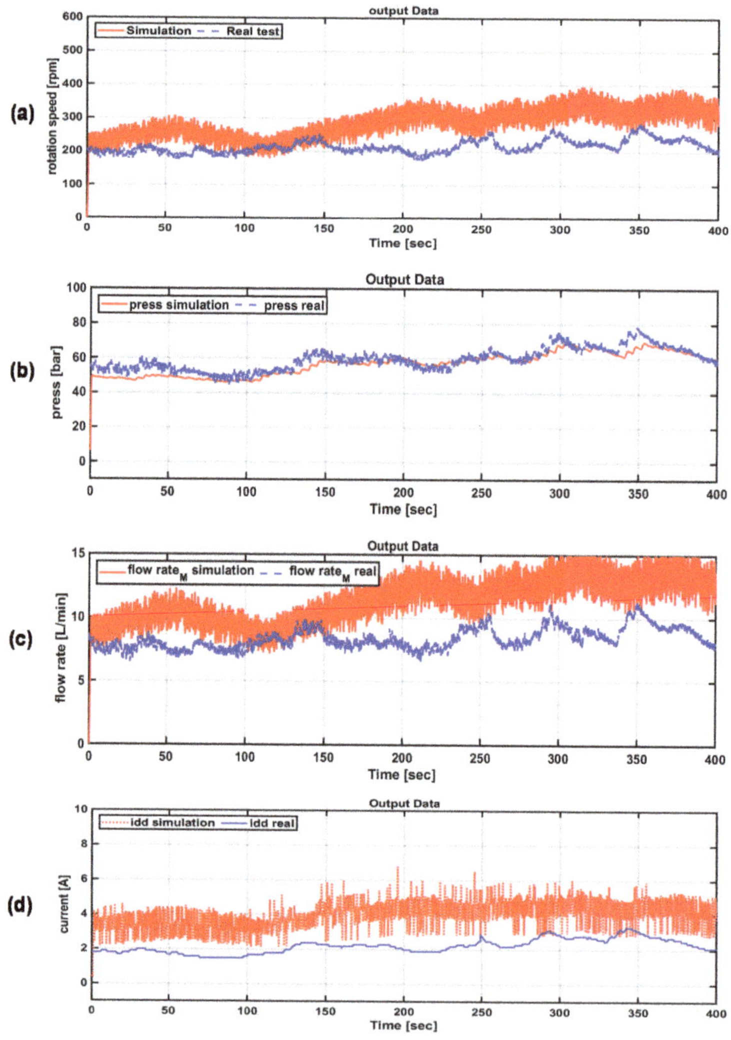

Figure 28. Comparison of PTO system characteristics according to real sea test results and P&O algorithm-based speed control algorithm of the simulator. (**a**) rotation speed; (**b**) mechanical pressure; (**c**) flow rate; (**d**) electrical load current.

Figure 29 shows the power generation results of the actual sea test and the P&O algorithm-based speed control algorithm. The amount of power generated was compared to the amounts of input power (first generation amount, P1), mechanical power (second generation amount, P2), and electrical power (third generation amount, P3) generation. The actual sea area test and the average input power generated in the simulation were 1.3021 kW and 1.3177 kW, respectively; hence, similar results were obtained. The same floating body motion was applied, and because the pressure of the hydraulic system was almost identical in the two cases, the PTO force was almost identical, and there was little difference in the input power generation. P2 and P3 appear similar to one another. Because the simulator cannot consider the loss that occurs physically, the amount of power generated appeared slightly larger. The average output in the actual sea test and simulation were 0.766 kW and 1.067 kW, and the PTO system efficiencies were 42% and 81%, respectively. In addition, the amount of input power generated rapidly changes with time in the PTO system using the hydraulic system, but the variability of the mechanical and electrical outputs is reduced.

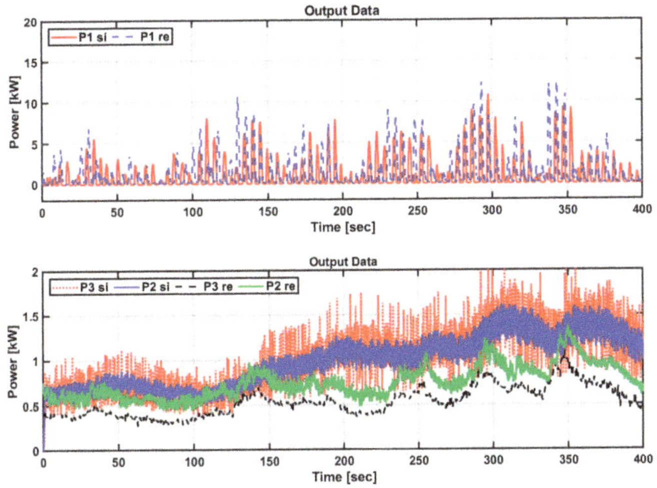

Figure 29. Comparison of the amounts of power generated in the actual sea test and with the speed control algorithm based on the P&O algorithm of the simulator.

5.2. Comparison of Torque Control Algorithm

Similar to the verification of the speed control algorithm based on the P&O algorithm, the input data shown in Figure 27 were used as the input data to verify the optimal torque control algorithm. Unlike the P&O algorithm, which monitors the output power amount, the optimal torque control algorithm adjusts the electrical torque based on the rpm change, according to the input energy required to obtain the maximum amount of power generation.

Figure 30 shows the results of the actual sea test and the characteristics of the PTO system according to the optimal torque control simulation. The pressures obtained using the simulated PTO system to which the optimal torque control algorithm was applied appear to be almost identical to the actual sea data. In addition, the changes in the rpm and the trend of the load change according to the input energy change are similar between the actual test and the simulation. Under the same input conditions as the P&O algorithm-based speed control algorithm, the PTO system and electrical load characteristics are similar. These findings were also verified by comparing the results of the optimal torque control algorithm with actual sea test data.

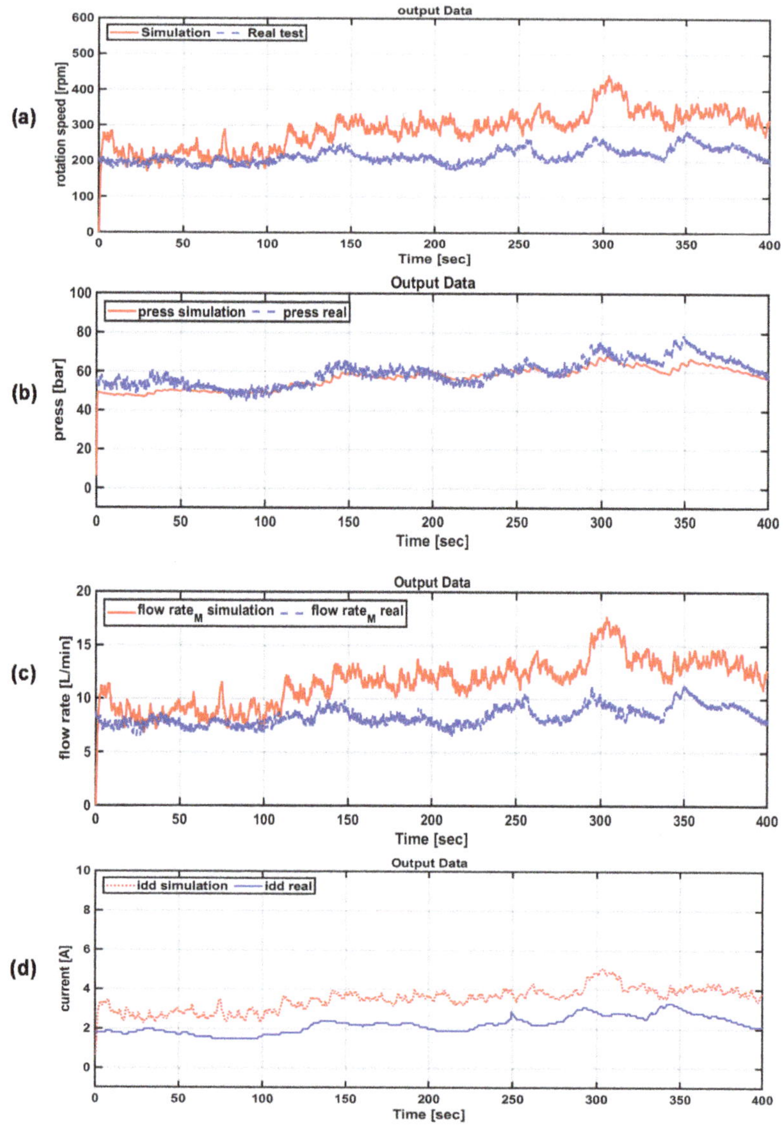

Figure 30. Comparison of the PTO system characteristics according to the actual sea test and optimal torque control simulation. (**a**) rotation speed; (**b**) mechanical pressure; (**c**) flow rate; (**d**) electrical load current.

Figure 31 shows the amounts of power generated in the actual sea test according to the optimal torque control algorithm. Specifically, the input power generation and mechanical and electrical power generation were compared. The amounts of input power generated in the actual sea test and simulation were 1.3021 kW and 1.3107 kW, respectively. Because the simulator does not consider loss, the amount of power generation appears slightly larger, but the trend of the generation data is similar. The actual sea test result and average output of the simulator were 0.7661 kW and 1.061 kW, respectively, and the efficiency of the PTO system was 42% and 81%, respectively. In addition, it can be confirmed that the output power generation amount is constant, even though the input energy changes rapidly.

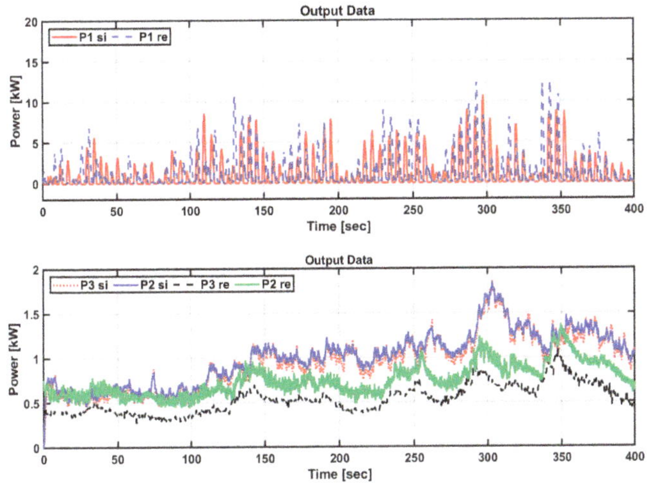

Figure 31. Comparison of power generation in actual sea test and simulation results.

In conclusion, by applying the speed and torque control algorithms, output performance similar to that in the actual sea test could be confirmed under the same input energy conditions. That is, it was confirmed that the maximum power control algorithms proposed in this paper can be applied to a hydraulic-based PTO system. However, it was not possible to compare the performances of the algorithms under various conditions, as in the previous simulation. As a future study, we will analyze the power generation performance considering changes in each algorithm and control variables.

6. Conclusions

Among the PTO systems for FWEC, the hydraulic system type has already been widely used because it facilitates the achievement of low speed and high torque. Therefore, in this study, modeling was conducted using a hydraulic system generator power converter to analyze the performance of a PTO system using a hydraulic system. Using this approach, the characteristics and power generation performance of the PTO system to which a load control algorithm for maximum power control was applied were analyzed. Unlike previous studies on input power generation using a hydraulic-based PTO system, the characteristics of a hydraulic-based PTO system were analyzed through electrical load control in this study. Based on this, the input power, output power, and PTO system efficiency were analyzed. The maximum power control algorithm was analyzed by applying the P&O algorithm-based speed control algorithm and optimal torque control algorithm. By analyzing the characteristics and power generation performance of the PTO system according to the control variables of each algorithm, a maximum power control algorithm suitable for load control of a hydraulic-based PTO system was proposed. Although the power generation performance and PTO system efficiency characteristics differed depending on the algorithm and control variable, input power control was possible by controlling the pressure of the hydraulic circuit through electrical load control.

In conclusion, the optimal torque control algorithm proposed in this paper proved to be suitable for load control of a PTO system based on a hydraulic system. This simulator was verified based on actual sea test results. In addition, the simulation results can be used as basic data for load control in actual sea tests of the PTO system based on a hydraulic system for wave power generation in the future.

Funding: This research received no external funding.

Conflicts of Interest: The author declares no conflict of interest.

Abbreviations

The following abbreviations are used in this manuscript:

ω_{pitch}	Pitch angular velocity of floating body
θ_{pitch}	Pitch angle of the floating body
Q_{pump}	Rotary vane pump flow
D_{pump}	Rotary vane pump displacement
F_{PTO}	FTO force on floating body
F_{opt}	Force of the FTO to obtain the maximum power generation
P_H	High pressure in hydraulic piping
P_L	Low pressure in hydraulic piping
Q_{check}	Flow of check valve
Q_{accu}	Accumulator flow
Q_{motor}	Hydraulic motor flow
V_{accu}	Accumulator volume
P_{H_pre}	Initial high pressure of hydraulic pipe
V_{accu_pre}	Accumulator initial volume
γ	Accumulator specific weight
D_{motor}	Hydraulic motor displacement
T_m	Mechanical torque
T_e	Electrical torque
ω_{motor}	Hydraulic motor angular speed
L_{dq}	D-Q axis inductance of generator
ω_e	Electrical angular frequency
i_{dq}	D-Q axis generator current
V_{dq}	D-Q axis generator voltage
Ψ_{pm}	Generator flux linkage
R_s	Generator armature resistance
P_e	Electrical power
N_p	Number of generator poles
D	DC/DC Converter duty ratio
T_{sp}	DC/DC Converter controlled sampling
ω_{wave}	Period of input wave energy
ω_{reson}	Resonance period of input wave energy
B_{pto}	Power take-off damping factor
P_{abs}	Input absorption power
P_m	Mechanical power
k_{opt}	Torque damping factor

References

1. Noble, D.R.; O'Shea, M.; Judge, F.; Robles, E.; Martinez, R.; Khalid, F.; Thies, P.R.; Johanning, L.; Corlay, Y.; Gabl, R.; et al. Standardising Marine Renewable Energy Testing: Gap Analysis and Recommendations for Development of Standards. *J. Mar. Sci. Eng.* **2021**, *9*, 971. [CrossRef]
2. Davey, T.; Sarmiento, J.; Ohana, J.; Thiebaut, F.; Haquin, S.; Weber, M.; Gueydon, S.; Judge, F.; Lyden, E.; O'Shea, M.; et al. Round Robin Testing: Exploring Experimental Uncertainties through a Multifacility Comparison of a Hinged Raft Wave Energy Converter. *J. Mar. Sci. Eng.* **2021**, *9*, 946. [CrossRef]
3. Xu, R.; Wang, H.; Xi, Z.; Wang, W.; Xu, M. Recent Progress on Wave Energy Marine Buoys. *J. Mar. Sci. Eng.* **2022**, *10*, 566. [CrossRef]
4. Martins, J.C.; Fragassa, C.; Goulart, M.M.; dos Santos, E.D.; Isoldi, L.A.; Gomes, M.D.N.; Rocha, L.A.O. Constructal Design of an Overtopping Wave Energy Converter Incorporated in a Breakwater. *J. Mar. Sci. Eng.* **2022**, *10*, 471. [CrossRef]
5. Drew, B.; Plummer, A.R.; Sahinkaya, M.N. A review of wave energy converter technology. *Proc. Inst. Mech. Eng. A* **2009**, *223*, 887–902. [CrossRef]
6. Cruz, J. *Ocean Wave Energy—Current Status and Future Perspectives*; Springer: Berlin, Germany, 2008; p. 434.
7. Salter, S.H.; Taylor, J.R.M.; Caldwell, N.J. Power conversion mechanisms for wave energy. *Proc. Inst. Mech. Eng. M* **2002**, *216*, 1–27. [CrossRef]
8. Rico, H.H.; Morten, M.K.; Enrique, V. Discrete displacement hydraulic power take-off system for the Wavestar wave energy converter. *Energies* **2013**, *6*, 4001–4044.
9. Markel, P.; John, V.R. A review of wave-to-wire models for wave energy converters. *Energies* **2016**, *9*, 506.

10. Mueller, M.; Baker, N. Direct Drive Wave Energy Converters. Available online: http://www.cder.dz/download/upec-1.pdf (accessed on 23 July 2013).
11. Polinder, H.; Mueller, M.; Scuotto, M.; Prado, M. Linear generator systems for wave energy conversion. In Proceedings of the 7th European Wave and Tidal Energy Conference, Porto, Portugal, 11–14 September 2007.
12. Li, B.; Macpherson, D.E.; Shek, J.K.H. Direct drive wave energy converter control in irregular waves. In Proceedings of the IET Conference on Renewable Power Generation (RPG 2011) IET, Edinburgh, UK, 6–8 September 2011; p. 64.
13. Poullikkas, A. Technology Prospects of Wave Power Systems. Electron. *J. Energy Environ.* **2014**, 2, 47–69.
14. Arof, H.; Nor, W.; Nor, K.M. Linear generator: Design and simulation. In Proceedings of the National Power Engineering Conference IEEE, Bangi, Malaysia, 15–16 December 2003; pp. 306–311.
15. Hansen, R.; Andersen, T.; Pedersen, H. Model based design of efficient power take-off systems for wave energy converters. In Proceedings of the 12th Scandinavian International Conference on Fluid Power (SICFP 2011), Tampere, Finland, 18–20 May 2011; Volume 2, pp. 35–49.
16. António, F.D.O. Modeling and control of oscillating-body wave energy converters with hydraulic power take-off and gas accumulator. *Ocean Eng.* **2007**, *34*, 2021–2032.
17. Babarit, A.; Guglielmi, M.; Clément, A.H. Declutching control of a wave energy converter. *Ocean Eng.* **2009**, *36*, 1015–1024. [CrossRef]
18. Hansen, R.; Andersen, T.; Pedersen, H. Comparison of reactive and non-reactive control strategies for wave energy converters with non-ideal power take-off systems. *Renew. Energy* **2013**. submitted for publication.
19. Skinner, N. Wave Energy Converter Device. European Patent EP2284386A2, 16 February 2011.
20. Linjama, M.; Vihtanen, H.P.; Sipola, A.; Vilenius, M. Secondary controlled multi-chamber hydraulic cylinder. In Proceedings of the 11th Scandinavian International Conference on Fluid Power (SICFP09), Linkoeping, Sweden, 2–4 June 2009.
21. Henderson, R. Design, simulation, and testing of a novel hydraulic power take-off system for the Pelamis wave energy converter. *Renew. Energy* **2006**, *31*, 271–283. [CrossRef]
22. Yemm, R.; Pizer, D.; Retzler, C.; Henderson, R. Pelamis: Experience from concept to connection. *Phil. Trans. R Soc. A* **2012**, *370*, 365–380. [CrossRef] [PubMed]
23. Joseba, L.A.; Carlos, A.U.J.; Jose, E.A.F. Power Take-Off Device for Wave Energy Transformation. European Patent EP2466118A1, 20 June 2012.
24. Lasa, J.; Antolin, J.C.; Angulo, C.; Estensoro, P.; Santos, M.; Ricci, P. Design, construction and testing of a hydraulic power take-off for wave energy converters. *Energies* **2012**, *5*, 2030–2052. [CrossRef]
25. Hansen, R.; Andersen, T.; Pedersen, H. Determining required valve performance for discrete control of PTO cylinders for wave energy. In Proceedings of the ASME Symposium on Fluid Power and Motion Control (FPMC 2012), Bath, UK, 12–14 September 2012; American Society of Mechanical Engineers: New York, NY, USA, 2012; pp. 565–578.
26. Ratanak, S.; Sean, C.; Sam, K.; Asher, S.; Ted, K.A. PTO-Sim: Development of a power take off modeling tool for ocean wave energy conversion. In Proceedings of the 2015 IEEE Power & Energy Society General Meeting (IEEE-PES 2015), Denver, CO, USA, 26–30 July 2015.
27. Ratanaks, S.; Asher, S.; Ted, B.; Kelley, R.; Carlos, M. Development of PTO-SIM: A power performance module for the open-source wave energy converter code WEC-SIM. In Proceedings of the ASME 2015 34th International Conference on Ocean, Offshore and Arctic Engineering (OMAE2015), St. John's, NL, Canada, 31 May–5 June 2015.
28. Abdelsalam, K.; Massoud, A.M.; Ahmed, S.; Enjeti, P.N. High-performance adaptive perturb and observe MPPT technique for photovoltaic based microgrids. *IEEE Trans. Power Electron.* **2011**, *26*, 1010–1021. [CrossRef]
29. Femia, N.; Petrone, G.; Spagnuolo, G.; Vitelli, M. Optimization of perturb and observe maximum power point tracking method. *IEEE Trans. Power Electron.* **2005**, *20*, 963–973. [CrossRef]
30. Xu, J.; Yang, Y.; Hu, Y.; Xu, T.; Zhan, Y. MPPT Control of Hydraulic Power Take-Off for Wave Energy Converter on Artificial Breakwater. *J. Mar. Sci. Eng.* **2020**, *8*, 304. [CrossRef]
31. Amon, E.A.; Brekken, T.K.A.; Schacher, A.A. Maximum Power Point Tracking for Ocean Wave Energy Conversion. *IEEE Trans. Ind. Appl.* **2012**, *48*, 1079–1086. [CrossRef]
32. Hardy, P.; Cazzolato, B.S.; Ding, B.; Prime, Z. A maximum capture width tracking controller for ocean wave energy converters in irregular waves. *Ocean Eng.* **2016**, *121*, 516–529. [CrossRef]
33. Mendes, R.; Calado, M.D.R.; Mariano, S.J.P.S. Maximum Power Point Tracking for a Point Absorber Device with a Tubular Linear Switched Reluctance Generator. *Energies* **2018**, *11*, 2192. [CrossRef]
34. Ding, B.; Cazzolato, B.S.; Arjomandi, M.; Hardy, P.; Mills, B. Sea-state based maximum power point tracking damping control of a fully submerged oscillating buoy. *Ocean Eng.* **2016**, *126*, 299–312. [CrossRef]
35. Li, W.; Chau, K.T.; Hua, F.; Ching, T.W. Maximum power point tracking control of a linear magnetic-geared generator for direct-drive wave energy conversion. In Proceedings of the 10th International Conference on Advances in Power System Control, Operation & Management (APSCOM 2015), Hong Kong, China, 8–12 November 2015; pp. 1–6.
36. Salter, S.H.; Jeffrey, D.C.; Taylor, J.R.M. The Architecture of Nodding Duck Wave Power Generators. 1976. Available online: https://hdl.handle.net/1842/37407 (accessed on 21 March 2022).
37. Ha, Y.J.; Park, J.Y.; Shin, S.H. Numerical Study of Non-Linear Dynamic Behavior of an Asymmetric Rotor for Wave Energy Converter in Regular Waves. *Processes* **2021**, *9*, 1477. [CrossRef]

Journal of
Marine Science and Engineering

MDPI

Article

An Improved Hydraulic Power Take-Off Unit Based on Dual Fluid Energy Storage for Reducing the Power Fluctuation Problem in the Wave Energy Conversion System

Mohd Afifi Jusoh, Zulkifli Mohd Yusop, Aliashim Albani, Muhamad Zalani Daud and Mohd Zamri Ibrahim *

Renewable Energy & Power Research Interest Group (REPRIG), Eastern Corridor Renewable Energy Special Interest Group, Faculty of Ocean Engineering Technology and Informatics, Universiti Malaysia Terengganu, Kuala Nerus 21030, Terengganu, Malaysia

* Correspondence: zam@umt.edu.my; Tel.: +60-966-833-28

Abstract: The power take-off (PTO) stability is one of the most important concerns for wave energy converters (WECs). The PTO unit converts the mechanical energy produced by the wave absorber (WA) unit into useful electrical energy. Due to the drastic input energy variation of real wave motions, the generated electrical power from the PTO unit significantly fluctuates and is potentially harmful to electrical and electronic appliances. This paper proposes an improved hydraulic PTO (HPTO) for the WECs. An improved HPTO unit comprises a dual high-pressure accumulator (HPA) module and fluid energy control (FEC) module, which significantly enhances the generated electrical power from the generator under irregular wave circumstances. A complete model of wave absorber device with conventional and improved HPTO units was built in MATLAB/Simulink using a Simscape fluids toolbox. The parameters of the FEC control strategy were optimized using a genetic algorithm. The improved HPTO unit model was simulated with five irregular wave inputs to evaluate its performance in irregular conditions. The effects of the HPA pressure constraints on the improved HPTO unit performance were also investigated. Overall, the simulation results indicate that the improved HPTO unit was able to generate a stable power up to 87.3% of WECs in an irregular sea state.

Keywords: wave energy conversion system; hydraulic power take-off; power fluctuation mitigation; control strategy

Citation: Jusoh, M.A.; Yusop, Z.M.; Albani, A.; Daud, M.Z.; Ibrahim, M.Z. An Improved Hydraulic Power Take-Off Unit Based on Dual Fluid Energy Storage for Reducing the Power Fluctuation Problem in the Wave Energy Conversion System. *J. Mar. Sci. Eng.* **2022**, *10*, 1160. https://doi.org/10.3390/jmse10081160

Academic Editors: Giuseppe Giorgi and Mauro Bonfanti

Received: 30 June 2022
Accepted: 16 August 2022
Published: 21 August 2022

1. Introduction

Renewable energy sources have a vast potential to reduce the negative impacts caused by the combustion of fossil fuels. Ocean waves are one of the emerging renewable energy sources that have a great potential to contribute to global green energy production. Over the past several decades, the generation of electricity from the ocean waves has received much interest among researchers [1,2]. As a result, several wave energy converter system (WECs) concepts through various harnessing techniques have been designed, developed, tested, and patented worldwide [3–5]. These existing WECs can be categorized according to their operational principles, such as overtopping, oscillating water column, and wave-activated-body (WAB) devices [6–8].

Basically, WECs are made up of three primary parts: the wave absorber (WA) unit, power take-off (PTO) unit, and control system (CS) unit. The WA unit is a front-end device that captures energy from ocean wave motions. The PTO unit is a device that converts the mechanical energy produced by the WA into useful electrical energy. The CS unit is used to monitor and control the operation of the WA and PTO units. The PTO unit is one of the most important element of the WECs. The performances of the PTO unit directly correlate to the amount of generated electricity from the absorbed ocean wave energy. In addition, the size and mass of the PTO unit can directly contribute to the overall design and behavior of the WA unit. From the literature, the various concepts of the PTO unit have been invented for different types of WA [9,10]. These PTO concepts can generally be classified into four main categories: direct-mechanical drive, direct-electrical drive, direct-turbine drive, and mechanical-hydraulic drive [10,11]. From the recent study by the authors in [10], the hydraulic PTO (HPTO) is among the most efficient PTO for various kinds of WA devices due to its significant advantages, such as being well-adapted to any ocean wave condition, high efficiency, and high controllability [12]. The HPTO unit is easy to build using standard hydraulic components that are widely applied in different hydraulic system applications [13].

From the literature, many kinds of HPTO concepts have been developed for different types of WECs [9,10,14]. Commonly, the HPTO unit combines several modules, such as actuation, rectification, fluid energy storage, power generation, and reservoir [10]. The actuation module, which commonly comprises single or multiple hydraulic actuators, is used to absorb the mechanical energy produced from the interaction of the ocean waves and WA device into fluid energy. From previous studies, several different types of hydraulic actuators, such as single-acting cylinder [15], double-acting cylinder [16,17], and multi-chamber cylinder [18–20], have been utilized in the actuation module of the HPTO unit. Aside from that, the rectification module is used to ensure the unidirectional flow through the generation module. The configuration of the rectification module directly depends on the actuation module configuration. Normally, the rectification module is a combination of two or four non-return valves, and its structure depends on the configuration of the hydraulic actuation module [10]. In a few studies [18–21], the rectification modules based on the combination of some electronic directional valves were proposed to increase the efficiency of the HPTO unit. Moreover, the fluid energy storage module consists of a single or multi hydraulic accumulator. The fluid energy storage module is used to smooth the overall pressure and prevent cavitation incidents in the HPTO unit. Finally, the power generation module converts the fluid energy to usable electrical energy. The power generation module combines the hydraulic motor and electrical generator.

The electrical power generated from the WECs majorly depends on the efficiency of the HPTO unit. According to [22], two major issues of the HPTO unit need to be emphasized in order to improve the efficiency of the HPTO unit, as follows: (1) an appropriate selection of the HPTO unit architecture and (2) the accurate selection of the parameters of the HPTO unit components. Several works have been conducted by researchers to increase the efficiency of the HPTO unit by optimizing the architecture of the HPTO unit. In [20], the use of a multi-chamber or hydraulic actuator in the actuation module and the digital rectification module were proposed in the HPTO unit architecture. In [23], several incremental modifications to the HPTO unit architecture were carried out, such as adding more accumulators in the energy storage module, using two kinds of rectification modules, and adding bypass circuits in the generation module. In [24,25], the hydraulic transformer unit was applied in the generation module to solve the variable pressure of the HPTO unit. In [26], several numbers of hydraulic accumulators have been added to the energy storage module of the HPTO unit, and the experiment result shows the HPTO unit was able to generate electricity at smaller wave conditions. In [27], multi hydraulic actuators in the actuator module sharing a single generation module were presented, and the experiment results indicated that the WECs could generate more stable energy.

Moreover, the HPTO unit efficiency can also be enhanced by optimizing its important parameters. In [28–33], several important HPTO unit parameters were identified, including the hydraulic actuator size, hydraulic accumulator capacity, hydraulic motor displacement, etc. In [28–30], the influences of the important HPTO unit parameters on the WECs effectiveness in various sea conditions were investigated. The overall result indicates that these important parameters highly influence the efficiency of the HPTO unit. For this reason, several studies have been conducted to enhance the efficiency of the HPTO unit by optimizing its important parameters [22,34–38]. Authors in [22] employed the genetic algorithm (GA) and non-linear programming by quadratic Lagrangian (NLPQL) for determining the optimal parameters of the HPTO unit. While in [38], the HPTO unit optimization was conducted for different sea states using the optimization toolbox in MATLAB software. Similarly in [36], the optimization of the hydraulic energy storage module was conducted, and the result indicated that optimization of the hydraulic energy storage module helps to increase the system efficiency and the electrical energy generated from the HPTO unit.

Apart from that, the voltage fluctuations generated from the electrical sources are potentially harmful to electrical and electronic appliances [39,40]. Thus, the stability of the electrical energy generated from the HPTO unit is a crucial issue that should be emphasized; however, it is challenging to stabilize the electrical energy generated from the HPTO unit. The instability of the generated electrical energy from the HPTO unit is due to the unstable speed of the electrical generator and the hydraulic motor, which depends on the internal pressure of the HPTO unit. Authors in [10], suggest that the HPTO unit pressure can be regulated by controlling several components of the HPTO unit, such as hydraulic motor, hydraulic accumulator, hydraulic rectifier, battery energy storage, etc. Several solutions have been proposed to improve the stability of the generated electrical energy from the HPTO unit as addressed in [16,41–48]. In [41], the control strategy based on the fuzzy controller was designed to regulate the displacement of the hydraulic motor. The simulation results for the irregular wave condition verified that the proposed control strategy could mitigate the fluctuations in the output of the HPTO unit; however, the variable-displacement hydraulic motor requires more expense and more attention than the fixed-displacement hydraulic motor, as reported in [10]. In [42], a control strategy for the HPTO unit was proposed to realize the flexible rectification and hydraulic motor speed control simultaneously. From the results, the control strategy was able to accelerate the hydraulic motor slowly during the starting operation of the HPTO unit to reduce the negative impact on the power grid; however, the regular wave input was in used instead of irregular wave input in [42]; thus, the efficiency of the control strategy in irregular wave condition was unidentified.

The primary goal of this paper is to improve the stability of the generated electrical power output from the HPTO unit of WECs. The main contributions of the present paper are given as follows: (1) a novel fluid energy control (FEC) to regulate the flow of fluid entering the hydraulic motor is proposed; (2) a novel control strategy for FEC module for improving the stability of the generated electrical power from the HPTO unit is proposed. This paper is organized in four sections. Section 2 presents the technical descriptions of the considered WECs, including the configuration and the operation of the improved HPTO unit. Section 3 describes the dynamic modeling of the WA device and improved HPTO unit. The findings and discussions are presented in Section 4. Finally, Section 5 discusses the conclusion and future work.

2. WECs Concept

A similar concept of WECs in [22] was considered in the present study; however, several modifications to the HPTO unit are proposed to improve the quality of the electrical power generated from the WECs. In this section, the descriptions of WA device are firstly presented. Then, the configuration of the improved HPTO unit concept are described.

2.1. Descriptions of WA Concept

The simplified layout of the considered WECs in the present study is given in Figure 1. From the figure, the WA device consists of a floater and hingeable arm attached to automatic floater level control and the static platform facing the dominant ocean wave direction. An automatic floater level control is designed in the wave absorber device to ensure the wave absorber device can operates at any sea level. Thus, the effect of the sea tide level can be significantly reduced. The WA device is then connected to the HPTO and the CS units, separately located in the HPTO house through a hydraulic actuator (HA) module. A single-rod double-acting type of hydraulic cylinder is utilized as a hydraulic actuator to ensure the mechanical energy produced during the bi-directional pitch motion of the

WA device can be maximally absorbed, as suggested in [10]. The hydraulic actuator barrel is attached to the fixed platform, while the piston rod is connected to the floater's arm. The hydraulic actuator is then connected to the rest of the HPTO unit components using two hydraulic hoses. The working principle of the considered WA device is almost similar to the WA device concept used in [19,20]. The passing ocean waves through the floater initially cause the WA device to pitch up and down. The WA device directly drives the hydraulic actuator and operates the HPTO unit.

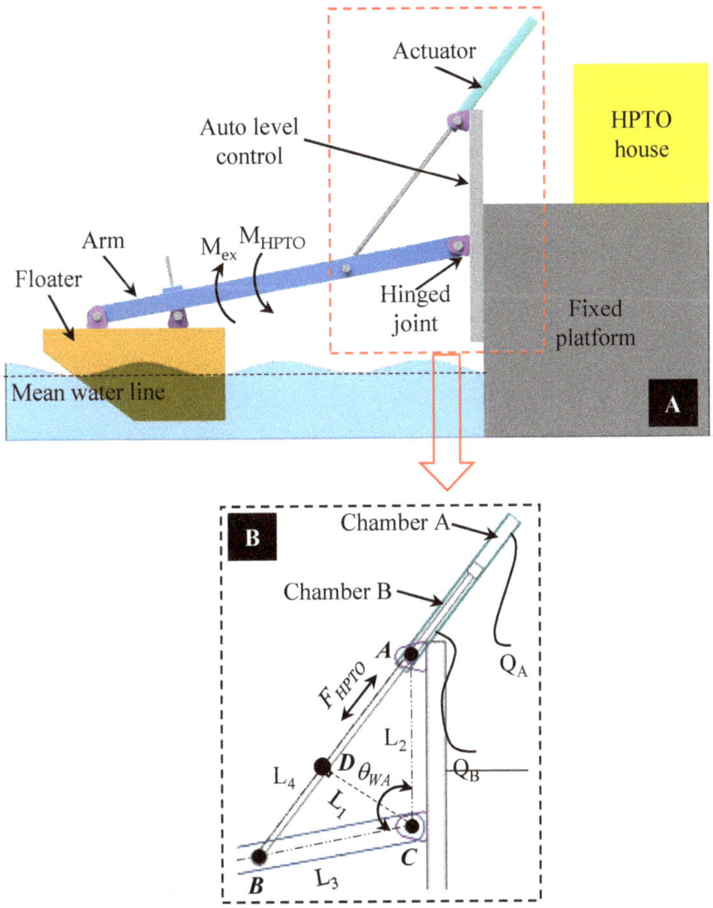

Figure 1. Schematic diagram of the considered WECs. (**A**) A complete system diagram, and (**B**) the enlarged diagram of the interconnection between the WA device and the HPTO unit.

2.2. Descriptions of the Improved HPTO Concept

Figure 2 illustrates the simplification of the conventional HPTO concept previously was used in [22] and the improved HPTO unit concept proposed in the present study. In the conventional HPTO concept, the hydraulic actuator module is connected to the rectification module, which consists of the four non-return check valves (CV_1, CV_2, CV_3, and CV_4) through hydraulic hoses. By relating Figures 1B and 2A, the CV_1 inlet and CV_4 outlet terminals are connected to the top side of the hydraulic actuator barrel. The inlet and outlet terminals of CV_3 and CV_2 are connected to the bottom side of the hydraulic actuator barrel. The rectification module is then connected to the generation module, which includes a fixed displacement hydraulic motor (HM) coupled with the permanent magnet synchronous generator (G). The fluid energy storage module comprises two bladder-type hydraulic accumulators, namely a high-pressure accumulator (HPA) and a low-pressure accumulator (LPA), placed between the rectification module and generation module. Additionally, the pressure relief valve (RV) is installed on the high-pressure line to avoid the over-pressurized event in the HPTO unit.

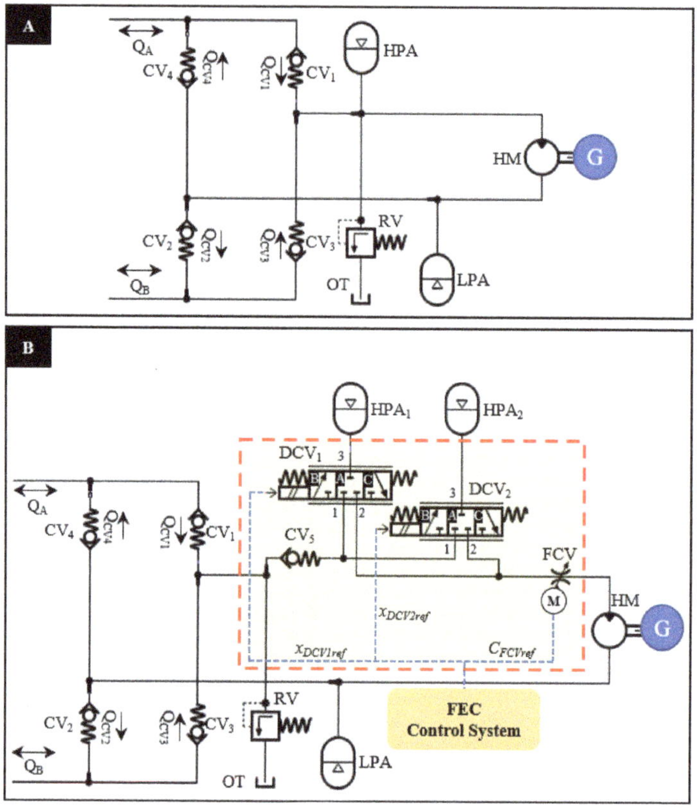

Figure 2. Schematic diagram of HPTO unit concepts. (**A**) Conventional HPTO unit from [22] and (**B**) an improved HPTO unit proposed in the present study.

During the operation, the back-and-forth motions of the hydraulic actuator due to the WA device motion generate high-pressure fluid in the hydraulic actuator barrel. The high-pressure fluid from the top side of the hydraulic actuator barrel flows through CV_1, HPA, HM, LPA, and CV_2 to the bottom side of the hydraulic actuator barrel during the upward pitch motion. Conversely, during the downward pitch motion, the high-pressure fluid from the bottom side of the hydraulic actuator barrel flows through CV_3, HPA, HM, LPA, and CV_4 to the top side of the hydraulic actuator barrel. Finally, the high-pressure fluid flowing through the hydraulic motor forces the hydraulic motor and generator to rotate synchronously in the same direction, resulting in useful electrical power; thus, the generated electricity from the generator significantly depends on the hydraulic motor speed, which is subjected to the ocean wave input characteristics.

In [22], the operation of the conventional HPTO unit was optimized using a genetic algorithm. The optimization process was intended to maximize the generated electrical power from the HPTO unit. Seven important parameters of the conventional HPTO unit, for example, hydraulic motor displacement, hydraulic cylinder size, hydraulic accumulator capacity, etc., were considered during optimization process. Overall, the findings demonstrated that the conventional HPTO unit could generate electrical power reached up to 62% of the rated capacity under irregular wave circumstances; however, the generated electrical power from the conventional HPTO unit highly fluctuated under irregular wave circumstances; thus, as suggested by the authors, some improvements should be implemented on the conventional HPTO unit to ensure it can generate a better quality of electrical power at the maximum capability.

The output power of the generator, as well as the frequency of the electricity, can be maintained if the hydraulic motor speed remains consistent. For the variable-displacement hydraulic motor case, the hydraulic motor speed can be controlled either by adjusting the volumetric displacement or the volumetric flow rate of the hydraulic motor; however, the variable-displacement hydraulic motor requires more expense and more attention than the fixed-displacement hydraulic motor [10]. Since the fixed-displacement hydraulic motor was considered in the conventional HPTO concept, the hydraulic motor speed only can be controlled by adjusting the volumetric flow rate of the hydraulic motor. To reduce the hydraulic motor speed, mechanical fluctuation, and electrical power fluctuation, several modifications to the conventional HPTO concept were proposed, as presented in Figure 2B. First, a dual HPA module is used instead of a single HPA module was used in the conventional HPTO unit. Second, the FEC module is added to provide a less-fluctuated and more constant fluid flow to the hydraulic motor by managing the charge/discharge processes of the dual HPA module and controlling the volumetric flow rate.

In the modified HPTO unit, the dual HPA module is used to eliminate the fluctuations of the hydraulic oil flow to the hydraulic motor. HPA_1 is used as a primary fluid energy storage, while HPA_2 is used as a backup fluid energy storage. As indicated in Figure 2B, the HPA_1 and HPA_2 with similar volume capacity are connected in parallel to the FEC module. FEC module consists of a hydraulic check valve (CV_5), 3-way direction control valves (DCV_1 and DCV_2), an electrical motorized flow control valve (FCV), a control strategy, and other electronic components. CV_5 is added at the charging line to prevent the reverse hydraulic oil flow from the HPA_1 and HPA_2 during their high-pressure state. DCV_1 and DCV_2m which is operated by the electric solenoid valve are used to control the charging and discharging processes of the HPA_1 and HPA_2 by regulating their hydraulic flows directions. DCV_1 and DCV_2 are composed of three ports and three blocks. Ports 1 and 2 of DCV_1 and DCV_2 are connected to the charging and discharging lines. Ports 3 of DCV_1 and DCV_2 are connected to the HPA_1 and HPA_2, respectively. In addition, FCV is used to control the volumetric flow rate from the HPA_1 and HPA_2 to hydraulic motor. In the improved HPTO unit model, the FEC module was assumed to be powered by external power supply, such as solar power or local power sources.

The overall operation of the FEC module is driven by a proposed control strategy, as depicted in Figure 3. The control strategy is divided into two parts. Part A is designed to control the charging and discharging process of the dual HPA module within the pressure constraint. The minimum and maximum hydraulic oil pressure constraints of HPA_1 and HPA_2 are set to 47 Bar and 48.5 Bar, respectively. The feedback output of the HPA_1 pressure is used as input for the control strategy in part A. The reference positions of DCV_1 and DCV_2 ($x_{DCV1,ref}$ and $x_{DCV2,ref}$) are the outputs from the control strategy in part A. Part B of the control strategy is designed to regulate the volumetric flow from HPA_1 and HPA_2 to the hydraulic motor. Figure 4 presents a closed-loop PID controller that is used in part B of the control strategy. The generator rated power ($P_{G,Set}$) and the instantaneous electrical power generated output (P_G) are the inputs of the closed-loop PID controller. In this case, the $P_{G,Set}$ is set to 100 W. A proportional–integral–derivative (PID) controller automatically adjusts the FCV flow coefficient (C_{FCV}) with respect to $P_{G,Set}$. $C_{FCV,ref}$ is the output from the closed-loop PID controller.

The operation of the FEC module is started with analyzing the pressure state of HPA_1 (P_{HPA1}), whether it reaches the lowest ($P_{HPA1,min}$) or highest ($P_{HPA1,max}$) of considered constraints. When P_{HPA1} reaches $P_{HPA1,max}$, the FEC module controller automatically regulates DCV_1 and DCV_2 to positions C and B in order to change HPA_1 and HPA_2 to discharge and charge mode. Conversely, the FEC module controller automatically regulates DCV_1 and DCV_2 to positions B and C once P_{HPA1} reaches $P_{HPA1,min}$ in order to change

HPA_1 and HPA_2 to charge and discharge mode. During P_{HPA1} is within the constraint, DCV_1 and DCV_2 positions remain as previously mentioned. The charging and discharging processes of HPA_1 and HPA_2 are completed once P_{HPA1} reaches $P_{HPA1,min}$ and $P_{HPA1,max}$, respectively. Finally, the PID controller automatically regulates $C_{FCV,ref}$ according to the deviation of $P_{G,Set}$ and P_G. The overall FEC module process iteratively loops until the end of the HPTO unit operation.

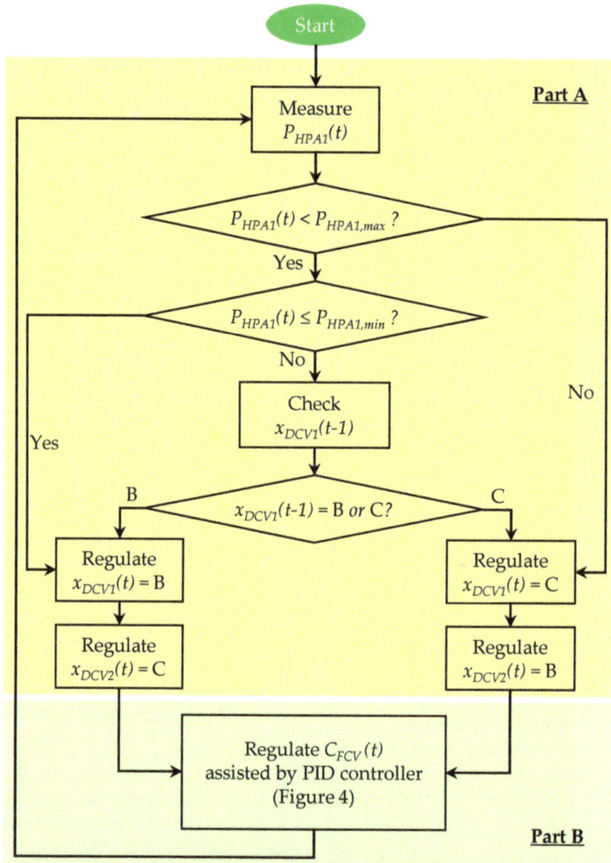

Figure 3. A proposed control strategy of FEC module.

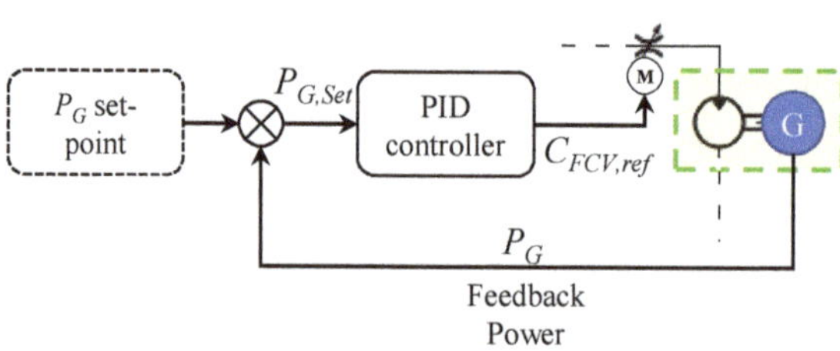

Figure 4. A closed-loop PID controller for flow control valve.

3. Wave-to-Wire Model of the Considered WECs

3.1. Irregular Wave Inputs

In the present study, five different irregular wave profiles are created in order to evaluate the robustness of the proposed control strategy in mitigating the power fluctuations of HPTO unit. The different irregular wave profiles are created based on the Pierson–Moskowitz spectrum using AQWA toolbox in ANSYS (Pennsylvania, United States). The characteristics of five different irregular wave profiles with different H_s and T_p are summarized in Table 1. Sea state A is considered as the nominal wave condition. Sea states B and C are used for the different H_s cases, while sea states D and E are considered for the different T_p cases. For sea state A, the values of H_s and T_p are set to 0.8 m and 4.5 s. For sea states B and C, the H_s is set to 0.6 m and 1.0 m (±0.2 m of nominal H_s), while T_p is maintained at the nominal value for both cases. Meanwhile, for sea states D and E, H_s is maintained at the nominal value while T_p is set to 2.5 s and 6.5 s (±2.0 s of nominal T_p). The wave profiles of sea states A to E are generated for a 500 s time period. The wave spectra and the example of generated irregular wave profiles for 50 s time period are presented in Figures 5 and 6.

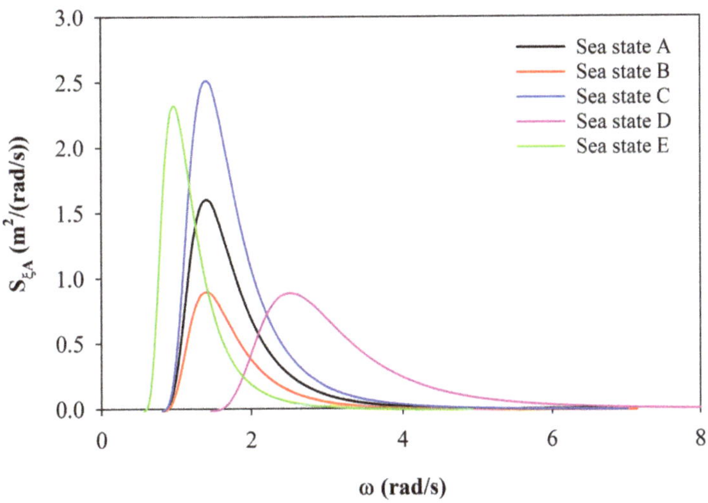

Figure 5. Wave spectra for different sea states.

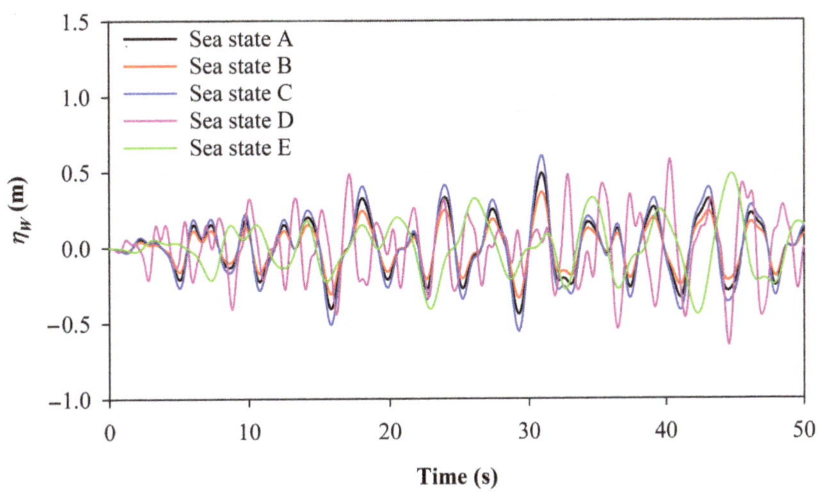

Figure 6. Examples of five different irregular wave profiles for 50 s time period.

Table 1. The characteristics of five different irregular wave profiles.

Sea State	H_s	T_p	Remarks
A	0.8	4.5	Nominal H_s & T_p
B	0.6	4.5	Lower H_s
C	1.0		Larger H_s
D	0.8	2.5	Lower T_p
E		6.5	Larger T_p

3.2. Dynamic Model of the WA Device

The hydrodynamic pitch motion of the WA device in the actual ocean waves are described based on the time-domain equation as given in Equation (1) [19,38]. J_{WA} and $J_{add,\infty}$ refers to the inertia moment of the WA device and the added mass at the infinite frequency. h_{ex} is the excitation force coefficient, k_{rad} is the radiation impulse response function coefficient, and k_{res} is the hydrostatic restoring coefficient. $\ddot{\theta}_{WA}$, $\dot{\theta}_{WA}$ and θ_{WA} are the instantaneous angular acceleration, velocity, and position of the WA device during the pitch motion, respectively. In the present study, the hydrodynamic diffraction analysis using ANSYS/AQWA computational fluid dynamics (CFD) software was used to determine the h_{ex}, k_{rad}, and k_{res} coefficients, as previously implemented in [49–51]. The results from the preliminary hydrodynamic diffraction simulation are presented in Figure 7.

$$(J_{WA} + J_{add,\infty})\ddot{\theta}_{WA}(t) + \int_0^t k_{rad}(t-\tau)\dot{\theta}_{WA}(t) + k_{res}\theta_{WA}(t) + M_{HPTO}(t) = \int_{-\infty}^{\infty} h_{ex}(t-\tau)\eta_W \, d\tau \tag{1}$$

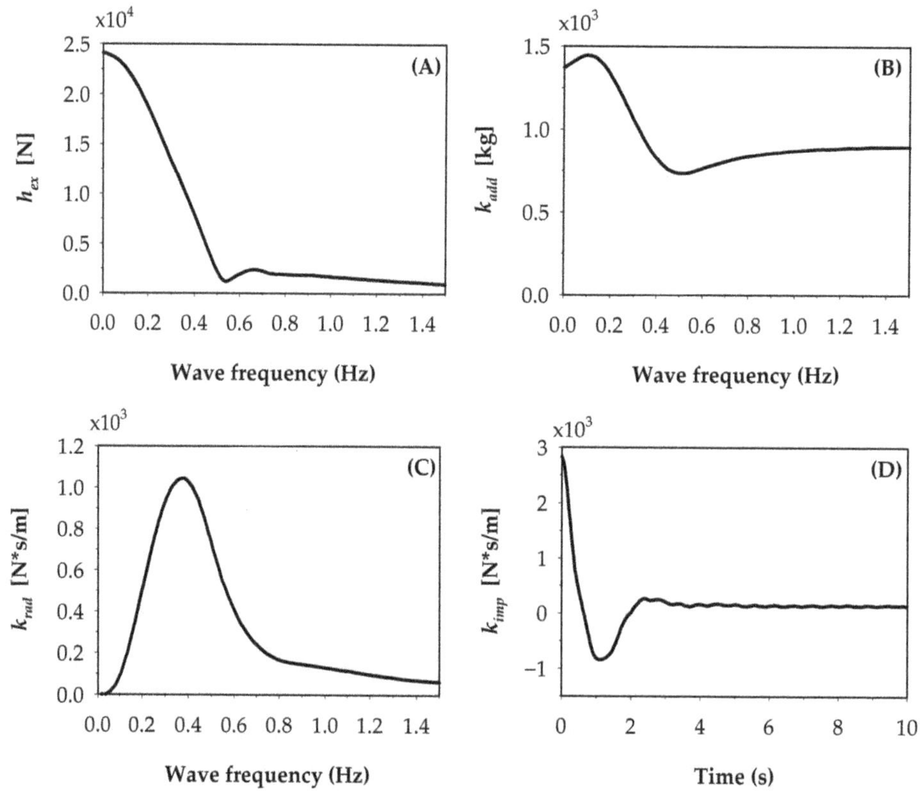

Figure 7. Hydrodynamic parameters of wave absorber device from preliminary simulation in ANSYS/AQWA. (**A**) Excitation force coefficient; (**B**) Added mass coefficient; (**C**) Radiation damping coefficient; (**D**) Impulse response function.

Meanwhile, the moment of due to the HPTO unit force (M_{HPTO}) in Equation (1) can be calculated using Equation (2), where F_{HPTO} refers to the feedback force from the hydraulic actuation module applied to the WA device during the pitch motion, as depicted in Figure 1. L_1 is the perpendicular length from point C to the hydraulic actuator rod. L_1 can be calculated using Equation (3), where L_2, L_3 and L_4 are the length from points A to C and B to C and A to B, respectively. $\theta_{WA,0}$ is the angular position of the WA device during the default position. x_{HA} refers to the linear displacement of the hydraulic actuator, which can be described using Equation (4).

$$M_{HPTO} = F_{HPTO} L_1 \tag{2}$$

$$L_1 = \frac{L_2 L_3 sin(\theta_{WA,0} - \theta_{WA})}{L_4 + x_{HA}} \tag{3}$$

$$x_{HA} = L_4 - \sqrt{L_2^2 + L_3^2 - 2L_2 L_3 cos(\theta_{WA,0} - \theta_{WA})} \tag{4}$$

The WA device submodel was built based on the hydrodynamic pitch motion functions in Equations (1)–(4) using a mathematical function block, as depicted in Figure 8A. The inputs of the WA device submodel are the irregular wave profile (η_W), and the feedback force from the hydraulic actuation module of the HPTO unit, F_{HPTO}. The linear displacement of the hydraulic actuator, x_{HA} is the output of the WA device submodel. The hydrodynamic coefficients of the WA device, such as h_{ex}, k_{rad}, and k_{res} were determined from the hydrodynamic diffraction analysis conducted using ANSYS/AQWA software (version 2019 R3).

3.3. HPTO Unit Models

The HPTO unit submodels were built using the Simscape Fluids toolbox provided in the MATLAB/Simulink based on the conventional and improved concepts in Figure 2A,B. The HPTO unit submodels were built using the available hydraulic components in the Simscape Fluids toolbox. The detail descriptions of each components in the Simscape Fluids can be obtained in [52]. For hydraulic actuation module, a single rod double chamber with a fixed body actuator was used. This actuator model includes pressure dynamics in the volumes either side of the piston and viscous friction. The piston diameter, rod diameter, stroke length, and initial piston displacement are some examples of the important parameters used in the hydraulic actuator. The ideal translational velocity source block was used to convert x_{HA} input from the WA device submodel into velocity dimension ($\dot{x}_{HA}$).

For rectification module, a combinations of four spring check valve blocks (CV_1 to CV_4) were used. In these check valve models, a functional hysteresis was specified to the model in order to take into account dry friction effects. The maximum cross-sectional area and flow discharge coefficient are the main parameters of the check valve model. For the fluid energy storage, the blocks model of a gas-charged accumulator (HPA_1, HPA_2 and LPA) were used. The accumulator blocks consists of a precharged gas chamber and a fluid chamber. The accumulators gas were assumed to obey an adiabatic polytropic gas law, where the hydraulic fluid within the accumulators were assumed to have the same pressure as the gas. The gas precharge pressure, accumulator volume, and adiabatic index are the vital parameters of the accumulator blocks.

For the generation module, the fixed displacement hydraulic motor block was coupled with a rotational damper with a variable damping. A rotational damper with a variable damping was used to replicate the electric generator behavior of the generation module in order to reduce the HPTO submodel complexity. The damping coefficient of the considered generator (d_G) was used in the rotational damper to represent the actual resistive torque of the electric generator imposed on the hydraulic motor. The instantaneous electrical power produced by the generator (P_G) was calculated using Equation (5), where ω_G is generator speed, τ_G is generator torque and η_G is generator efficiency.

$$P_G = 2\pi\omega_G \tau_G \eta_G \tag{5}$$

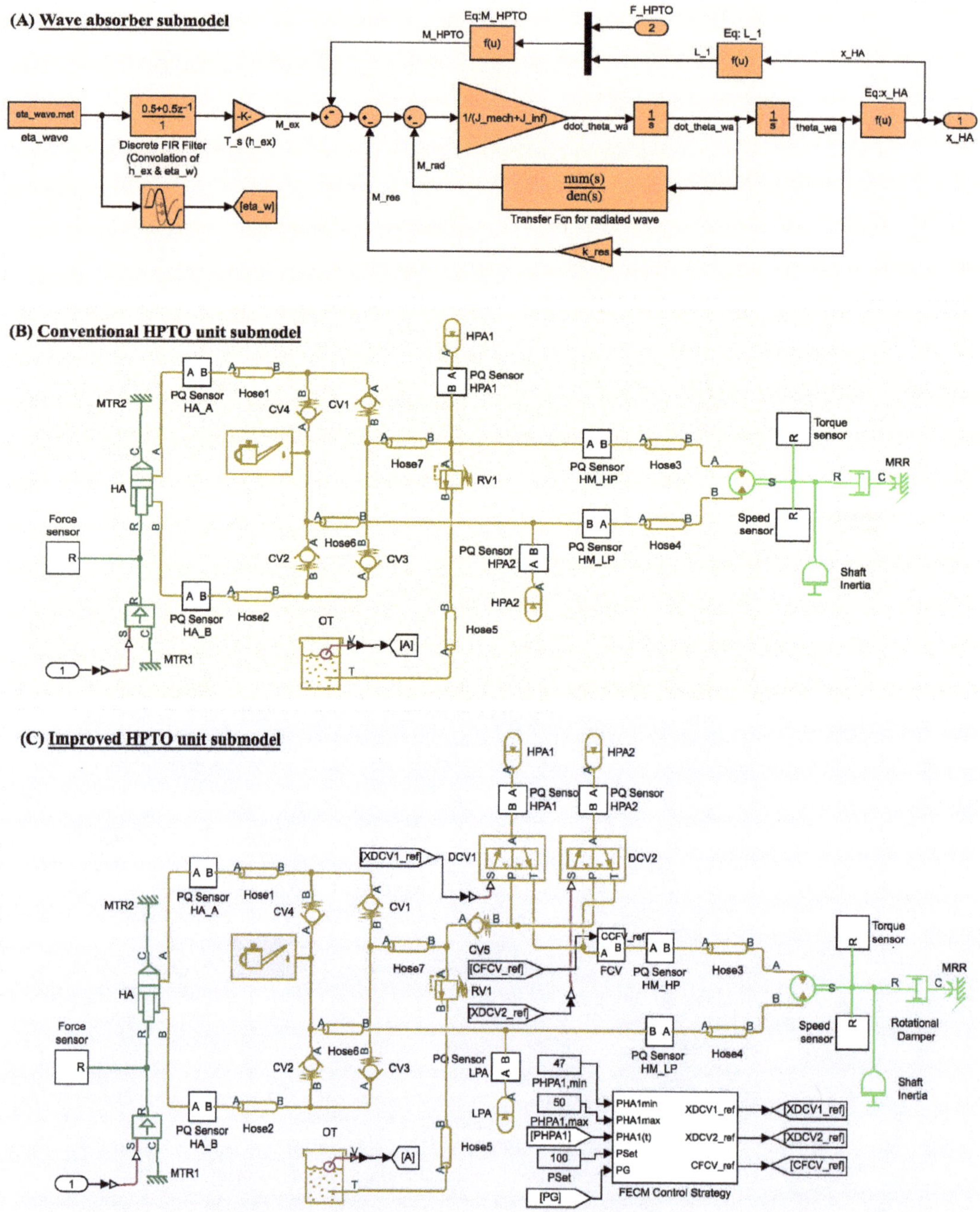

Figure 8. A developed WECs simulation model in MATLAB/Simulink. (**A**) WA device submodel; (**B**) Conventional HPTO unit submodel; (**C**) Improved HPTO unit submodel.

For the improved HPTO unit model, the FEC module was modeled using 3-way directional valve (DCV_1 and DCV_2) and variable orifice valve (FCV) blocks. The 3-way directional valve block represents a directional control valve with three ports and two positions, or flow paths. The dynamic behavior of the 3-way directional valve blocks were set based on the second-order system between input signal and valve opening. The model parameterization of DCV and FCV blocks were set based on the maximum cross-sectional area and valve opening factor. The detail descriptions of the DCV and FCV blocks parameters setting can be obtained in [52]. The flow charge coefficient, leakage area, and maximum opening area are the important parameters of the DCV and FCV blocks. Furthermore, the control strategy of the FEC module was modeled using mathematical blocks and PID controller block based on the concept in Figure 3.

In addition, the hydraulic pipeline block was used to connect all the modules in the HTPO unit models. The hydraulic pipeline block accounts for friction loss along the pipe length and for fluid compressibility. In addition, several sensors such as force, pressure, flow rate, etc., were included in both HPTO unit models. Finally, all the parameters setting of both HPTO unit models are given in the Table 2.

Table 2. HPTO models parameters.

Descriptions	Value	Unit
Actuation module		
Piston diameter	0.035	m
Rod diameter	0.022	m
Stroke length	0.3	m
Initial displacement of piston	0.15	m
Rectification module		
Flow discharge coefficient	0.75	-
Maximum cross-sectional area	0.003	m^2
Fluid energy storage module		
HPA_1 & HPA_2 pre-charge pressure	47	bar
LPA pre-charge pressure	3.2	bar
HPA_1 & HPA_2 volume capacity	2.8	L
LPA volume capacity	4	L
Adiabatic index	1.4	-
Generation module		
Rated power of generator	100	W
Rated speed of generator	200	rpm
Rated torque of generator	6	Nm
Damping coefficient of generator	0.024	Nm/rpm
Efficiency of the generator	0.95	-
Volumetric displacement of hydraulic motor	8	cc/rev
Volumetric efficiency of hydraulic motor	0.95	-
Mechanical efficiency of hydraulic motor	0.95	-
Fluid energy control module *		
Flow discharge coefficient of DCV_1, DCV_2 & FCV	0.7	-
Leakage area of DCV_1, DCV_2 & FCV	1×10^{-12}	m^2
Maximum opening area of DCV_1 & DCV_2 & FCV	0.5×10^{-4}	m^2

* Applicable for an improved HPTO unit only.

3.4. Determination of Optimal Parameters of FEC Control Strategy Using Genetic Algorithm

A FEC module control strategy consists of four important parameters, such as proportional gain (k_p), integral gain (k_i), derivative gain (k_d), and $P_{HPA1,max}$. In this study, the control strategy parameters were preliminarily determined using a genetic algorithm. Figure 9 illustrates the generalized simulation set-up diagram for parametric estimation using the genetic algorithm. In this estimation process, the genetic algorithm program based on Figure 10 was separately built in the MATLAB/M-file. The parameter settings of the genetic algorithm are gathered in Table 3. The genetic algorithm in the MATLAB/M-file was then linked to the WECs model in MATLAB/Simulink. The irregular wave profile based on the nominal condition (sea state A) was used as the wave input of WECs model. A specific objective function ($OF(x_{CS})$) was created for genetic algorithm based on the FEC module control strategy in Figure 3. $OF(x_{CS})$ can be described using Equations (6)–(8). $MAAPD$ is the moving average of the absolute power deviation (PD). T_s is the simulation period. Using $OF(x_{CS})$, the P_G generated from the HPTO unit is maximized by genetic algorithm through selecting the best values for the FEC module control strategy parameters.

The parameters determination process using genetic algorithm was initially started by randomly generating a population of chromosomes for k_p, k_i, k_d, and $P_{HPA1,max}$. Thereafter, for the first iteration, the random values from the generated population were chosen for each study parameter. The WECs model was then evaluated based on the $OF(x_{CS})$. The chromosomes of the population were then sorted according to the least cost or highest fitness. Some percentages of the best chromosomes were transferred directly to the next generation based on their merit. Then, three genetic algorithm operators named as selection, crossover, and mutation were implemented to manipulate the rest of the chromosomes for the next generation. During the selection rule, the parent's chromosome that contributed to the current population was selected for the next generation process. Then, pairs of selected parents were recombined by a crossover operator to produce new chromosomes. A mutation rule was then applied to the new chromosomes to avoid the genetic algorithm converging to the local optimum. This process was iterated until the satisfactory fitness level was reached.

Table 3. Parameters setting of parameters optimization using genetic algorithm.

Descriptions		**Value**	
Genetic algorithm setting			
Population size		100	
Reproduction ratio (%)		80	
Maximum number of generations		20	
Mutation probability (%)		10	
Mutation amplitude		0.2	
Seed		1	
Final accuracy		0.01	
Parameters constraints	**Min**		**Max**
k_p	0.01		1
k_i	0.01		1
k_d	0.01		1
$P_{HPA1,max}$	47		52

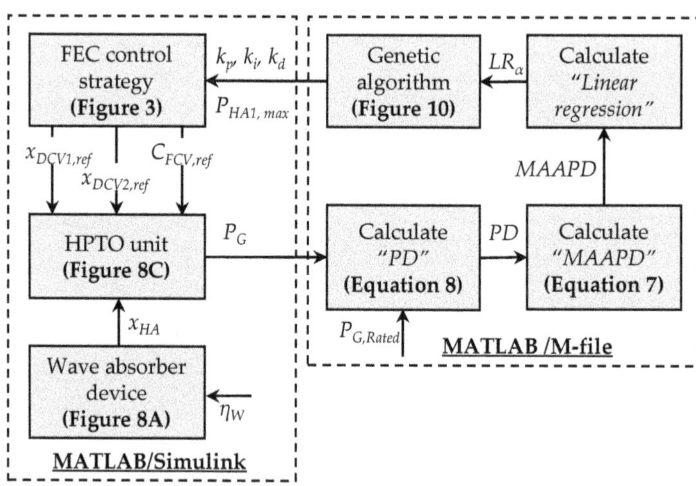

Figure 9. Illustration of simulation set-up for optimization of control strategy using genetic algorithm in MATLAB software.

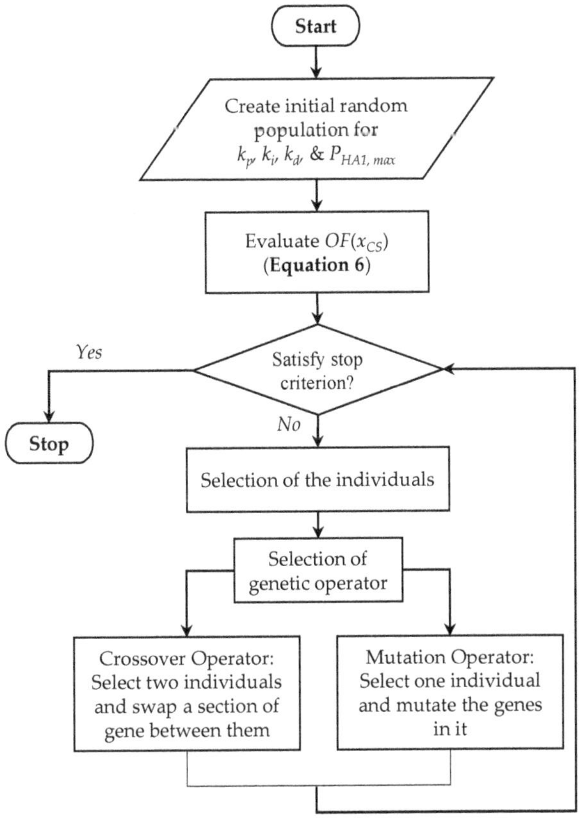

Figure 10. Procedure of parameters optimization using genetic algorithm.

$$OF(x_{CS}) = \min(LR_\alpha) \tag{6}$$

$$MAAPD = \frac{1}{T_s}\int_0^t |PD(t)|\ dt \tag{7}$$

$$PD = P_{G,Rated} - P_G \tag{8}$$

3.5. Simulation Analysis of the Improved HPTO Unit

The simulation analysis of the improved HPTO unit consists of two main parts. The first part of the simulation study was aimed to compare the performance of the conventional and improved HPTO unit under irregular wave circumstances; thus, the conventional and improved HPTO unit models were simulated using five different sea state inputs with different significant wave heights (H_s) and the peak wave periods (T_p). For the second part, the simulation study was aimed to investigate the effects of the $P_{HPA1,max}$ on the generated electrical power from the improved HPTO units performance; thus, the simulation of the improved HPTO unit model was repeatedly performed by adjusting the $P_{HPA1,max}$ value from 47.5 to 51.5 bar, respectively. In addition, three evaluation criteria, as derived in Equations (7)–(9), were considered to evaluate the generated electrical power in the simulation studies in part 1 and 2. N_{PD} in Equation (9) represents the number of PD at a certain level.

$$OP_{PD} = \frac{N_{PD}}{\sum_0^{T_s} N_{PD}} \times 100 \tag{9}$$

4. Results and Discussions

The simulation results are presented and discussed in this section: Section 4.1 presents the optimal parameters of FEC control strategy obtained using genetic algorithm. Sections 4.2 and 4.3 present the evaluation of an improved HPTO unit under different irregular sea states. Section 4.4 describes the effect of the $P_{HPA1,max}$ on the improved HPTO unit performance.

4.1. Optimal Parameters of FEC Control Strategy

Figure 11 depicts the chronological variation of the best variable for objective function, PID, and $P_{HPA1,max}$ parameters with the respect to the number of generations obtained from the optimization process performed by genetic algorithm. From the figure, the red vertical line indicates the termination point of the control strategy parameters optimization process by the genetic algorithm. Figure 11A shows that the overall optimization process of the best control strategy parameters was completed after 15 number of iterations once the genetic algorithm had satisfied its accuracy requirement. The overall parameters optimization process by genetic algorithm was carried out for 7 h 30 min. The lowest objective function was found equal to 0.05601, as depicted in Figure 11A. Figure 11B illustrates the chronological variation of the best variable for PID controller parameters throughout the optimization process. From the figure, the best values for PID controller, k_p, k_i, and k_d, were found to be equal to 0.989, 0.869, and 0.112, respectively. Figure 11C depicts the chronological variation of the best variable for $P_{HPA1,max}$ parameter and the optimal $P_{HPA1,max}$ was equal to 48.5 bar.

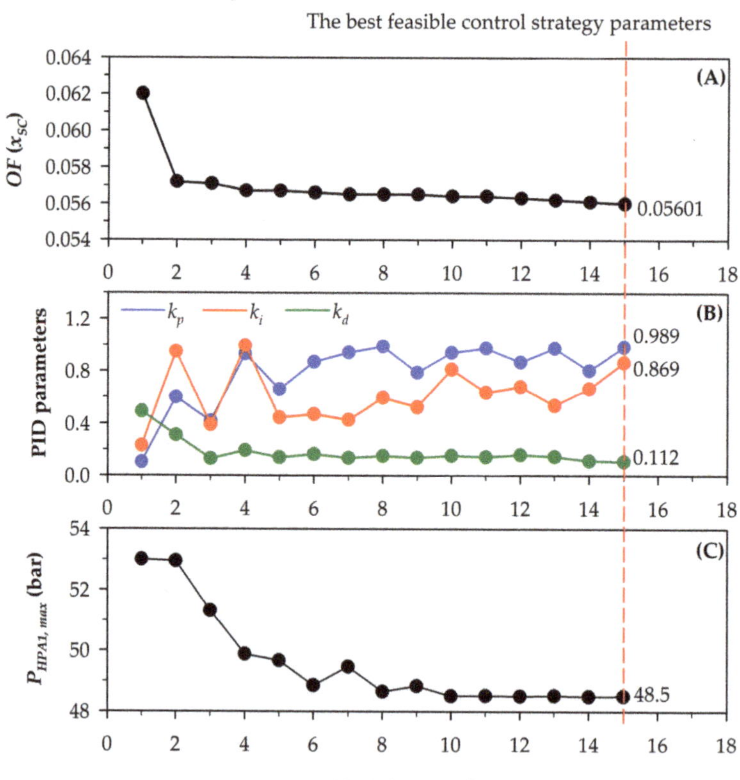

Figure 11. Chronological variation of the best value obtained in each generation for (**A**) objective function, (**B**) PID controller, and (**C**) $P_{HPA1,max}$ parameters.

4.2. Evaluation of an Improved HPTO Unit Nominal Sea State (Sea State A)

This section presents the comparison of the conventional and improved HPTO unit performances in the nominal sea state (sea state A). The performance evaluations were conducted by comparing the quality and the amount of the generated electrical power, P_G from both HPTO unit. Figure 12 presents the profiles of the electrical power generated from the conventional ($P_{G,conv.}$) and improved ($P_{G,imp.}$) HPTO units for a 150 s of a 500 s total time simulation period (for better visualization). The figure shows that the $P_{G,conv.}$ frequently fluctuated below the rated capacity (100 W) and potentially fluctuates within a second. The fluctuation of the generated electrical power is due to the irregular behavior of the ocean wave input, which produces inconsistent high-pressure flow to the power generation module.

Technically, the fluctuation occurrences in the mechanical and electrical side of the HPTO unit may produce several negative impacts. For the mechanical side, the fluctuation of high-pressure flow may cause sudden vibrations on HPTO unit components and directly produce unnecessary noise and heat. The vibration also may increase the leakage potential on the components involved [10]. For the electrical side, the fluctuation of the generated electrical power from the HPTO unit may cause a disastrous impact on the electrical appliances. The lower voltage generated from the HPTO unit causes the electrical equipment to not function at the expected load. The higher voltage generated from the HPTO unit can damage the equipment insulation.

The unnecessary power fluctuations in the generated electrical power were significantly reduced by employing the improved HPTO unit, as illustrated in Figure 12. The figure indicates that the improved HPTO unit, which consists of the FEC module, can transform the high-fluctuation electrical power to a more stable electrical power. The stability power generated by the improved HPTO unit depends on the pressure level of HPA_1 and HPA_2. During the high-pressure state of HPA_1 and HPA_2, the improved HPTO unit can generate electricity at its rated capacity. During a low-pressure state, the generation module of improved HPTO unit is deactivated from the power generation process.

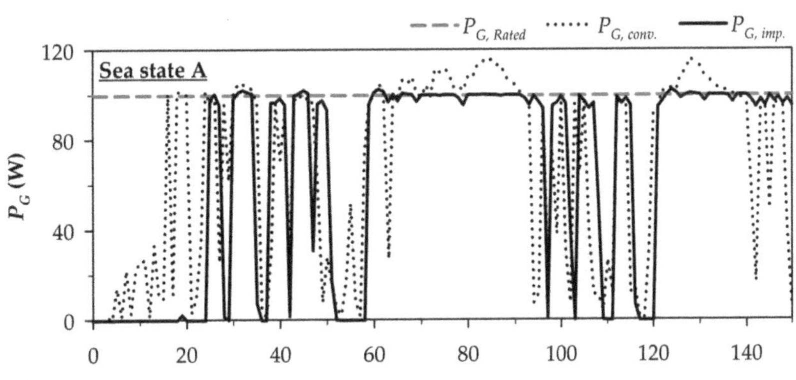

Figure 12. Comparison of P_G generated from conventional and improved HPTO unit outputs in the sea state A.

PD and OP_{PD} evaluation criteria, as described in Equations (8) and (9), were used to measure the quality of the generated electrical power in Figure 12. Figure 13 illustrates the OP_{PD} results of both HPTO unit cases for 500 s of WECs operation. PD represents the power deviation between P_G and $P_{G,Rated}$; OP_{PD} represents the percentage of PD occurrences. The negative and positive value of PD refers to the higher and lower P_G rather than the $P_{G,Rated}$. PD equal to 0 W means the P_G is generated at the $P_{G,Rated}$ level. PD equal to 100 W means the electrical power is not generated from the HPTO unit. Figure 13 depicts that the PD occurrences for the conventional HPTO unit case were found between

−20 W to 100 W. The $P_{G,conv.}$ was above the rated capacity during 12.8% of WECs operation. For 38.1% of WECs operation, the $P_{G,conv.}$ was below the rated capacity. The $P_{G,conv.}$ was at the rated capacity only during 10.2% of WECs operation.

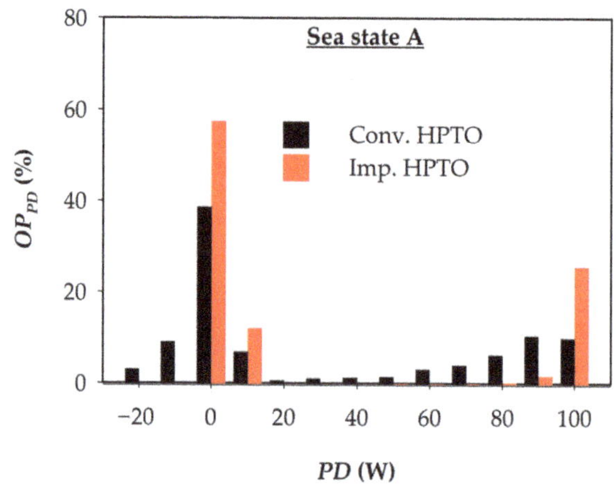

Figure 13. A comparison of OP_{PD} for conventional and improved HPTO unit models in sea state A.

Meanwhile, for improved HPTO unit case, PD was significantly found at 0 W, 10 W, 90 W, and 100 W only; therefore, the over-generated electrical power apparently does not exist in the improved HPTO unit case. $P_{G,imp.}$ was at the rated capacity during 57.6% of WECs operation. For 16.4% of WECs operation, $P_{G,imp.}$ was lower than the rated capacity. Figure 13 also indicates that the generation module in both HPTO units was not operated (off-time) several times. The figure shows that the generation module off-time of the improved HPTO unit was longer than the conventional HPTO unit. The longer off-time may be due to the dual HPA module used in the improved HPTO unit instead of the single HPA module used in the conventional HPTO unit. The occurrences of generation module off-time for the improved HPTO unit was 25.9% of WECs operation. In contrast, the occurrences of generation module off-time for the conventional HPTO unit was 10.2% of WECs operation. Technically, a longer off-time can increase the lifespan of the components in the generation module, such as the hydraulic motor and generator, due to the higher heat reduction.

Figure 14 presents a comparison of moving average of absolute power deviation, $MAAPD$ for both HPTO units in sea state A. R_1 and R_2 represent the linear regression lines of $MAAPD$ for conventional and improved HPTO units ($MAAPD_{conv.}$ and $MAAPD_{imp.}$), respectively. In overall, the variation of $MAAPD_{conv.}$ & $MAAPD_{imp.}$ can be linearized based on the R_1 and R_2 functions. The slope coefficients of R_1 and R_2 indicate the average change rate of $MAAPD_{conv.}$ and $MAAPD_{imp.}$, where the greater slope coefficient indicates the greater the rate of change. The slope coefficient of the R_1 and R_2 were obtained equal to 0.066 and 0.057, as shown in Figure 14A. From the figure, the R_2 was higher than R_1 during the early state, which means that the $MAAPD_{imp.}$ was greater than $MAAPD_{conv.}$ during the early state. This due to a longer off-time of generation module (up to 20 s) in improved HPTO unit during early state operation, as depicted in Figure 12. A longer off-time of generation module at early state operation is due to the high charging process of the dual HPA module; however, the R_2 was lower than R_1 after 320 s, which means that the $MAAPD_{imp.}$ was smaller than $MAAPD_{conv.}$, as depicted in Figure 14A. A smaller $MAAPD_{imp.}$ was due to the efficient control of FEC module used in the improved HPTO unit. The intercept point between R_1 and R_2 was found at 177.5 s with the magnitude of 12.5 W, as depicted in Figure 14B.

Figure 15 compares the moving average of $P_{G,conv.}$ and $P_{G,imp.}$. To be fair, in comparison, the $P_{G,conv.}$ and $P_{G,imp.}$ were limited up to the rated capacity during the moving average. Based on the figure, a longer off-time of the generation module of the improved HPTO unit during the early stage operation does not affect the amount of $P_{G,imp.}$. The figure shows that the final average of $P_{G,imp.}$ was higher than that of $P_{G,conv.}$. The final average of $P_{G,imp.}$ and $P_{G,conv.}$ were equal to 69.77 W and 67.44 W, respectively.

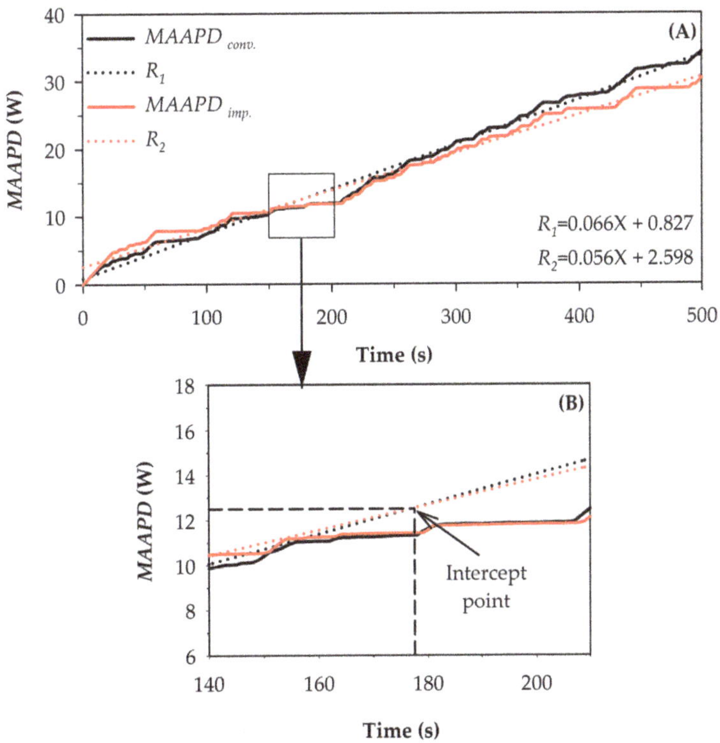

Figure 14. A comparison of $MAAPD$ for conventional and improved HPTO unit models in sea state A, (**A**) $MAAPD$ profile for 500 s and (**B**) enlarged image focused on the intercept point between R_1 and R_2.

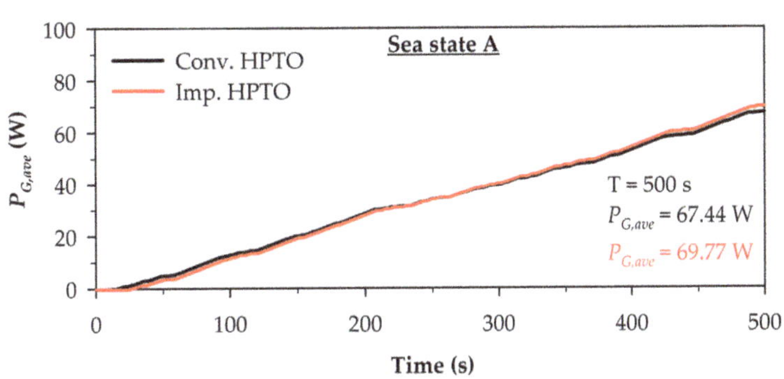

Figure 15. A comparison of $P_{G,ave}$ for conventional and improved HPTO unit models in sea state A.

The behavior profiles of selected components in the improved HPTO unit are depicted in Figure 16. Figure 16A presents the active position of DCV_1 and DCV_2 profiles during the improved HPTO unit operation. At positions B and C, HPA_1 and HPA_2 are, respectively, connected to the charging and discharging line, as depicted in Figure 2B. Figure 16A shows that the HPA_1 and HPA_2 operations were interchangeable throughout the HPTO unit operation with respect to the HPA_1 pressure constraint. The switching periods of DCV_1 and DCV_2 depend on the charging and discharging period of HPA_1. Figure 16B illustrates that the initial fluid pressures of HPA_1 and HPA_2 were set to 0 bar. At the initial operation of the improved HPTO unit, the HPA_1 was first connected to the charging line since it was assumed as a primary fluid energy storage, as depicted in Figure 16A. At the same time, HPA_2 was connected to the discharging line. The charging and discharging process were interchangeable after 25 s after P_{HPA1} reached $P_{HPA1,max}$ constraint, as depicted in Figure 16B.

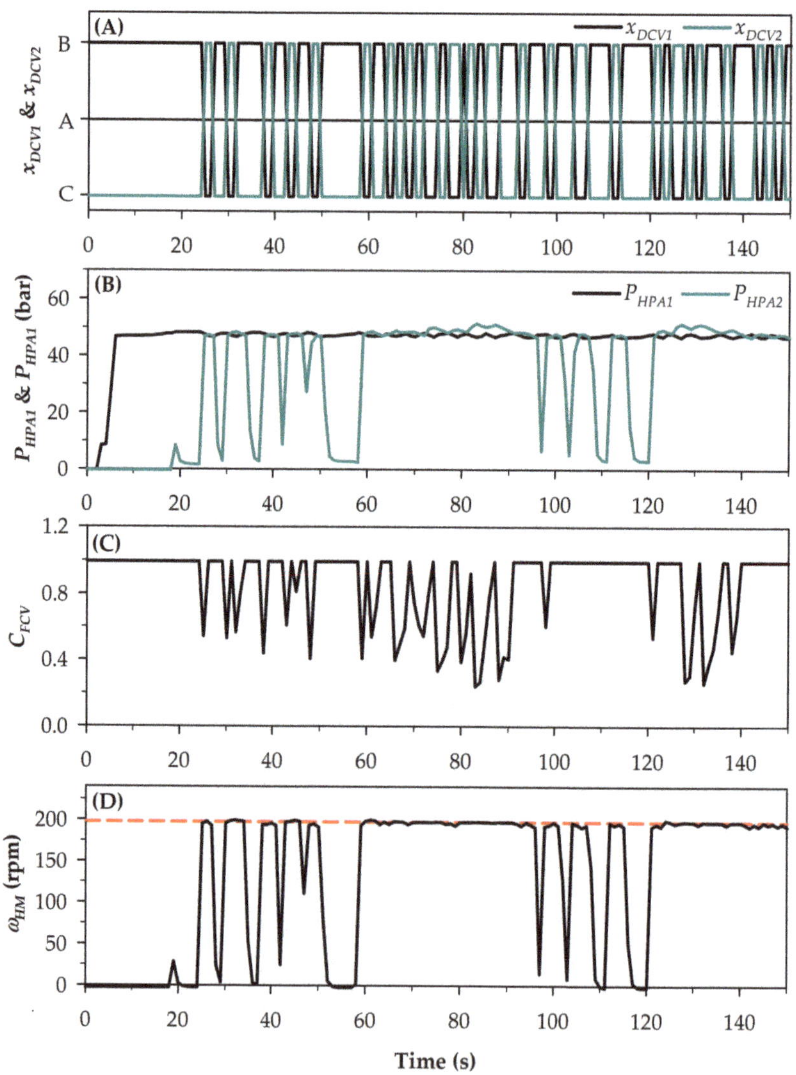

Figure 16. The behavior profiles of selected components in the improved HPTO unit. (**A**) The active position of DCV_1 and DCV_2. (**B**) The oil pressure in HPA_1 and HPA_2. (**C**) The flow coefficient of FCV. (**D**) Hydraulic motor speed.

Figure 16C shows the variation of FCV flow coefficient, C_{FCV}, throughout the improved HPTO unit operation. The FCV operation depends on the hydraulic motor speed as well as the generated electrical power from the generator, as depicted in Figure 4. Technically, the C_{FCV} can be varied between 0 to 1; however, in this case, the C_{FCV} was approximately varied from 0.2 to 1 during the high-pressure to low-pressure states of the P_{HPA1}, as presented in Figure 16C. In this regard, the fluid flow through the hydraulic motor was approximately regulated to 1.68 L/min during the highest pressure state to ensure the hydraulic motor operates at its rated speed, as depicted in Figure 16D. In contrast, the fluid flow from the dual HPA module to the hydraulic motor was unregulated during the low-pressure state of P_{HPA1}.

4.3. Evaluation of an Improved HPTO Unit in Different Significant Wave Height and Peak Wave Period Inputs (Sea State B, C, D, and E)

The simulations of conventional and improved HPTO units were repeated using four different sea states to ensure the robustness of the improved HPTO unit at the different significant wave heights and periods. A similar power quality analysis used in Section 4.2 was carried out on the generated electrical power from both HPTO units in different significant wave heights and wave periods. Figures 17 and 18 present the $P_{G,conv.}$ and $P_{G,imp.}$ profiles for different significant wave height and wave period cases, respectively. By comparing Figure 12 to Figures 17A and 18A, the $P_{G,conv.}$ and $P_{G,imp.}$ in the lower significant wave height and period (sea states B and D) cases fluctuated more than in the nominal sea state. The $P_{G,conv.}$ in the sea states B and D mostly fluctuated below the rated capacity, as indicated in Figures 17A and 18A. The conventional HPTO unit could not generate reliable electricity during the lower significant wave height and wave period; however, the fluctuation problem in the electrical power generation can be reduced by using the improved HPTO unit. Figures 17A and 18A show that the improved HPTO unit could generate more reliable electricity than the conventional HPTO unit. From the figure, the improved HPTO unit generates less fluctuated electrical power after 30 s (sea state B) and 32 s (sea state D) of WECs operation, respectively.

Meanwhile, for higher significant significant wave height and wave peak period cases (sea states C and E), Figures 17B and 18B show that the $P_{G,conv.}$ and $P_{G,imp.}$ fluctuated less than $P_{G,conv.}$ and $P_{G,imp.}$ in the nominal sea state case, particularly from improved HPTO unit. The improved HPTO unit generates better-quality electrical power in the sea states C and E after 18 s (sea state C) and 25 s (sea state E) of WECs operation. The generator off-time of the improved HPTO unit was reduced, particularly in sea state E. Moreover, $P_{G,conv.}$ fluctuated more than the $P_{G,imp.}$. The $P_{G,conv.}$ was frequently fluctuated below and above the rated capacity, which is potentially up to 120 W, as indicated in Figures 17B and 18B.

Figures 19 and 20 illustrate the PD occurrences of conventional and improved HPTO units for different significant wave height and wave peak period cases. Overall, the comparison of the PD occurrences in Figures 19 and 20 show that the PD occurrences of the improved HPTO unit case in sea states B to E were lower than the conventional HPTO unit case. The figures show that the PD occurrences of the conventional HPTO unit were found between -20 W and 100 W in all sea states. For improved HPTO unit, the PD occurrences mainly were significantly found at 0 W, 10 W, 90 W, and 100 W for all sea states. Figure 19A,B indicate that the improved HPTO unit generates electrical power at the rated capacity in sea states B and C during 34.3% and 78.6% of WECs operation. The conventional HPTO unit generates electrical power at the rated capacity in sea states B and C only during 21.4% and 32.1% of WECs operation. Meanwhile, the improved HPTO unit generates electrical power at the rated capacity for sea states D and E during 37.3% and 53.7% of WECs operation, as indicated in Figure 20A,B. For the conventional HPTO unit, the generated electrical power at the rated capacity in sea states B and C only during 25.4% and 36.7% of WECs operation, respectively.

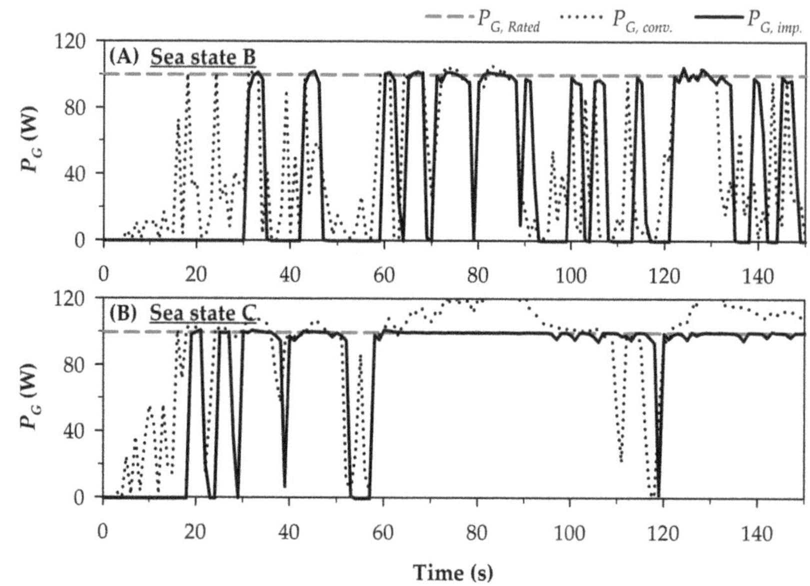

Figure 17. Comparison of the $P_{G,conv.}$ and $P_{G,imp.}$ profiles for the different significant wave height cases.

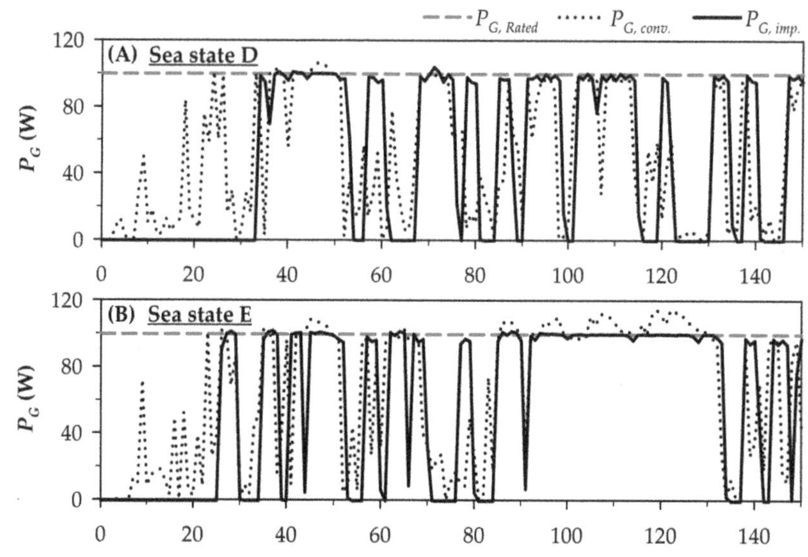

Figure 18. Comparison of the $P_{G,conv.}$ and $P_{G,imp.}$ profiles for the different wave peak period cases.

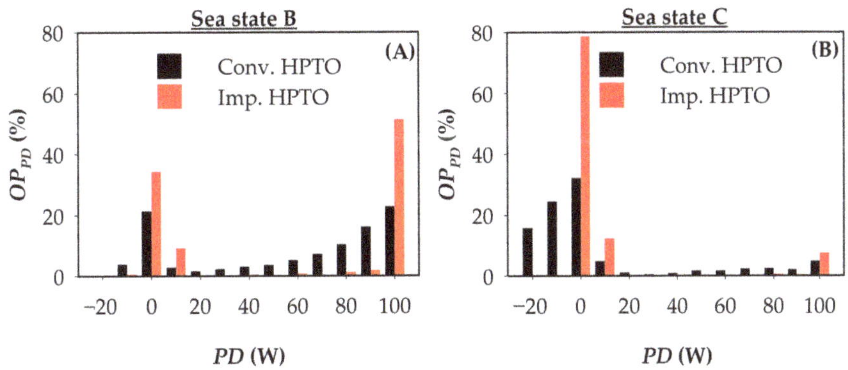

Figure 19. The *PD* occurrences of conventional and improved HPTO units' outputs for the significant wave height cases.

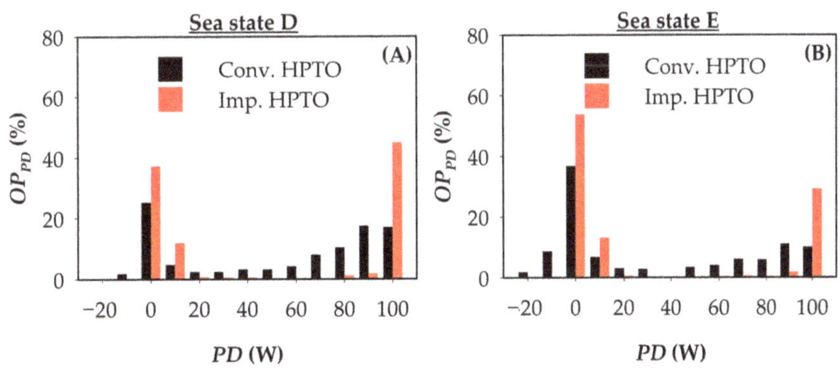

Figure 20. The *PD* occurrences of conventional and improved HPTO units' outputs for the significant wave period cases.

In overall, the generation module off-time of the improved HPTO unit was significantly higher than the generation module off-time of the conventional HPTO unit, particularly in sea state B, D, and E, as depicted in Figures 19 and 20. The longer off-time of generation module did not affect the overall power generated from the improved HPTO unit. Figures 21 and 22 present the comparison of $MAAPD$ and LR_{α} for conventional and improved HPTO unit models for the significant wave height and wave peak period cases. The figures depict that the $MAAPD_{imp.}$ was lower than the $MAAPD_{conv.}$ in all sea states. For example, in sea state B, the final $MAAPD$ and LR_{α} of the improved HPTO unit were equal to 55.77 W and 0.107, respectively, whereas the final $MAAPD$ and LR_{α} of the conventional HPTO unit were equal to 57.52 W and 0.114, respectively. In sea state C, the $MAAPD$ and LR_{α} of the improved HPTO unit were significantly lower than the $MAAPD$ and LR_{α} of the conventional HPTO unit. This is due to the shorter generation module off-time, as depicted in Figure 19B. In sea state C, the final $MAAPD$ and LR_{α} of improved HPTO unit were equal to 9.8 W and 0.012, respectively, whereas the final $MAAPD$ and LR_{α} of conventional HPTO unit were equal to 20.11 W and 0.036, respectively, as depicted in Figure 21.

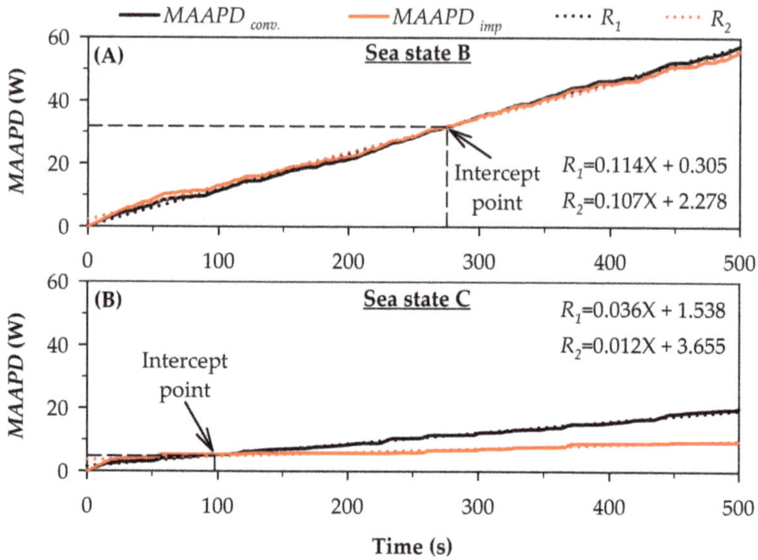

Figure 21. A comparison of *MAAPD* for conventional and improved HPTO unit models for the different significant wave height cases.

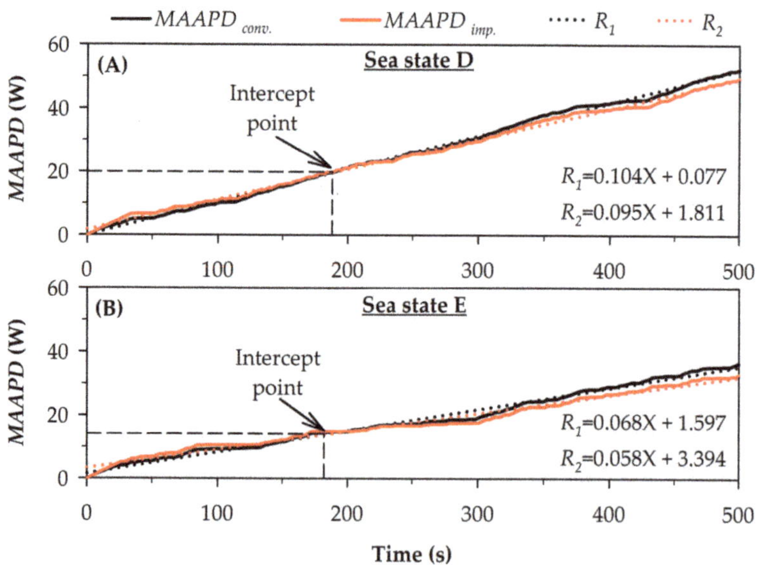

Figure 22. A comparison of *MAAPD* for conventional and improved HPTO unit models for the different wave peak period cases.

Further analysis of the generated power from the improved HPTO unit in different sea states based on the overall average power was performed to ensure the average of $P_{G,imp.}$ is not significantly lower than the average of $P_{G,conv.}$. Figures 23 and 24 compare the average of $P_{G,conv.}$ and $P_{G,imp.}$ for the different significant wave height and wave period cases. The overall results in Figures 23 and 24 indicate that the average of the $P_{G,imp.}$ was higher than the average of the $P_{G,conv.}$ in all sea states after 500 s of WECs operation. For example, during the lower significant wave height case (sea state B), the average of $P_{G,imp.}$ reached up to 44.36 W, while the average of the $P_{G,conv.}$ only reached up to

42.86 W, as illustrated in Figure 23A. In the higher significant wave height case (sea state C), the average of $P_{G,imp.}$ reached up to 90.48 W, while the average of the $P_{G,conv.}$ only reached up to 87.29 W, as indicated in Figure 23B. During lower and higher wave peak period sea states, the average of the $P_{G,imp.}$ reached up to 50.44 W and 67.04 W, as indicated in Figure 24. The average of the $P_{G,conv.}$ was only equal to 47.86 W and 64.81 W, as shown in Figure 24, respectively. In summary, Figures 23 and 24 verified that the improved HPTO unit could generate a better quality and amount of electrical power compared to the conventional HPTO unit.

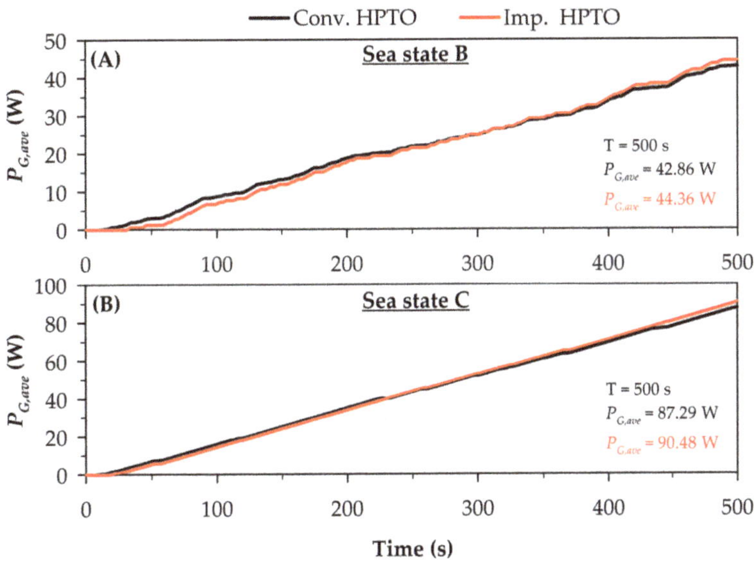

Figure 23. The average $P_{G,conv.}$ and $P_{G,imp.}$ for the different significant wave height cases.

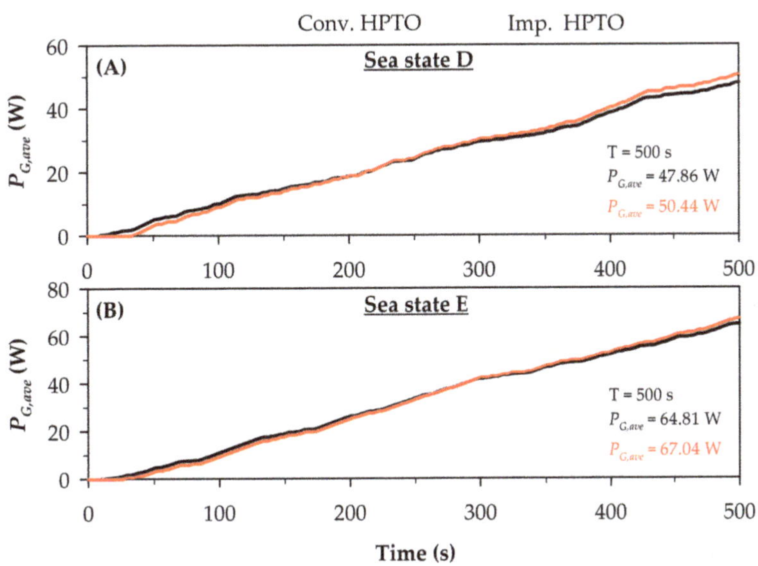

Figure 24. The average $P_{G,conv.}$ and $P_{G,imp.}$ for the different significant wave period cases.

4.4. Effect of HPA_1 Pressure Constraint on the Improved HPTO Unit Output

Further simulations of the improved HPTO unit with the different HPA_1 pressure constraint were conducted. These simulations were intended to investigate the effect of the upper-pressure constraint of HPA_1 ($P_{HPA1,max}$) on the generated electrical power from the improved HPTO unit. Technically, $P_{HPA1,max}$ is one of the parameters influencing the accumulation period and the accumulated energy in HPA_1. The lower and higher value of $P_{HPA1,max}$ affects the accumulation period and the accumulated energy in the HPA_1. At the lower $P_{HPA1,max}$ setting, the accumulation period is short, and the accumulated energy is low. Conversely, the accumulation period is long and the accumulated energy is high at the higher $P_{HPA1,max}$ setting. Subsequently, the accumulation period and accumulated energy in HPA_1 influence the hydraulic motor speed as well as the electrical power generated from the generator. In short, the $P_{HPA1,max}$ setting influences the generated electrical power from the HPTO unit. Figure 25A,B present the example of $P_{G,imp.}$ profiles for the lower (47.5 bar) and higher (49.5 bar) $P_{HPA1,max}$ cases. The comparison of results in Figure 25A indicates that, at the lower $P_{HPA1,max}$ setting, the $P_{G,imp.}$ can be generated earlier (within 18 s) than the optimal $P_{HPA1,max}$ setting (within 25 s). On the contrary, $P_{G,imp.}$ was slightly delayed (within 30 s) than the optimal case using the higher $P_{HPA1,max}$ setting, as indicated in Figure 25B.

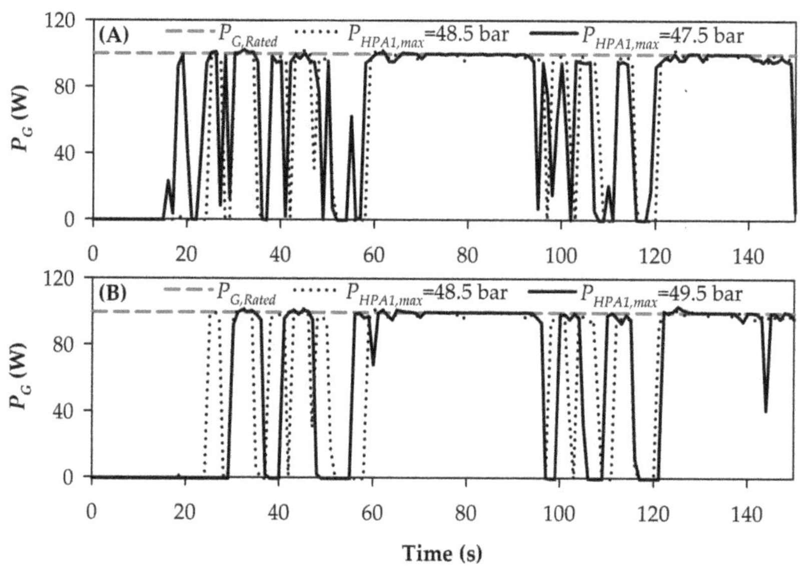

Figure 25. The profiles of the generated electrical power from the improved HPTO unit for different $P_{HPA1,max}$ cases. (**A**) $P_{HPA1,max}$ = 47.5 bar; (**B**) $P_{HPA1,max}$ = 49.5 bar.

Figure 25 illustrates the effect of the $P_{HPA1,max}$ setting on the PD occurrences in the generated electrical power from the improved HPTO unit. Overall, the PD occurrences in the generated electrical power were reduced by increasing the $P_{HPA1,max}$ setting. As indicated in Figure 25, $P_{G,imp.}$ more fluctuated during the lower $P_{HPA1,max}$ setting while less fluctuated during the higher $P_{HPA1,max}$ setting. Figure 26 compares the PD occurrences of $P_{G,imp.}$ for three different $P_{HPA1,max}$ settings (47.5 bar, 48.5 bar, and 49.5 bar). The $P_{HPA1,max}$ setting effects on the PD occurrences of $P_{G,imp.}$ can be clearly observed from the figure. The improved HPTO unit with the higher $P_{HPA1,max}$ setting could generate more stable electrical power (PD = 0 W). As indicated in Figure 26, the improved HPTO unit with a higher $P_{HPA1,max}$ setting could generate electrical power at the rated capacity during 60.7% of WECs operation. The generated electrical power from improved HPTO unit with lower $P_{HPA1,max}$ setting only up to 44.7% of WECs operation. The figure also depicts that the generated electrical power with 10% power fluctuation (PD = 10 W) also was lower during the higher $P_{HPA1,max}$ setting, which only up to 8% of WECs operation. A more stable electrical power from the improved HPTO unit with a higher $P_{HPA1,max}$ setting was due to the higher fluid energy accumulated in the HPA_1. Furthermore, the higher $P_{HPA1,max}$ setting also increased the off-time of the generation module in the improved HPTO unit (PD = 100 W). As indicated in Figure 26, the off-time of the generation module was longer, up to 28.1% of WECs operation. A longer off-time of the generation module was due to the larger pressure constraint during the higher $P_{HPA1,max}$ setting. A larger pressure constraint required a more extended energy accumulation period of HPA_1. Figure 27 depicts the average of $P_{G,imp.}$ and LR_α corresponds to different $P_{HPA1,max}$. The dotted-red line indicates the optimal $P_{HPA1,max}$ setting, which corresponds to the average of $P_{G,imp.}$ and LR_α. From the figure, the lower and the higher $P_{HPA1,max}$ setting significantly influences the average of $P_{G,imp.}$ and LR_α. The increasing and reducing of the $P_{HPA1,max}$ setting significantly increased the overall LR_α of the improved HPTO unit. As shown in Figure 27, the LR_α was increased from 0.05601 (optimal) to 0.05915 and 0.05855. An increase in LR_α has reduced the average of $P_{G,imp.}$, as presented in Figure 27. The figure shows that the average of $P_{G,imp.}$ was reduced up to 69 W and 67.4 W when the $P_{HPA1,max}$ setting was set to 47.5 bar and 51.5 bar, respectively. The detailed findings of the $P_{HPA1,max}$ effects on the $P_{G,imp.}$ are summarized in Table 4.

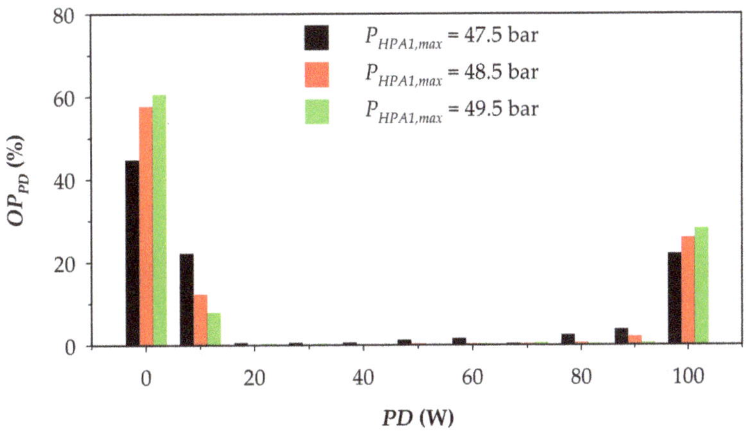

Figure 26. The PD occurrences of the generated electrical power from the improved HPTO unit for different $P_{HPA1,max}$ cases.

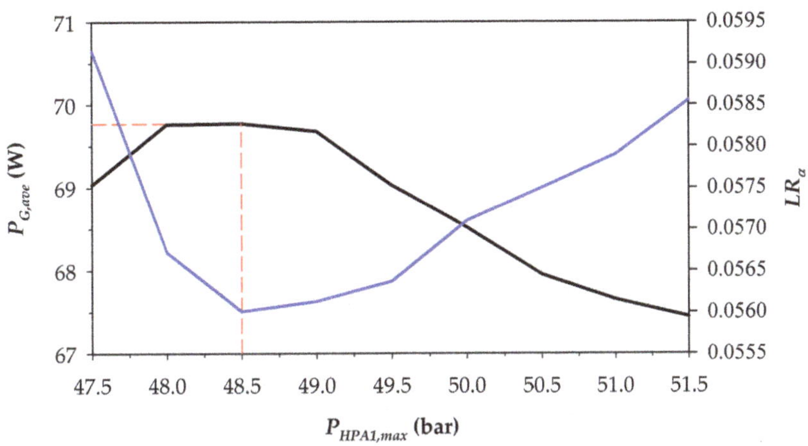

Figure 27. The average of $P_{G,imp.}$ and LR_α corresponds to different $P_{HPA1,max}$.

Table 4. The detailed findings of the $P_{HPA1,max}$ effects on the the $P_{G,imp}$.

Descriptions	Unit	$P_{HPA1,max}$								
		47.5	48.0	48.5	49.0	49.5	50.0	50.5	51.0	51.5
OP_{PD}										
100 W	%	21.96	24.75	25.95	26.95	28.14	28.14	29.54	30.54	29.94
90 W	%	3.79	2.59	2.00	1.40	0.60	1.20	0.80	1.00	1.00
80 W	%	2.40	1.00	0.60	0.40	0.40	0.40	1.00	0.60	0.40
70 W	%	0.40	0.80	0.40	0.40	0.60	0.20	0.40	0.00	0.00
60 W	%	1.60	0.60	0.40	0.40	0.40	0.00	0.20	0.00	0.40
50 W	%	1.20	0.20	0.40	0.40	0.20	0.80	0.00	0.20	0.00
40 W	%	0.60	0.00	0.00	0.20	0.20	0.20	0.00	0.20	0.20
30 W	%	0.60	0.20	0.20	0.00	0.40	0.20	0.20	0.00	0.20
20 W	%	0.60	0.20	0.00	0.40	0.40	0.20	0.40	0.20	0.20
10 W	%	22.16	18.36	12.38	10.38	7.98	8.18	6.99	6.39	7.98
0 W	%	44.71	51.30	57.68	59.08	60.68	60.48	60.48	60.88	59.68
−10 W	%	0.00	0.00	0.00	0.00	0.00	0.00	0.00	0.00	0.00
−20 W	%	0.00	0.00	0.00	0.00	0.00	0.00	0.00	0.00	0.00
MAAPD										
Final value	W	30.9986	30.2751	30.3207	30.4898	31.1239	31.6452	32.2186	32.4980	32.7173
Slope coefficient	-	0.05915	0.05672	0.05601	0.05613	0.05637	0.05710	0.05751	0.05792	0.05855
y-axis intercept point	W	1.74041	2.24836	2.59752	2.85415	3.29133	3.60071	4.30470	4.55947	4.80109
$P_{G,ave}$										
Final value	W	69.0355	69.7628	69.7719	69.6776	69.0286	68.5179	67.9496	67.6495	67.4428

5. Conclusions

This study proposed an improved HPTO unit model for the WECs to improve the quality of the generated electrical power. A novel FEC module was introduced to mitigate the fluctuation issues in the generated electrical power from the HPTO unit. A comprehensive simulation analysis was carried out using MATLAB/Simulink software. In the first part, the improved HPTO unit model was simulated using five irregular wave inputs to evaluate its performance in irregular circumstances. In the second part, the improved HPTO unit model was simulated using the different HPA_1 pressure constraints in order to investigate the effects of the pressure constraints on the generated electrical power from the improved HPTO unit. From the simulation investigations, the following conclusions can be drawn:

1. The genetic algorithm was introduced to obtain the optimal parameters of the control strategy for the improved HPTo unit. The optimization process of the control strategy parameters was carried out for 7 h 30 min.
2. By employing the improved HPTO unit, the unnecessary fluctuations in the generated electrical power from the WECs were significantly reduced. The integration between the dual HPA module and FEC module in the improved HPTO unit transformed the high-fluctuation electrical power to a more stable electrical power. The improved HPTO unit could generate electricity at the rated capacity during 57.6% of WECs operation during nominal sea state.
3. The performances of the improved HPTO unit at the different sea states were also higher than the conventional HPTO unit performance. At the lower (sea state B) and higher (sea state C) wave height sea states, the improved HPTO unit could generate electricity at the rated capacity during 34.3% and 78.6% of WECs operation. At the lower (sea state D) and higher (sea state E) wave period sea states, the improved HPTO unit could generate the electricity at the rated capacity during 37.3% and 53.7% of WECs operation, respectively.
4. Overall, the lower and the higher $P_{HPA1,max}$ setting significantly influences the average of $P_{G,imp.}$ and the power fluctuation occurrences. The average of $P_{G,imp.}$ was reduced up to 69 W and 67.4 W when the $P_{HPA1,max}$ setting was set to 47.5 bar and 51.5 bar, respectively.

The presented improved HPTO unit model may help designers, engineers, and researchers to reduce the fluctuation problems in WECs applications. The introduced optimization method for the control strategy parameters can be applied to solve the other optimization problem in order to increase the performance WECs. The experiment validation of the improved HPTO unit is compulsory to ensure its effectiveness in real application; therefore, further study on the experiment validation of the improved HPTO unit should be explored. Finally, the implementation of the proposed control strategy in different types of HPTO unit is suggested to be conducted in future.

Author Contributions: M.A.J., conceptualization, methodology, software, data curation, analysis, writing—original draft; M.Z.I., writing—review and editing and supervision, project administration, funding acquisition; M.Z.D., conceptualization, methodology, writing—review and editing, supervision; Z.M.Y., software, data curation, analysis, writing—original draft; A.A. data curation, analysis, writing—review and editing. All authors have read and agreed to the published version of the manuscript.

Funding: This project was funded by the Ministry of Higher Education (MOHE) under the Fundamental Research Grant Scheme (FRGS/1/2019/TK07/UMT/01/1).

Institutional Review Board Statement: Not applicable.

Informed Consent Statement: Not applicable.

Data Availability Statement: Data available on request due to restrictions of privacy.

Acknowledgments: The authors would like to thank the Ministry of Higher Education (MOHE) and Universiti Malaysia Terengganu (UMT) for financial support for this research.

Conflicts of Interest: The authors declare no conflict of interest.

Abbreviations

The following abbreviations are used in this manuscript:

CS	Control System
CV	Check Valve
DCV	Directional Control Valve
FCV	Flow Control Valve
FEC	Fluid Energy Control
GA	Genetic Algorithm
HA	Hydraulic Actuator
HM	Hydraulic Motor
HPA	High Pressure Accumulator
HPTO	Hydraulic Power Take-Off
LPA	Low Pressure Accumulator
LVQNN	Learning Vector Quantitative Neural Network
NLPQL	Non-Linear Programming by Quadratic Lagrangian
PTO	Power Take-Off
RV	Relief Valve
WA	Wave Absorber
WAB	Wave-Activated-Body
WECs	Wave Energy Conversion System

References

1. Melikoglu, M. Current status and future of ocean energy sources: A global review. *Ocean Eng.* **2018**, *148*, 563–573. [CrossRef]
2. Rusu, E.; Onea, F. A review of the technologies for wave energy extraction. *Clean Energy* **2018**, *2*, 10–19. [CrossRef]
3. Titah-Benbouzid, H.; Benbouzid, M. An up-to-date technologies review and evaluation of wave energy converters. *Int. Rev. Electr. Eng.* **2015**, *10*, 52–61. [CrossRef]
4. Farrok, O.; Ahmed, K.; Tahlil, A.D.; Farah, M.M.; Kiran, M.R.; Islam, M.R. Electrical power generation from the oceanic wave for sustainable advancement in renewable energy technologies. *Sustainability* **2020**, *12*, 2178. [CrossRef]

5. Ilyas, A.; Kashif, S.A.; Saqib, M.A.; Asad, M.M. Wave electrical energy systems: Implementation, challenges and environmental issues. *Renew. Sustain. Energy Rev.* **2014**, *40*, 260–268. [CrossRef]
6. Yusop, Z.M.; Ibrahim, M.Z.; Jusoh, M.A.; Albani, A.; Rahman, S.J.A. Wave-Activated Body Energy Converter Technologies: A Review. *J. Adv. Res. Fluid Mech. Therm. Sci.* **2020**, *76*, 76–104. [CrossRef]
7. Wang, L.; Kolios, A.; Cui, L.; Sheng, Q. Flexible multibody dynamics modelling of point-absorber wave energy converters. *Renew. Energy* **2018**, *127*, 790–801. [CrossRef]
8. Ozkop, E.; Altas, I.H. Control, power and electrical components in wave energy conversion systems: A review of the technologies. *Renew. Sustain. Energy Rev.* **2017**, *67*, 106–115. [CrossRef]
9. He, X.; Xiao, G.; Hu, B.; Tan, L.; Tang, H.; He, S.; He, Z. The applications of energy regeneration and conversion technologies based on hydraulic transmission systems: A review. *Energy Convers. Manag.* **2020**, *205*, 112413. [CrossRef]
10. Jusoh, M.A.; Ibrahim, M.Z.; Daud, M.Z.; Albani, A.; Yusop, Z.M. Hydraulic power take-off concepts for wave energy conversion system: A review. *Energies* **2019**, *12*, 4510. [CrossRef]
11. Wang, L.; Isberg, J.; Tedeschi, E. Review of control strategies for wave energy conversion systems and their validation: The wave-to-wire approach. *Renew. Sustain. Energy Rev.* **2018**, *81*, 366–379. [CrossRef]
12. Penalba, M.; Ringwood, J.V.; Penalba, M.; Ringwood, J.V. A Review of Wave-to-Wire Models for Wave Energy Converters. *Energies* **2016**, *9*, 506. [CrossRef]
13. Zhang, D.; Li, W.; Lin, Y.; Bao, J. An overview of hydraulic systems in wave energy application in China. *Renew. Sustain. Energy Rev.* **2012**, *16*, 4522–4526. [CrossRef]
14. Lin, Y.; Bao, J.; Liu, H.; Li, W.; Tu, L.; Zhang, D. Review of hydraulic transmission technologies for wave power generation. *Renew. Sustain. Energy Rev.* **2015**, *50*, 194–203. [CrossRef]
15. Kurniawan, A.; Pedersen, E.; Moan, T. Bond graph modelling of a wave energy conversion system with hydraulic power take-off. *Renew. Energy* **2012**, *38*, 234–244. [CrossRef]
16. Tri, N.M.; Truong, D.Q.; Thinh, D.H.; Binh, P.C.; Dung, D.T.; Lee, S.; Park, H.G.; Ahn, K.K. A novel control method to maximize the energy-harvesting capability of an adjustable slope angle wave energy converter. *Renew. Energy* **2016**, *97*, 518–531. [CrossRef]
17. Ding, B.; Cazzolato, B.S.; Arjomandi, M.; Hardy, P.; Mills, B. Sea-state based maximum power point tracking damping control of a fully submerged oscillating buoy. *Ocean Eng.* **2016**, *126*, 299–312. [CrossRef]
18. Pedersen, H.C.; Hansen, R.H.; Hansen, A.H.; Andersen, T.O.; Bech, M.M. Design of full scale wave simulator for testing Power Take Off systems for wave energy converters. *Int. J. Mar. Energy* **2016**, *13*, 130–156. [CrossRef]
19. Hansen, A.H.; Asmussen, M.F.; Bech, M.M. Model predictive control of a wave energy converter with discrete fluid power power take-off system. *Energies* **2018**, *11*, 635. [CrossRef]
20. Hansen, R.H.; Kramer, M.M.; Vidal, E.; Hansen, R.H.; Kramer, M.M.; Vidal, E. Discrete displacement hydraulic power take-off system for the wavestar wave energy converter. *Energies* **2013**, *6*, 4001–4044. [CrossRef]
21. Gaspar, J.F.; Calvário, M.; Kamarlouei, M.; Guedes Soares, C. Power take-off concept for wave energy converters based on oil-hydraulic transformer units. *Renew. Energy* **2016**, *86*, 1232–1246. [CrossRef]
22. Jusoh, M.A.; Ibrahim, M.Z.; Daud, M.Z.; Yusop, Z.M.; Albani, A. An Estimation of Hydraulic Power Take-off Unit Parameters for Wave Energy Converter Device Using Non-Evolutionary NLPQL and Evolutionary GA Approaches. *Energies* **2020**, *14*, 79. [CrossRef]
23. Gaspar, J.F.; Calvário, M.; Kamarlouei, M.; Soares, C.G. Design tradeoffs of an oil-hydraulic power take-off for wave energy converters. *Renew. Energy* **2018**, *129*, 245–259. [CrossRef]
24. Othman, F.; Sadeghian, M.S.; Ebrahimi, F.; Heydari, M. A Study on Sedimentation in Sefidroud Dam by Using Depth Evaluation and Comparing the Results with USBR and FAO Methods. In Proceedings of the International Proceedings of Chemical, Biological and Environmental Engineering, Kuala Lumpur, Malaysia, 8–9 Jun 2013; Volume 51, p. 6. [CrossRef]
25. Gaspar, J.F.; Kamarlouei, M.; Sinha, A.; Xu, H.; Calvário, M.; Faÿ, F.X.; Robles, E.; Soares, C.G. Speed control of oil-hydraulic power take-off system for oscillating body type wave energy converters. *Renew. Energy* **2016**, *97*, 769–783. [CrossRef]
26. Liu, Z.; Qu, N.; Han, Z.; Zhang, J.; Zhang, S.; Li, M.; Shi, H. Study on energy conversion and storage system for a prototype buoys-array wave energy converter. *Energy Sustain. Dev.* **2016**, *34*, 100–110. [CrossRef]
27. Do, H.T.; Dang, T.D.; Ahn, K.K. A multi-point-absorber wave-energy converter for the stabilization of output power. *Ocean Eng.* **2018**, *161*, 337–349. [CrossRef]
28. Jusoh, M.A.; Yusop, Z.M.; Albani, A.; Daud, M.Z.; Ibrahim, M.Z. Investigations of hydraulic power take-off unit parameters effects on the performance of the wab-wecs in the different irregular sea states. *J. Mar. Sci. Eng.* **2021**, *9*, 897. [CrossRef]
29. Liu, C.; Yang, Q.; Bao, G. Influence of hydraulic power take-off unit parameters on power capture ability of a two-raft-type wave energy converter. *Ocean Eng.* **2018**, *150*, 69–80. [CrossRef]
30. Gao, H.; Xiao, J. Effects of power take-off parameters and harvester shape on wave energy extraction and output of a hydraulic conversion system. *Appl. Energy* **2021**, *299*, 117278. [CrossRef]
31. Jusoh, M.A.; Ibrahim, M.Z.; Daud, M.Z.; Yusop, Z.M.; Albani, A.; Rahman, S.J.; Mohad, S. Parameters estimation of hydraulic power take-off system for wave energy conversion system using genetic algorithm. In Proceedings of the IOP Conference Series: Earth and Environmental Science, Bangkok, Thailand, 11–14 December 2019; Institute of Physics Publishing: Bristol, UK, 2020; Volume 463, p. 12129. [CrossRef]

32. Cargo, C.J.; Plummer, A.R.; Hillis, A.J.; Schlotter, M. Determination of optimal parameters for a hydraulic power take-off unit of a wave energy converter in regular waves. *Proc. Inst. Mech. Eng. Part J. Power Energy* **2012**, *226*, 98–111. [CrossRef]
33. Cargo, C.J.; Hillis, A.J.; Plummer, A.R. Optimisation and control of a hydraulic power take-off unit for a wave energy converter in irregular waves. *Proc. Inst. Mech. Eng. Part J. Power Energy* **2014**, *228*, 462–479. [CrossRef]
34. Cargo, C.J.; Hillis, A.J.; Plummer, A.R. Strategies for active tuning of Wave Energy Converter hydraulic power take-off mechanisms. *Renew. Energy* **2016**, *94*, 32–47. [CrossRef]
35. Sun, P.; Li, Q.; He, H.; Chen, H.; Zhang, J.; Li, H.; Liu, D. Design and optimization investigation on hydraulic transmission and energy storage system for a floating-array-buoys wave energy converter. *Energy Convers. Manag.* **2021**, *235*, 113998. [CrossRef]
36. Wang, D.; Lu, K. Design optimization of hydraulic energy storage and conversion system for wave energy converters. *Prot. Control Mod. Power Syst.* **2018**, *3*, 7. [CrossRef]
37. Chen, Q.; Yue, X.; Geng, D.; Yan, D.; Jiang, W. Integrated characteristic curves of the constant-pressure hydraulic power take-off in wave energy conversion. *Int. J. Electr. Power Energy Syst.* **2020**, *117*, 105730. [CrossRef]
38. Calvário, M.; Gaspar, J.F.; Kamarlouei, M.; Hallak, T.S.; Guedes Soares, C. Oil-hydraulic power take-off concept for an oscillating wave surge converter. *Renew. Energy* **2020**, *159*, 1297–1309. [CrossRef]
39. Jusoh, M.A.; Daud, M.Z. Control strategy of a grid-connected photovoltaic with battery energy storage system for hourly power dispatch. *Int. J. Power Electron. Drive Syst.* **2017**, *8*, 1830–1840. [CrossRef]
40. Jusoh, M.A.; Daud, M.Z. Particle swarm optimisation-based optimal photovoltaic system of hourly output power dispatch using Lithium-ion batteries. *J. Mech. Eng. Sci.* **2017**, *11*, 2780–2793. [CrossRef]
41. Fan, Y.; Mu, A.; Ma, T. Design and control of a point absorber wave energy converter with an open loop hydraulic transmission. *Energy Convers. Manag.* **2016**, *121*, 13–21. [CrossRef]
42. Wang, K.; Sheng, S.; Zhang, Y.; Ye, Y.; Jiang, J.; Lin, H.; Huang, Z.; Wang, Z.; You, Y. Principle and control strategy of pulse width modulation rectifier for hydraulic power generation system. *Renew. Energy* **2019**, *135*, 1200–1206. [CrossRef]
43. Ricci, P.; Lopez, J.; Santos, M.; Ruiz-Minguela, P.; Villate, J.L.; Salcedo, F.; Falcão, A.; Falcão, A.F. Control strategies for a wave energy converter connected to a hydraulic power take-off. *IET Renew. Power Gener.* **2011**, *5*, 234. [CrossRef]
44. Jianan, X.; Tao, X. MPPT Control of Hydraulic Power Take-Off for Wave Energy Converter on Artificial Breakwater. *J. Mar. Sci. Eng.* **2019**, *8*, 304. [CrossRef]
45. Zou, S.; Abdelkhalik, O. Control of Wave Energy Converters with Discrete Displacement Hydraulic Power Take-Off Units. *J. Mar. Sci. Eng.* **2018**, *6*, 31. [CrossRef]
46. Song, R.; Dai, Y.M.; Qian, X. Intermittent wave energy generation system with hydraulic energy storage and pressure control for stable power output. *J. Mar. Sci. Technol.* **2018**, *23*, 802–813. [CrossRef]
47. Gaspar, J.F.; Kamarlouei, M.; Sinha, A.; Xu, H.; Calvário, M.; Faÿ, F.X.; Robles, E.; Guedes Soares, C. Analysis of electrical drive speed control limitations of a power take-off system for wave energy converters. *Renew. Energy* **2017**, *113*, 335–346. [CrossRef]
48. Truong, D.Q.; Ahn, K.K. Development of a novel point absorber in heave for wave energy conversion. *Renew. Energy* **2014**, *65*, 183–191. [CrossRef]
49. Shadman, M.; Estefen, S.F.; Rodriguez, C.A.; Nogueira, I.C. A geometrical optimization method applied to a heaving point absorber wave energy converter. *Renew. Energy* **2018**, *115*, 533–546. [CrossRef]
50. Coiro, D.P.; Troise, G.; Calise, G.; Bizzarrini, N. Wave energy conversion through a point pivoted absorber: Numerical and experimental tests on a scaled model. *Renew. Energy* **2016**, *87*, 317–325. [CrossRef]
51. Chen, Z.; Zhou, B.; Zhang, L.; Sun, L.; Zhang, X. Performance evaluation of a dual resonance wave-energy convertor in irregular waves. *Appl. Ocean Res.* **2018**, *77*, 78–88. [CrossRef]
52. Mathworks. Hydraulics (Isothermal) Library—MATLAB & Simulink. Available online: https://www.mathworks.com/help/physmod/hydro/hydraulics-modeling.html (accessed on 2 December 2021).

Journal of
Marine Science and Engineering

MDPI

Article

Downsizing the Linear PM Generator in Wave Energy Conversion for Improved Economic Feasibility

Jian Tan *, Xuezhou Wang, Henk Polinder, Antonio Jarquin Laguna and Sape A. Miedema

Department of Maritime & Transport Technology, Delft University of Technology, 2268 CD Delft, The Netherlands
* Correspondence: j.tan-2@tudelft.nl

Abstract: A crucial part of wave energy converters (WECs) is the power take-off (PTO) mechanism, and PTO sizing has been shown to have a considerable impact on the levelized cost of energy (LCOE). However, as a dominating type of PTO system in WECs, previous research pertinent to PTO sizing did not take modeling and optimization of the linear permanent magnet (PM) generator into consideration. To fill this gap, this paper provides an insight into how PTO sizing affects the performance of linear permanent magnet (PM) generators, and further the techno-economic performance of WECs. To thoroughly reveal the power production of the WEC, both hydrodynamic modeling and generator modeling are incorporated. In addition, three different methods for sizing the linear generator are applied and compared. The effect of the selection of the sizing method on the techno-economic performance of the WEC is identified. Furthermore, to realistically reflect the relevance of PTO sizing, wave resources from three European sea sites are considered in the techno-economic analysis. The dependence of PTO sizing on wave resources is demonstrated.

Keywords: wave energy converter; linear PM generator; downsizing

Citation: Tan, J.; Wang, X.; Polinder, H.; Laguna, A.J.; Miedema, S.A. Downsizing the Linear PM Generator in Wave Energy Conversion for Improved Economic Feasibility. *J. Mar. Sci. Eng.* **2022**, *10*, 1316. https://doi.org/10.3390/jmse10091316

Academic Editor: Giuseppe Giorgi and Mauro Bonfanti

Received: 18 August 2022
Accepted: 13 September 2022
Published: 17 September 2022

1. Introduction

Ocean waves contain substantial energy, which can be exploited as an important supplementary of clean energy next to wind and solar energy. Wave energy has been investigated over decades as a renewable energy resource [1–3]. Over 200 different types of wave energy converters (WECs) have been reported, but they are still far away from large-scale commercialization [4–6]. This mainly results from the fact that WECs correspond to a higher value of the levelized cost of energy (LCOE) than other technologies of renewable energy [4,7]. Therefore, it is significant to bring the LCOE down for the further development of WECs.

In the operation of WECs, wave energy is first absorbed as a form of mechanical energy by the interaction between the buoy and incoming waves [8]. The absorbed mechanical energy is then converted into usable electricity by a power take-off (PTO) system. As a core component, the PTO capacity is highly influential to the techno-economic performance of WECs. According to recent scientific research, reducing the PTO capacity to an appropriate level helps to lower the LCOE of WECs [9–11]. On the one hand, the PTO capacity is directly related to the cost. As stated in [12], PTO systems normally account for over 20% of total capital expenditure (CAPEX), and oversized PTO systems are therefore not economically favorable. On the other hand, the PTO capacity indicates the physical constraints, such as force, peak power and displacement limit. In order to comply with them, the PTO parameters are usually required to be adapted, which consequently confines the mechanical energy absorption of WECs. A collective sizing method considering both buoy and PTO sizing has been proposed to improve the techno-economic performance of WECs [9]. It has been applied to a point absorber WEC in three different wave sites, leading to the conclusion that suitably sizing WECs could reduce the LCOE. More specifically, downsizing the PTO capacity is rather effective. For instance, it has the potential to decrease the LCOE by over 30% in particular circumstances. However, some assumptions in the reference [9] need to be further studied when evaluating the power performance of WECs with differently sized PTO capacities. The influence of PTO sizing could play a role at two conversion stages in the power extraction. Firstly, the PTO capacity is highly related to the physical limits, which determines the maximum absorbed power. Secondly, the PTO sizing inevitably brings changes to its design, which might make a difference to the conversion efficiency from the absorbed power to the electrical power. The second aspect was not taken into account in previous papers [9–11] discussing the PTO downsizing, in which the PTO size on the power conversion efficiency was neglected. Hence, a better understanding of the effect of the sizing on PTO conversion efficiencies is expected to improve the size optimization methods of WECs.

Among various types of PTO systems in WECs, linear generators have received significant attention because of their high efficiencies and low maintenance demands [2,13]. However, the effect of sizing on the performance of linear generators of WECs is rarely discussed in literature even though downsizing the PTO capacity has shown the potential to reduce the LCOE. A preliminary work of this study has been reported in [1], in which the sizing of the generator was roughly performed without optimizing the main geometric parameters. In addition, the variation of the wave resource was not taken into account even though it is of significance to the size determination of WECs [14]. To address these issues, the present paper is thus extended to further demonstrate how the sizing affects the techno-economic performance of WECs equipped with linear permanent magnet (PM) generators as the PTO system.

This paper begins with a literature review of the application of the linear generator in wave energy conversion. Then, the WEC concept is described, and the specification of the linear PM generator reference is given. Next, the method to calculate the power performance of the WEC is presented, in which both hydrodynamic modeling and electrical modeling are considered. The main design parameters of the generator are resized to minimize the LCOE for a realistic sea site. For comparison, three different sizing methods of the electrical machines are applied to determine the optimal PTO capacity. The influence of sizing methods on the resized generator performance, the AEP and the LCOE of the WEC are analyzed. Moreover, the dependence of the generator efficiency and PTO size determination on the characteristics of wave resources is discussed. Finally, conclusions are drawn.

2. Review of the Application of Linear Generator in Wave Energy Conversion

The commonly used PTO systems in WECs are divided into five types by their operating principles, and they are briefly introduced below.

- Hydraulic PTO: In hydraulic PTO, the motion of a buoy drives the hydraulic piston to increase the pressure of the working medium. Then, the pressure of the medium is raised sufficiently high in an accumulator to rotate the hydraulic motor. The hydraulic PTO is robust and able to provide large forces at low frequencies, which highly matches the dynamic characteristics of WECs. Hydraulic PTO systems were widely used in WECs, such as Pelamis and Edinburgh Duck [15–17]. However, hydraulic PTO systems contain plenty of moving parts, which results in their complex structure. So, regular system maintenance and inspection are normally required, which is time-consuming and costly [2]. In addition, the conversion efficiencies of hydraulic PTO systems are relatively low [18].
- Hydro PTO: The hydro PTO mainly refers to the hydro turbine. Hydro turbines transfer the fluid flow to electricity. This type of PTO systems is commonly employed in overtopping devices, such as WaveDragon [19].
- Pneumatic PTO: The pneumatic turbine generally refers to the air turbine, which is driven by the oscillating air pressure. This technology is usually utilized in oscillating water column converters [20].

- Mechanical PTO: In the mechanical PTO system, a gearbox is used to convert the linear movement of the buoy of WECs to rotary motion for fitting conventional rotary generators. The oscillation of the buoy of WECs is of low speed due to the characteristics of ocean waves. For improving efficiency, another important function of the gearbox is to increase the speed of motion. In [21], a point absorber equipped with a bidirectional gearbox and rotary generator was introduced. More recently, WECs with mechanical PTO mechanism have been further investigated by numerical modeling [22] and small-scale testing [23].
- Direct-drive linear generator: The linear generator could be used as the direct-drive PTO system in the oscillating body WECs, and they are usually used in point absorber wave energy converters, such as AWS [24]. In direct-drive linear generators, intermediate transmission interfaces, such as gearbox and hydraulic motors, are not necessary. Instead, the oscillating buoy is directly coupled with the translator of the linear generator. The linear generators are commonly associated with higher efficiencies compared with other PTO systems. This is because there are less transmission losses resulting from a reduced number of energy conversion steps [2,24]. In addition, the reduced number of components in the PTO system increases the reliability of the whole WEC system [25].

The linear generator has been thought of as an appealing solution for wave energy technologies among all types of PTO systems because of the mentioned advantages. Hence, this paper concentrates on the WECs with a linear generator. Linear generators have been investigated and developed for application in WECs over years. For various WEC prototypes that have been successfully tested in the real oceanic environment, linear generators were serving as the PTO system [26,27]. So far, three full-scale WEC technologies equipped with the linear generator, excluding their updated versions, have been successfully tested in real oceanic climates. The key information of their prototypes is presented in Table 1.

Table 1. WECs applying the linear generator tested in the oceanic environment.

Parameters	AWS	Uppsala Concept	SeaBeavI
Rated power	2 MW	10 kW	10 kW
WEC type	Submerged point absorber	Floating point absorber	Floating point absorber
Generator structure	Bottom founded	Bottom founded	Floating
Generator topology	Flat and double sided	Flat and four sided	Tubular
First tested time	2004	2006	2007
Testing site	Portugal	Sweden	USA

The Archimedes Wave Swing (AWS) was the first large-scale WEC project using the linear generator [24]. It is a fully submerged WEC device, and it produces energy relying on the pressure difference on the moving part of the device when the ocean wave passes above. The linear generator in the AWS was designed to be double-sided for avoiding the huge force of attraction between the translator and stator. The maximum peak power of the tested prototype was 2 MW, which is remarkably high in the field of wave energy conversion [27]. Uppsala university developed a floating point absorber WEC concept, and the first full-scale prototype was deployed at the Lysekil research site in Sweden in 2006. Since then, thirteen follow-up prototypes have been deployed [28]. This WEC consists of a floating buoy and a linear generator founded on the seabed. The buoy is connected to the translator of the generator through a connection line. In the generator, the translator moves vertically along the rail, and four stator packages are mounted on a pillar fixed at the foundation structure [26]. Oregon University developed a dual-body WEC in which a linear generator was incorporated, called SeaBeavI Project [29]. The ocean testing of its 10 kW prototype was performed in Yaquina Bay in 2007. This WEC includes a moored cylindrical spar and a floating buoy. The heave motion of the spar is restrained by the mooring system, but the buoy is free to move vertically relative to the spar. In this concept, the spar and buoy act as the translator and stator of the linear generator, respectively. The coils are wound on the interior surface of the spar and permanent magnets are placed on the buoy. Besides, this generator was configured as a tubular geometry [25]. These three linear generators share some common features. First, all of them are permanent-magnet (PM) synchronous machines. Second, they are all longitudinal flux machines. Since the successful tests of these three concepts, a series of extensive studies have been performed. They were mainly intended to address the negative characteristics of linear generators when serving WECs.

The application of linear generators in wave energy conversion also faces some challenges. Firstly, the dimensions of linear generators are much larger than standard electrical machines for sustaining a similar power level. This results from the low typical speed of WECs. One solution to reduce the dimension is the use of transverse flux linear machines. The advantage of the transverse flux machines is the high force density, which results from the flux concentration and decoupled magnetic and electrical circuits [30]. In [31–33], WECs with transverse flux linear generators were investigated, in which a lower cost of permanent magnets and higher force/power density were indicated. However, their fabrication is much more complex than the longitudinal flux machines, and their leakage paths also lead to low power factors [30]. As a consequence, they have not been as of much interest as the longitudinal flux machines in the field of WECs. Another possible solution is to increase the movement speed of WECs. For example, a speed amplified linear generator was proposed for a WEC concept in [34], and the relative speed between the stator and translator was therefore increased. The second challenge is that the conversion efficiency of linear generators in WECs is still limited compared to other applications of direct-drive systems. This is because the induced voltage is decreased by the low speed and then a large current is required to produce the generator force. In this case, the resistive losses on the conductors are high. In [35–37], superconducting linear generators were studied to improve it. The results showed that superconducting generators could clearly contribute to the high efficiency and force density. This is because they are associated with less resistivity and a stronger magnetic field compared to conventional machines. However, the difficulty and high expense of making the cryogenic temperature in an oceanic environment are still hurdles in practice.

3. Description of WEC Concept

In this section, the WEC concept and the design of the reference linear PM generator are described. Figure 1 illustrates the concept of the investigated heaving point absorber, and the PTO mechanism of the concept is considered to be a linear PM generator. Figure 1 also depicts the connection scheme of the PM generator with the electrical converter and the grid. A three-phase back-to-back voltage source converter is applied here to regulate the stator current and terminal voltage because of its high efficiency [27]. The converter is connected to the output side of the linear generator. In the concept studied in the present paper, the translator of the bottom-founded machine is directly connected to the floating buoy. The radius of the spherical floater was designed to be 2.5 m, and the buoy draft is 2.5 m as well in still water. During operation, the incoming wave force excites the floater to oscillate in the heaving direction. The stroke of the point absorber is set as 4 m, and the maximum allowed velocity u_s is 1.25 m/s. To avoid excessive motion, the generator has end stop mechanisms mounted on both the top and bottom. More details about modeling, design and the sea trial test of the reference generator can be found in [27,38].

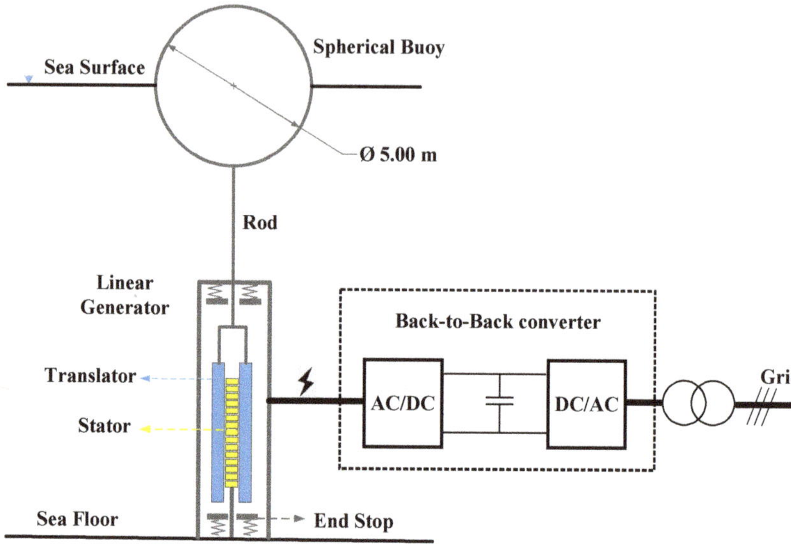

Figure 1. Schematic of the heaving point absorber WEC, and the connection scheme of the PM linear generator with the back-to-back converter and the grid.

In Figure 2a, a photo of the reference linear PM machine applied in the prototype of the AWS wave energy converter is shown. Figure 2b illustrates the cross-section of the reference linear PM generator, which indicates that it is a three-phase machine. The machine was designed to be double-sided so as to counteract the force of attraction between the translator and stator [39]. Magnets are placed on the translator segments, and coils are wound on the stator segments. Different from rotary electrical machines, the partial overlap between the translator and stator could happen to the linear generator. This phenomenon makes a part of the material of the machine unproductive during operation, which is negative to the conversion efficiency. It is known that a longer translator is able to reduce the occurrence of the partial overlap, but it could clearly increase the cost. Hence, a compromise always has to be made, and the translator was 3 m longer than the stator in the design of the reference machine.

(a)

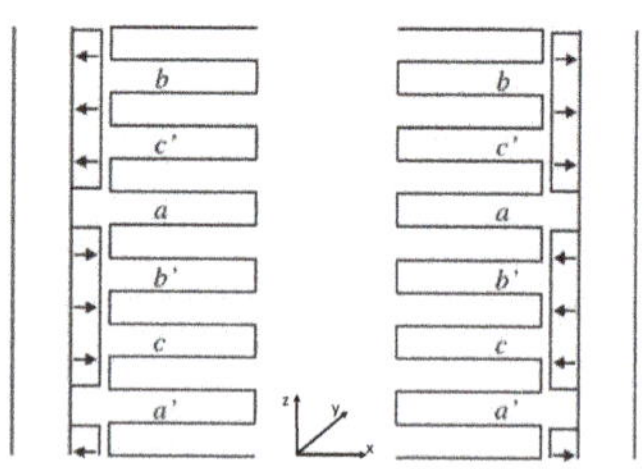

(b)

Figure 2. The photo of the reference linear electrical machine in the pilot of the AWS WEC, and the cross-section of the linear PM generator of the AWS plant. In figure (**b**), a/a′, b/b′ and c/c′ are the current directions [39], and the "z" axis in the coordinate is perpendicular to the ground plane. (**a**) Photo of the reference machine. (**b**) Cross-Section of the machine.

The key design parameters of the reference linear PM generator are specified in Table 2. As the design parameters of the original reference machine were not fully disclosed, some parameters (including the air gap length, slot width, pole width, tooth width and magnet thickness) were referred to the design of a direct-drive wind turbine generator design [40]. The referred generator has a similar power rating and maximum designed force to the reference machine studied in this paper. It is considered fair since this work is focused on the effect of the generator sizing instead of its design phases.

Table 2. Design parameters of the reference linear generator.

Parameters	Symbol	Value
Maximum average power	P_{rared}	1 MW
Maximum force	F_{max}	933 kN
Maximum velocity	u_{max}	2.2 m/s
Stroke	S	7 m
Translator length	L_t	8 m
Stator length	L_s	5 m
Air gap length	g	5 mm
Slot width	b_s	15 m
Stack length	l_s	1 m
Magnet pole width	b_p	79 mm
Tooth width	b_t	19 mm
Magnet thickness	l_m	15 mm
Number of conductors per slot	N_s	6

4. Methodology

In this section, a frequency domain model is established to calculate the hydrodynamic responses of the buoy. Next, an analytical model is presented to calculate the generator performance. Furthermore, the principle and optimization for resizing the generator are explained. Finally, an economic model necessary for analyzing the techno-economic performance is built.

4.1. Hydrodynamic Modeling

Based on linear wave theory, a frequency domain model is derived for the WEC. The buoy is constrained to oscillate in the heaving direction, thus only this degree of freedom is considered in the modeling. According to Newton's second law, the motion of the WEC as a rigid body can be described as

$$m\frac{\mathrm{d}\boldsymbol{u}(t)}{\mathrm{d}t} = \boldsymbol{F_{hs}}(t) + \boldsymbol{F_e}(t) + \boldsymbol{F_{pto}}(t) + \boldsymbol{F_r}(t) \tag{1}$$

where m is the mass of the oscillating body,

$\boldsymbol{F_{hs}}$ is the hydrostatic force,

$\boldsymbol{F_e}$ is the wave excitation force,

$\boldsymbol{F_r}$ is the wave radiation force,

$\boldsymbol{F_{pto}}$ is the PTO force, and

$\boldsymbol{u}$ is the velocity of the rigid body. If the body is assumed to perform harmonic motion under regular waves and a linear PTO model is used to simulate the dynamic behaviour of the PTO system, (1) could be rewritten in the form of complex amplitudes [41] as,

$$\hat{F}_e(\omega) = [R_i(\omega) + R_{pto}]\hat{u} + \mathrm{i}\omega\hat{u}[m + M_r(\omega)] + \mathrm{i}\hat{u}(-\frac{K_{pto}}{\omega} - \frac{S_{wl}}{\omega}) \tag{2}$$

where $R_i(\omega)$ is the hydrodynamic damping coefficient, R_{pto} is the PTO damping coefficient, ω is the wave frequency, $M_r(\omega)$ is the added mass of the buoy, $\hat{u}$ is complex amplitude of the vertical velocity, K_{pto} is the PTO stiffness coefficient and S_{wl} is the hydrostatic stiffness.

The intrinsic impedance of the heaving buoy and PTO impedance can be introduced as,

$$Z_i(\omega) = R_i(\omega) + \mathrm{i}X_i(\omega) \tag{3}$$

$$X_i(\omega) = \omega[m + M_r(\omega)] - \frac{S_{wl}}{\omega} \tag{4}$$

where $Z_i(\omega)$ and $X_i(\omega)$ embody the intrinsic impedance and the intrinsic reactance of the buoy respectively. Similarly, the impedance of PTO $Z_{pto}(\omega)$ and reactance of $_{pto}(\omega)$ can be given as,

$$Z_{pto}(\omega) = R_{pto}(\omega) + \mathrm{i}X_{pto}(\omega) \tag{5}$$

$$X_{pto}(\omega) = -\frac{K_{pto}}{\omega} \tag{6}$$

In this work, only the passive control strategy is used, which implies that there is only PTO damping force exerted on the WEC. Thus, (2) can be rewritten as,

$$\hat{F}_e(\omega) = [Z_i(\omega) + R_{pto}(\omega)]\hat{u} \tag{7}$$

The hydrodynamic characteristics of floating bodies, including $M_r(\omega)$, $R_i(\omega)$ and $F_e(\omega)$, are calculated numerically using the Boundary Element Method through the open source software Nemoh [42]. Then, by solving (7), the complex amplitude of velocity $\hat{u}$ could be obtained as,

$$\hat{u} = \frac{\hat{F}_e}{Z_i + R_{pto}} \tag{8}$$

In regular wave conditions, the time-averaged absorbed mechanical power can be obtained and expressed as,

$$\overline{P} = \frac{1}{2}R_{pto}|\hat{u}|^2 \tag{9}$$

The complex amplitude of the required generator force (or PTO force) is derived as,

$$\hat{F}_{ge} = R_{pto}\hat{u} \tag{10}$$

The amplitude of the the required generator force can be derived by substituting (8) to (10), and it gives

$$|\hat{F}_{ge}(\omega)| = \left|R_{pto}\frac{\hat{F}_e(\omega)}{Z_i + R_{pto}}\right| = \frac{|R_{pto}|}{|Z_i + R_{pto}|}|\hat{F}_e(\omega)| \tag{11}$$

4.2. Generator Modeling

The performance of variously sized generators is calculated based on an analytical generator model. Validation of the model has been performed through a full-scale test [38]. As magnetic saturation only plays a very limited role in this linear generator design [43], and the effect is neglected in this study.

In the air gap, the magnets on the translator induce the magnetic flux density, and its fundamental space harmonic is calculated as,

$$\hat{B}_{gm} = \frac{l_m}{\mu_{rm}g_{eff}}B_{rm}\frac{4}{\pi}\sin\left(\frac{\pi b_p}{2\tau_p}\right) \tag{12}$$

where l_m, μ_{rm} and g_{eff} embody the magnet length in the magnetization direction, the recoil permeability of the magnets and the effective air gap, respectively. B_{rm} represents the remnant flux density of the permanent magnets. According to [38], when the translator is moving, the induced no-load phase voltage in the stator winding can be calculated as

$$E_p = \sqrt{2}pl_sN_sk_wu\hat{B}_{gm}\frac{l_{act}}{L_s} \tag{13}$$

where u is the floater velocity, p is the number of pole pairs, l_{act} is the actual length of the overlap between the stator and translator, and k_w is the winding factor. l_{act} is a function of translator displacement, and it can be calculated as

$$l_{act}(z) = \begin{cases} L_s, & for \quad |z| < 0.5(L_t - L_s) \\ 0, & for \quad |z| > 0.5(L_t + L_s) \\ 0.5(L_t + L_s) - |z|, & else \end{cases} \tag{14}$$

where z represents the displacement of the translator (or the floater). The stator phase resistance R_t is calculated from the machine dimensions, the number of turns in a slot and the cross-section of a slot:

$$R_t = \rho_{Cu}\frac{l_{Cus}}{A_{Cus}} = \rho_{Cu}\frac{2N_s^2(l_s + 2\tau_p)}{ph_s b_s k_{sfil}} \tag{15}$$

where ρ_{Cu} is the resistivity of copper, and k_{sfil} is the copper fill factor of the slots. The iron losses are calculated as

$$P_{Fes} = P_{Fe0}\left(m_{Fest}\left(\frac{\hat{B}_{st}}{B_0}\right)^2 + m_{Fesy}\left(\frac{\hat{B}_{sy}}{B_0}\right)^2\right)\frac{f_e}{f_0} \tag{16}$$

where P_{Fe0} is the iron loss per unit mass at a frequency f_0 and flux density B_0, m_{Fest} and m_{Fesy} are the mass of the stator teeth and the stator yoke, respectively, and f_e is the electrical generator frequency. $\hat{B}_{st}$ and $\hat{B}_{sy}$ embody the fundamental space harmonic of the magnetic flux density in the stator and yoke, respectively. This is a rough approximation of the iron losses, in which the iron losses are assumed to be only proportional to the frequency.

After the PTO damping coefficient is derived, the required generator force is calculated based on (10). Then, the power taken by the stator winding of the machine can be expressed as

$$P_{wd} = F_{ge}v - P_{Fes} \tag{17}$$

For the sake of higher system efficiency, the operating points of the generator are determined by the method demonstrated in [27]. The stator current I_s can be represented by two orthogonal components, namely quadrature (or force making) component I_{sq} and direct (or flux making) component I_{sd}. When the stator current I_s is below the maximum current of the converter I_{conm}, the direct component I_{sd} is regulated to be zero. In this case, the stator current I_s is in phase with the induced no-load voltage E_p, as shown in the first phasor diagram in Figure 3. Thus, the current is calculated as

$$I_s = I_{sq} = \frac{P_{wd}}{mE_p} \tag{18}$$

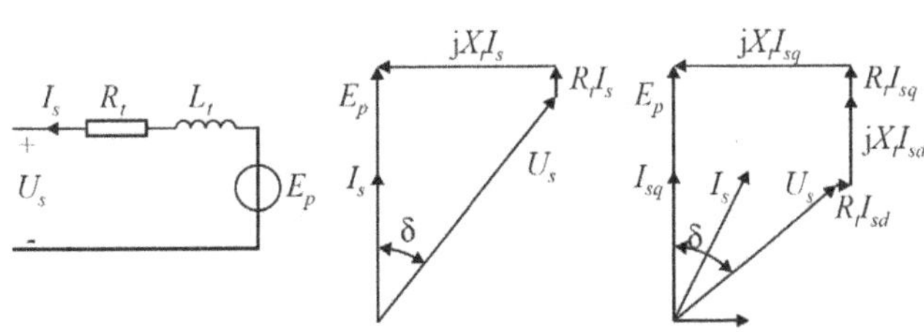

Figure 3. Equivalent circuit and the phasor diagram of the linear generator.

When I_s is greater than the maximum current of the converter I_{conm}, I_s is then limited to the defined maximum current. As a result, the actual force provided by the machine is decreased to lower than the required generator force. In this sense, the terminal voltage U_s can be calculated as

$$U_s = \sqrt{(E_p - I_{sq}R_t)^2 + (\omega_e L_t I_{sq})^2} \tag{19}$$

If U_s is about to violate the rated converter voltage U_{conm}, it would be limited by a direct component I_{sd} separated from the current, as illustrated in the second phasor diagram in Figure 3. In this condition, if the resulting current is still larger than the maximum converter current, the stator current and terminal voltage are limited to the maximum converter current and voltage. In these situations, the actual generator force would also be penalized.

The converter losses are dependent on three factors. The first one is the constant power dissipation, resulting from control, cooling and power supplies. The second is proportional to the current, representing the switching losses and conduction losses. The third is proportional to the current squared, which is related to conduction losses. Besides, only the losses from the generator side are considered in this model. Therefore, the converter losses can be calculated as

$$P_{conv} = \frac{P_{convm}}{31}[(1 + 20\frac{I_s}{I_{sm}} + 10(\frac{I_s}{I_{sm}})^2] \tag{20}$$

where P_{convm} represents the dissipated power of the converter at the rated condition. It is assumed to account for 3% of the rated power [44]. I_{sm} is the maximum generator side current of the converter. In the reference linear machine, the rated power, maximum line voltage and phase current of the converter were specified as 1 MW, 1500 V and 400 A, respectively [27]. The rated power was selected based on the maximum average power of the device.

4.3. Generator Sizing

As characteristics of wave climate vary with locations, sizing of WECs could result in devices better fitting the local wave resource. In the existing literature, PTO systems are commonly sized by the same scaling ratio as the buoy, in which PTO sizing is only playing a limited role. To allow for independent sizing of the PTO system, three sizing methods, different in complexity, are applied in this paper for comparison.

- Method 1: Scaling law
  The scaling principle of electrical generators is based on the fact that the force density per unit surface area of the machine remains rather constant when its dimension changes [40]. The force density is mainly related to two factors. The first one is the magnetic flux density in the air gap, and it is limited by the effect of magnetic saturation. The second one is the linear current density of the machine, and it is limited by the maximum allowed heat dissipation. These two factors could hardly be changed with the dimension if the topology design of the machine does not vary. In this paper, the machine is resized to suit various maximum forces. The force density of resized machines remains identical to the reference machine.
  For convenience, a scale factor λ is introduced here as,

$$\lambda = \frac{S_s}{S_o} \tag{21}$$

where S_o and S_s embody the geometrical lengths of the original and the scaled generator. In this method, stator, translator and stack length are sized based on an identical scale factor. However, in principle, they can be sized independently as implemented in method 2 introduced in the following text. Besides, only the stator, translator and stack lengths are considered in scaling, and other geometrical parameters of the machine are kept identical. For a fair comparison of differently sized generators, the stroke and speed limits are maintained to be the same. Then, the other parameters are scaled as follows:

$$\text{Force}: \quad \frac{F_{max_s}}{F_{max_o}} = \lambda^2 \tag{22}$$

$$\text{Velocity}: \quad \frac{u_{max_s}}{u_{max_o}} = 1 \tag{23}$$

$$\text{Current}: \quad \frac{I_{con_s}}{I_{con_o}} = 1 \tag{24}$$

where subscripts "s" and "o" indicate the scaled machine and original machine. The rating of the converter is accordingly scaled for each machine. The rated apparent power, current and terminal voltage should be selected during the rating. As the linear current density is assumed to remain identical for differently sized machines, the maximum phase current of the scaled converter is selected to be unchanged, namely I_{con} of 400 A. At the rated operating condition, the velocity of the buoy is u_{max} of 1.25 m/s, and then the resulting no-load voltage can be obtained by (13). Additionally, the no-load voltage at the rated point is assumed to be in phase with the current. As a consequence, the rated terminal voltage of the scaled converter, U_{con_s}, and the phase angle between the current and terminal voltage, namely δ, can be calculated according to the first phasor diagram in Figure 3. Therefore, the rated apparent power of the scaled converter is obtained as

$$S_{conv_s} = mI_{conm}U_{conm_s} \tag{25}$$

- Method 2: Scaling with optimization
In this method, the main parameters of the machine, including the translator length, stator length and stack length are optimized during sizing for the lowest LCOE. The optimization for each designed maximum generator force is expressed as

$$\begin{gathered} minimize\ f = \text{LCOE}(L_s, L_t, l_s) \\ subject\ to \begin{cases} 2\rho_{force\_ref}L_s l_s = F_{limit} : constraint\ 1 \\ 1.2L_s \leq L_t : constraint\ 2 \end{cases} \end{gathered} \tag{26}$$

where ρ_{force_ref} is the force density of the reference machine, and F_{limit} is the designed maximum generator force. In (26), constraint 1 is associated with the force density. The force density adopted here is considered as 46 kN/m^2, which is calculated based on the performance identification of the AWS generator in [27]. A factor of 2 is included on the left side of the constraint 1 because the machine is double-sided. Constraint 2 indicates that the translator length is required to be larger than 1.2 times the stator length, which is used to mitigate the effect of partial overlap between stator and translator. The maximum phase current, voltage and the rated apparent power of the scaled converter are calculated based on the same way as introduced in Method 1. As the objective function is computationally expensive in this case, the surrogate algorithm is adopted. In the algorithm, the objective function is approximated by a surrogate function, and therefore the optimization speed can be significantly improved [45]. The optimization is implemented in the Matlab environment (version 2020). The tolerance of the function and the maximum iteration steps are set to 1×10^{-5} and 3000, respectively, and the iteration process is terminated when any of the criteria is crossed.
- Method 3: Scaling with assuming a constant generator efficiency
For simplification, in studies discussing the effects of the sizing of WECs, the sizing of WECs was commonly implemented in the absence of the consideration of the variation of generator efficiencies. Instead, a constant energy conversion efficiency from the absorption stage to the grid was assumed. In this method, the generator size acts simply as a PTO force constraint in the hydrodynamic modeling. This method

is discussed in this paper for a comparison with the other two above methods to demonstrate the effects of this simplification on the generator size determination and the LCOE estimation. In this paper, this constant efficiency is considered to be 70%, as usually used in literature [9,46]. Moreover, as the design of the generator is not taken into account in this method, the cost could not be derived explicitly. To evaluate the techno-economic performance of WECs, the PTO related cost in this method is assumed to be the same as that estimated in method 2 for each designed maximum generator force limit.

4.4. Economic Modeling

The linear generator is resized to minimize the LCOE of WECs. To assess the LCOE of the WEC with differently sized generators, an economic model is presented in this part. The AEP of the WEC is calculated as

$$\mathrm{AEP} = A \cdot \sum_{x=1}^{x=n} P_{absorbed}(x) \cdot \eta(x) \cdot T(x) \tag{27}$$

where η is the generator efficiency; A is the availability of WECs to work, and it is set as 90% due to necessary operation and maintenance [47]; T represents the total hours of the occurrence of a certain sea state and x means the wave state. The sea site is considered as Yeu island situated in the oceanic territory of France, and its hours of occurrence of each wave state are shown in Appendix A. By equating the time-averaged power transport per unit length of the wavefront of the incoming waves, the irregular wave states are converted to regular wave states for the sake of simplification [9]. The LCOE of a WEC can be expressed as

$$\mathrm{LCOE} = \frac{\mathrm{CAPEX} + \sum_{Y=1}^{N} \frac{\mathrm{OPEX}_Y}{(1+r)^Y}}{\sum_{Y=1}^{n} \frac{\mathrm{AEP}_Y}{(1+r)^Y}} \tag{28}$$

in which CAPEX and OPEX represent the Capital Expenditure (CAPEX) and the Operational Expenditure (OPEX), respectively. The annual OPEX is assumed to be 8% of the CAPEX; r, N and Y embody the discount rate, the lifespan and the evaluated year. The CAPEX of a WEC can be derived as

$$\mathrm{CAPEX} = C_S + C_F + C_I + C_{PTO} + C_C \tag{29}$$

where C_S, C_F, C_I, C_{PTO} and C_C are the costs of the structure, foundation and mooring, installation, PTO and grid connection, respectively. C_S, C_F and C_I are related to the structural mass of the system. They are estimated using the method introduced in [9]. As for the linear generator, the costs are predicted as

$$C_{PTO} = 2(C_{Fe}M_{Fe} + C_{Cu}M_{Cu} + C_{pm}M_{pm}) \tag{30}$$

In Equation (30), a factor of 2 is included to cover the costs of manufacturing and converter; C_{Fe}, C_{Cu} and C_{pm} embody the cost of iron, copper and the permanent magnet per unit mass, and their values are referred to [48]; M_{Fe}, M_{Cu} and M_{pm} are the mass of iron, copper and permanent magnet, respectively. The relevant economic parameters applied in this work are listed in Table 3.

Table 3. Parameters in the economic modeling [14,48,49].

Parameter	Value	Unit
Iron price (C_{Fe})	3.3	Euros/kg
Copper price (C_{Cu})	15.2	Euros/kg
Permanent magnet price (C_{pm})	24.7	Euros/kg
Lifespan (N)	20	years
Discount rate (r)	8%	–
Availability of operation (A)	0.9	–

It should be acknowledged that only a simplified economic model is applied in this work. The economic parameters can be affected by a variety of factors, but the influence of economic modeling on the generator sizing of the WEC is not taken into consideration. The method used here is limited to this specific type of WEC and a single device. For instance, the economic parameters might be different for WEC farms. Because the operation and maintenance strategy could be optimized for WEC farms to better compromise the availability of the WEC and the total OPEX [50]. Then, the estimation of the OPEX has to be updated. Besides, the cost of the WEC structure and the PTO could be decreased when entering into the stage of series manufacture for WEC farms. Moreover, it can be expected that the lifespan and the operational demand of the WECs are correlated with the local wave climate [51]. For example, the frequent occurrence of harsh wave conditions has a negative influence on the lifespan of the WEC. This implies that the economic parameters could differ with the wave site where the WEC is considered to be deployed. Furthermore, a common challenge in modeling the economics of WECs results from the lack of information of the practical projects, which causes the disagreement of economic parameters in different references. Nevertheless, given the scope of this work is to identify the effect of the generator sizing on the performance, it is thought reasonable at the current stage to use this model for constructing an economic implication.

5. Results and Discussion

5.1. Generator Performance

The performance of the linear PM generator is calculated for a particular regular wave state, which is depicted in Figure 4. The concerned generator is sized by method 2 with a maximum designed generator force of 200 kN, and the resized parameters will be shown in Figure 5 and discussed in the following text. It is seen from Figure 4 that the root mean square (RMS) of the no-load voltage doesn't present a sinusoidal curve, which results from the partial overlap between the translator and stator. A negative effect of the partial overlap can be observed as that the stator current is driven relatively higher for maintaining the required generator force. Consequently, the copper and converter losses are also increased.

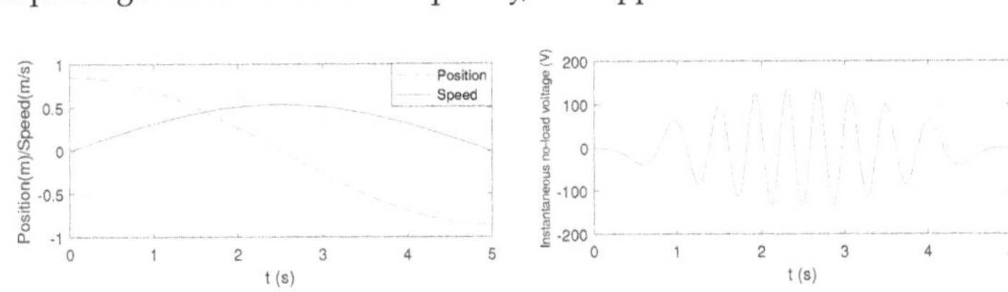

(a) (b)

Figure 4. *Cont.*

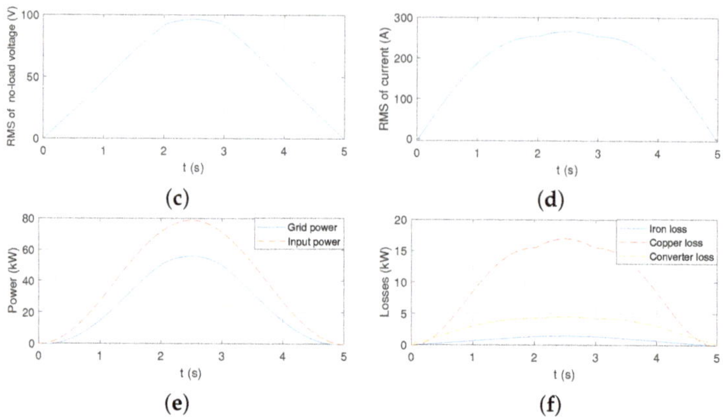

Figure 4. Generator performance over a half of wave period with a displacement of 0.85 m and a wave period of 10.0 s. The considered machine is sized with a maximum designed generator force of 200 kN based on sizing method 2. (**a**) Position and speed of the translator. (**b**) Instantaneous no-load voltage induced by the translator. (**c**) RMS value of the no-load voltage. (**d**) RMS value of the stator current. (**e**) Input mechanical power and electrical power to the grid. (**f**) Electrical losses.

5.2. Comparison of Sizing Methods

5.2.1. On the Generator Performance

The reference linear generator is resized corresponding to the designed maximum generator forces ranging from 60 kN to 280 kN. The machine parameters scaled by method 1 and method 2 are compared in Figure 5. It can be seen that all of the translator, stator and stack lengths show a trend to increase with the designed maximum generator force regardless of the sizing method. However, these two methods result in very different machine parameters. The stack lengths obtained using method 2 are clearly longer than those by method 1, but the translator and stator lengths resulting from method 2 are shorter. Noteworthy, based on method 2, the length difference between the translator and the stator is limited if compared to the cases using method 1. This indicates that the negative effect of the partial overlap is not significant for the combination of the machine and the considered wave resource. This can be interpreted as the fact the displacement of the WEC is relatively small in the wave states dominating energy production.

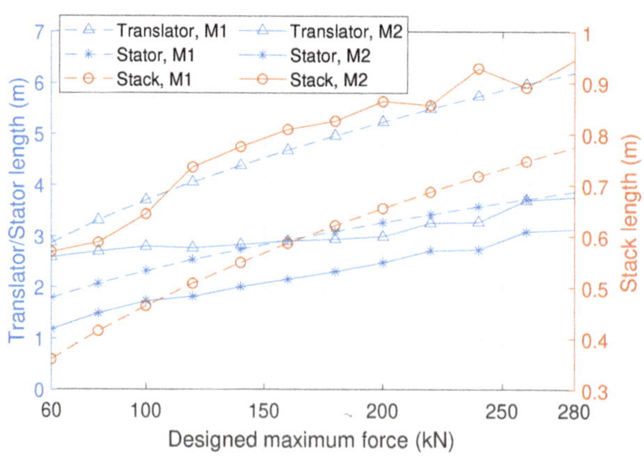

Figure 5. Scaled translator, stator and stack lengths for different designed maximum generator forces. "M1" and "M2" embody method 1 and method 2 for generator sizing.

Figure 6 shows the efficiency maps of linear generators with the designed maximum force of 100 kN, 160 kN and 200 kN, respectively, in which regular wave states are considered. Sizing method 1 and method 2 are both applied for comparison. The efficiency at each wave state is calculated as the ratio between the electrical power and the absorbed power of the floater. It is visible in Figure 6 that larger generators generally correspond to higher efficiencies. Besides, the efficiency tends to increase with the wave height and decline with the wave period. For the generator with a designed maximum force of 100 kN, a few empty cells can be noticed at low wave periods and very high wave heights. This is because the buoy velocity tends to be high in these operating conditions. Then, the PTO damping has to be increased to lower the velocity, but the larger PTO damping values are necessarily related to the larger PTO force. Therefore, with a relatively strict designed maximum force, it is not feasible to satisfy both the defined force constraint and velocity constraint.

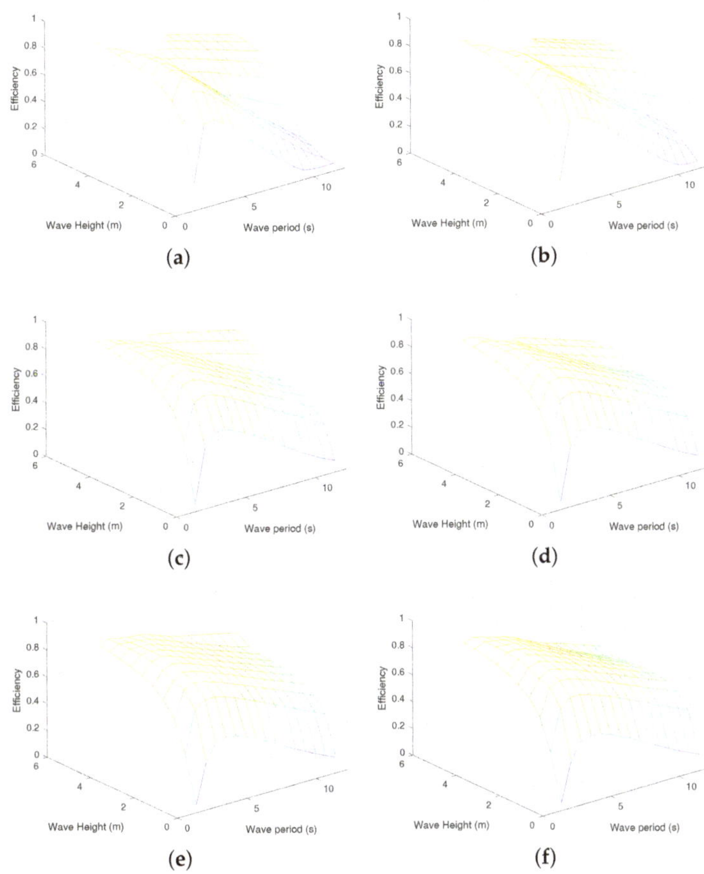

Figure 6. Comparison between efficiency maps of generators scaled based on different methods. (**a**) Designed maximum force is 100 kN, based on method 1. (**b**) Designed maximum force is 100 kN, based on method 2. (**c**) Designed maximum force is 160 kN, based on method 1. (**d**) Designed maximum force is 160 kN, based on method 2. (**e**) Designed maximum force is 200 kN, based on method 1. (**f**) Designed maximum force is 200 kN, based on method 2.

The overall conversion efficiencies of the generators operating in Yeu island are shown in Figure 7, which is defined as the ratio between the yearly electrical power production and the yearly energy absorbed by the buoy. First, the overall efficiency has a clear dependence on the generator size no matter which sizing method is applied. Generally, larger generators present higher conversion efficiencies. In this particular case, the overall conversion efficiencies are around 60% to 65% for the generator sized with a designed maximum force of 60 kN, while they increase to be higher than 75% for generators with a designed maximum force of 280 kN. This observation clearly reflects the inaccuracy of the assumption used in previous sizing studies [9,14,52] that the generator efficiency is not changing during PTO sizing. Secondly, even though the optimization procedure is incorporated in method 2, it does not necessarily lead to a higher efficiency. For instance, the overall efficiency of the generator scaled based on method 2 is associated with slightly lower efficiency compared to that using method 1 when the designed maximum force is larger than 250 kN. Nevertheless, the contribution of the optimization can be found in Figure 8 as method 2 gives a lower PTO cost when the designed maximum force is higher than 100 kN. For instance, the PTO cost of the generator with a designed maximum force of 280 kN is 160,000 Euros and 12,000 Euros for method 1 and 2, respectively, while their generator efficiencies are comparable.

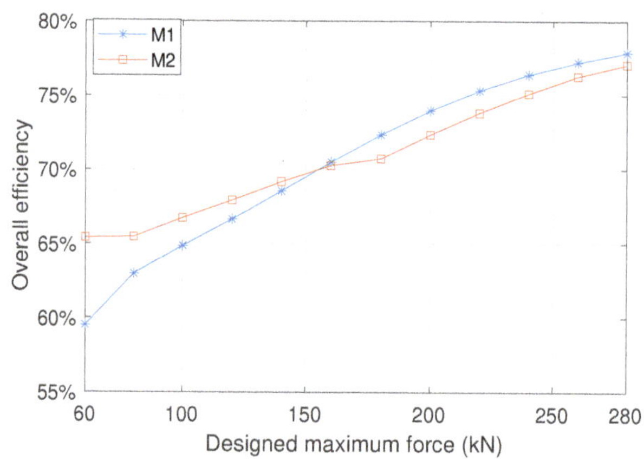

Figure 7. Scaled translator, stator and stack lengths for different designed maximum generator forces.

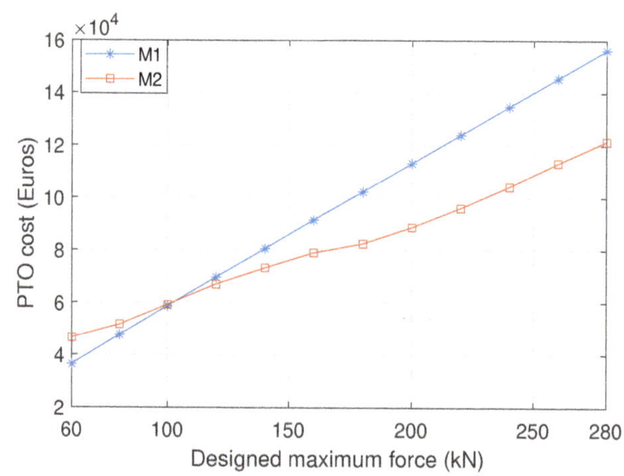

Figure 8. The PTO related cost of generators scaled by different methods.

5.2.2. On the Techno-Economics and the PTO Size Determination

The AEP and the LCOE of the WEC with differently sized generators are depicted in Figures 9 and 10, in which three different sizing methods are applied, respectively. From Figure 9, it is visible that the method 1 and method 2 result in similar values of the AEP, but the relative discrepancy of method 3 is noticeable. For example, when the designed maximum force is 280 kN, the AEP is approximately 130 MWh for method 1 and method 2, but the AEP is calculated to be around 117 MWh by using method 3. In this sense, the relative error reaches 10%. However, unlike the trends of the AEP, these three methods result in very different values of the LCOE, as shown in Figure 10. For instance, when the designed maximum force of the generator is 280 kN, the LCOE is estimated to be around 0.45 Euros, 0.38 Euros and 0.42 Euros for the method 1, 2 and 3, respectively. In this sense, the optimization of generator parameters leads to the reduction of the LCOE of around 15% and 10% with regard to method 1 and 3. In addition, as a consequence of the optimization of the main parameters, method 2 is always related to a lower value of the LCOE compared to method 1. Furthermore, the selection of the generator sizing method is significant to the PTO size determination. In this case, the optimal PTO force limit corresponding to the lowest LCOE is 120 kN, 140 kN and 100 kN for method 1, 2 and 3, respectively. In this sense, the relative error resulting from the assumption of a constant generator efficiency on the PTO size determination is around 29%. It can be concluded that method 3, which assumes the constant generator efficiency during sizing, could lead to a notable misestimate of the AEP, LCOE and the optimal PTO size. Thus, taking the influence of sizing on the generator efficiency into consideration is strongly suggested when conducting PTO sizing. Method 2, namely scaling with the optimization of the main parameters, could better reflect the techno-economic potential of the WEC, since it presents a lower LCOE compared to method 1, which is based on the scaling law. However, it has to be acknowledged that the optimization requires higher computational efforts, while the scaling law requires negligible computational loads with a given reference machine.

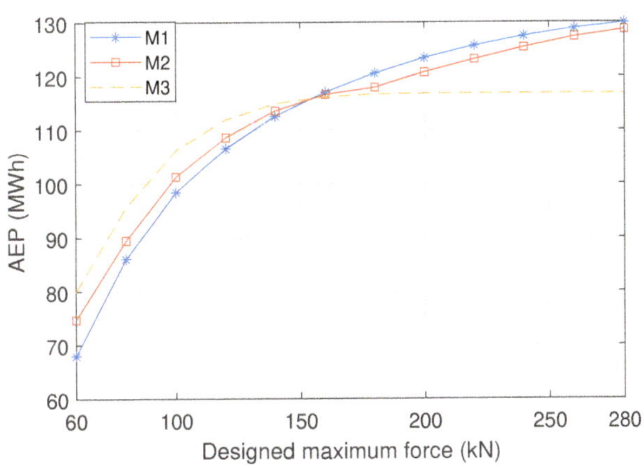

Figure 9. The AEP of the WEC as a function of the designed maximum generator force. "M3" embodies the method 3 for generator sizing.

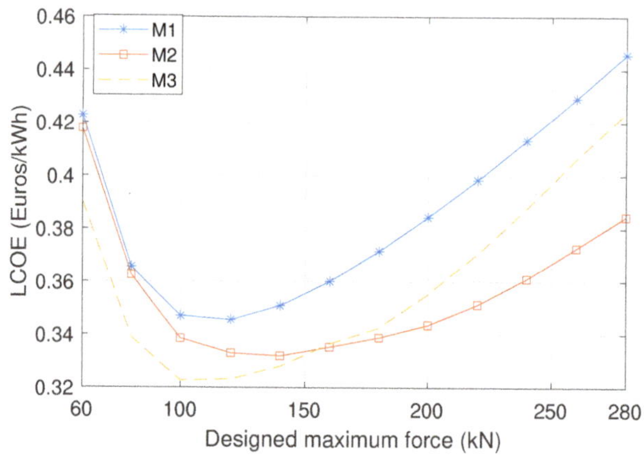

Figure 10. The LCOE of the WEC as a function of the designed maximum generator force.

5.3. Dependence of the Generator Sizing on Wave Resources

To demonstrate the dependence of the generator sizing and techno-economic performance of the WEC on different characteristics of wave resources, two more wave sites are considered as a comparison in this subsection. They are DK North Sea Point 2 (DK2) and Biscay Marine Energy Park (BIMEP), located in the oceanic territories of Denmark and Spain, respectively. Their scatter diagrams are shown in Appendix A. The mean wave power density of Yeu island, DK 2 and BIMEP are 26 kW/m, 12 kW/m and 21 kW/m, respectively. Only method 2 is applied in this subsection for PTO sizing. Figure 11a shows the optimized design parameters of the linear generator in different wave sites, and the parameters are normalized to the optimized values for Yeu island. It can be found that the wave resource makes a difference to the optimization of the design parameters.

Figure 11b shows the overall efficiencies of the differently sized linear generators in the three wave sites. It can be seen that the overall efficiency is highly related to the wave resource. The overall efficiencies in Yeu island and DK 2 are clearly higher than that in BIMEP. For instance, when the designed maximum generator force is 60 kN, the efficiencies in Yeu island and DK 2 are around 65% while it is only 45% in BIMEP. This can be explained by the fact that the wave states in BIMEP are concentrated in relatively long wave periods, which are associated with low translator velocities. Therefore, the generator efficiency is not only dependent on the generator size but also to the wave resource, and it is unrealistic to be represented by a constant value. Figure 11c depicts the LCOE of WECs in different wave sites. It is visible that the value of the LCOE clearly differs with the wave site. For instance, when the designed maximum force is 100 kN, the values of the LCOE in Yeu island, DK 2 and BIMEP are estimated to be around 0.33, 0.52 and 0.67 Euros/kWh, respectively. The reason is that the energy production of the WEC varies in different wave sites. The difference of wave resources in each site is reflected by its specific distribution of the occurrence of wave states. The distribution of the considered three wave sites are clearly different, as presented in Appendix A. However, the power absorption of a WEC is highly dependent on the wave state, since the wave period and height are crucial to the dynamics of the buoy [53,54]. In addition to the power absorption stage, the generator efficiency also presents a strong relation to the wave state because of the variation of the translator velocity and displacement, as shown in Figure 6. Therefore, it is reasonable that there appears a difference in the LCOE in the three wave sites. It also implies the importance of wave resource of deployment on the techno-economic performance of the WEC. In this case, the LCOE in Yeu island is the lowest among the three. The LCOE in DK 2 is lower than that in BIMEP while BIMEP corresponds to a higher mean wave power density. Hence, more powerful wave resources do not necessarily contribute to better techno-economic performance. Furthermore, the optimal generator size differs with wave resources. The lowest LCOE values in Yeu island, DK 2 and BIMEP correspond to the designed maximum generator force of 120 kN, 100 kN and 160 kN, respectively.

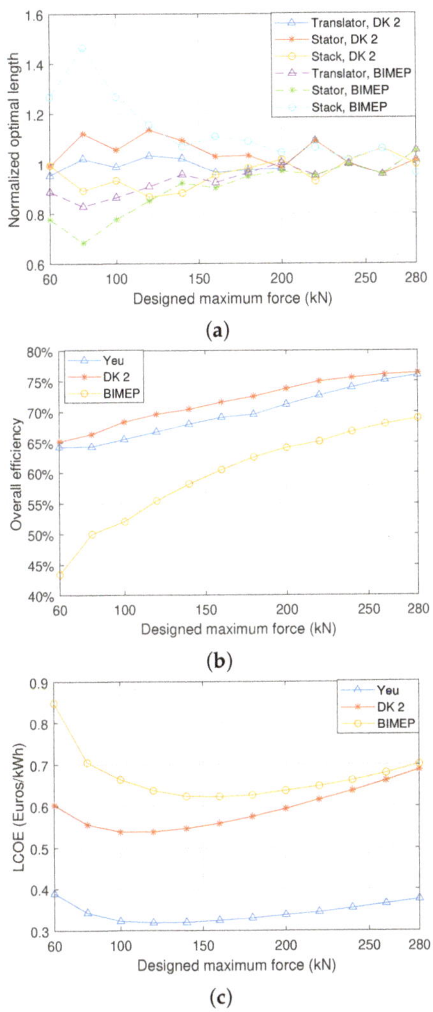

Figure 11. The optimized design parameters, the overall efficiency and the LCOE of the WEC with various designed maximum forces in three different wave sites. (**a**) The optimized design parameters for DK 2 and BIMEP, which are normalized to the corresponding values in Yeu island. (**b**) The overall efficiency of the linear generator as a function of the designed maximum generator force. (**c**) The LCOE of the WEC as a function of the designed maximum generator force.

6. Conclusions

The influence of PTO sizing on the power conversion efficiency of a linear generator in a point absorber WEC is investigated in this work. The relevance of PTO sizing to the techno-economic performance of WECs is also studied. To estimate the power production of WECs with the variously sized PTO systems, hydrodynamic modeling and generator modeling are both taken into consideration. Different methods are applied for sizing the linear PM generator for various designed maximum generator forces. Based on the results, it is concluded that the sizing of the generator is important to the LCOE of WECs. The overall conversion efficiency has a notable dependence on the size of generators. Thus, assuming a constant generator efficiency causes an obvious mis-estimate on the determination of the AEP, LCOE and the PTO capacity. In particular cases, the relative errors in the estimation of the AEP and the optimal PTO capacity reach 10% and 29%. Compared with the method based on the scaling law, the scaling with implementing the optimization of main machine parameters could better reflect the techno-economic potential of WECs when conducting PTO sizing. Moreover, the overall generator efficiency and the LCOE are notably related to the wave resource. The wave resource has a clear influence on the LCOE and overall efficiencies of the linear generator. As the overall efficiency of the generator is far from being constant, it is important to take the variation of the generator performance into consideration when sizing PTO systems of WECs for different sea sites.

Author Contributions: Conceptualization, J.T., X.W. and H.P.; methodology, J.T., X.W. and H.P.; software, J.T.; validation, H.P.; formal analysis, J.T., X.W., A.J.L., H.P. and S.A.M.; investigation, J.T., X.W., A.J.L., H.P. and S.A.M.; resources, J.T., X.W., A.J.L., H.P. and S.A.M.; data curation, J.T., X.W., A.J.L., H.P. and S.A.M.; writing—original draft preparation, J.T.; writing—review and editing, J.T., X.W., A.J.L., H.P. and S.A.M.; visualization, J.T.; supervision, A.J.L., H.P. and S.A.M.; project administration, A.J.L., H.P. and S.A.M.; funding acquisition, A.J.L., H.P. and S.A.M. All authors have read and agreed to the published version of the manuscript.

Funding: This research has received funding from China Scholarship Council under Grant: 20180695 0003.

Institutional Review Board Statement: Not applicable.

Informed Consent Statement: Not applicable.

Data Availability Statement: The data presented in this study are available on request from the corresponding author.

Conflicts of Interest: The authors declare no conflict of interest.

Appendix A

The scatter diagrams of wave states in Yeu island, DK 2 and BIMEP are shown below.

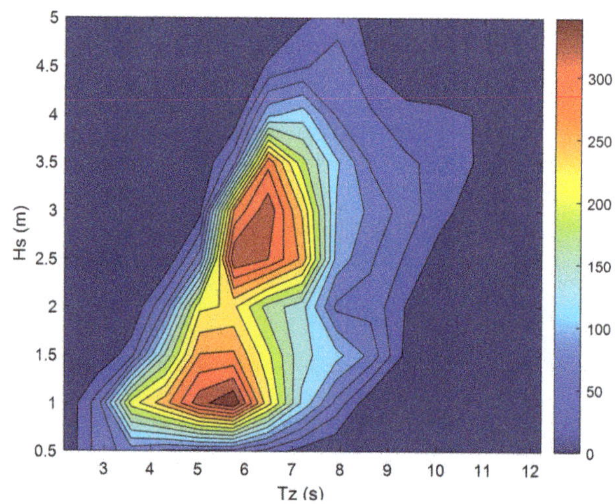

Figure A1. Hours of occurrence of each wave state of Yeu island [54].

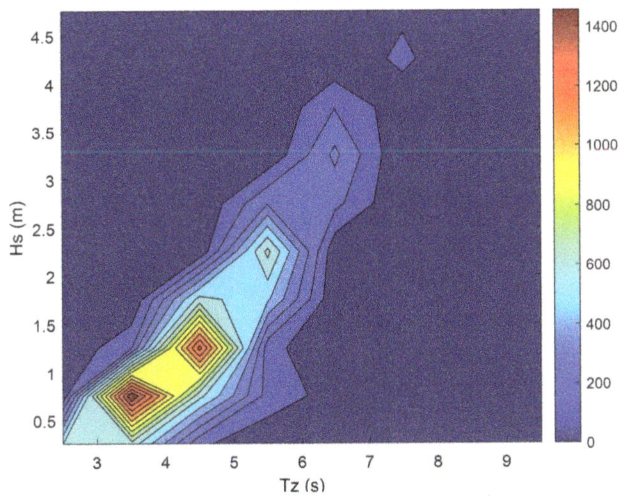

Figure A2. Hours of occurrence of each wave state of DK 2 [14].

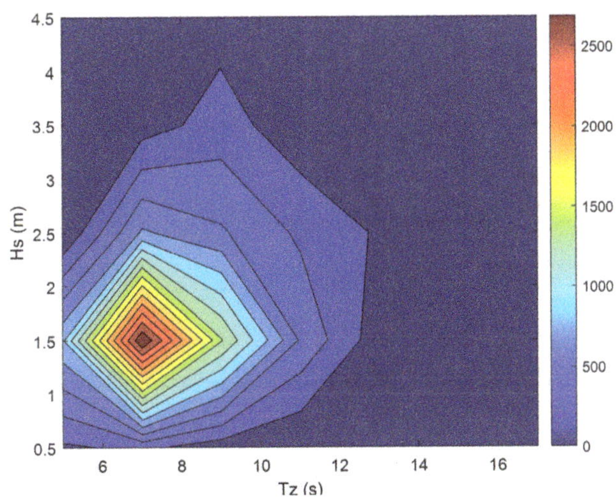

Figure A3. Hours of occurrence of each wave state of BIMEP [14].

References

1. Tan, J.; Wang, X.; Jarquin Laguna, A.; Polinder, H.; Miedema, S. The Influence of Linear Permanent Magnet Generator Sizing on the Techno-Economic Performance of a Wave Energy Converter. In Proceedings of the 2021 13th International Symposium on Linear Drives for Industry Applications (LDIA), Wuhan, China, 1 July 2021; pp. 1–6. [CrossRef]
2. Ahamed, R.; McKee, K.; Howard, I. Advancements of wave energy converters based on power take off (PTO) systems: A review. *Ocean. Eng.* **2020**, *204*, 107248. [CrossRef]
3. Aderinto, T.; Li, H. Ocean Wave energy converters: Status and challenges. *Energies* **2018**, *11*, 1250. [CrossRef]
4. Shen, W.; Chen, X.; Qiu, J.; Hayward, J.A.; Sayeef, S.; Osman, P.; Meng, K.; Dong, Z.Y. A comprehensive review of variable renewable energy levelized cost of electricity. *Renew. Sustain. Energy Rev.* **2020**, *133*, 110301. [CrossRef]
5. Lavidas, G.; Blok, K. Shifting wave energy perceptions: The case for wave energy converter (WEC) feasibility at milder resources. *Renew. Energy* **2021**, *170*, 1143–1155. [CrossRef]
6. Iglesias, G.; Astariz, S.; Vazquez, A. The economics of wave and tidal energy. *Wave Tidal Energy* **2018**, 513–532. [CrossRef]
7. Wang, L.; Zhao, T.; Lin, M.; Li, H. Towards realistic power performance and techno-economic performance of wave power farms: The impact of control strategies and wave climates. *Ocean Eng.* **2022**, *248*, 110754. [CrossRef]
8. Pecher, A. *Handbook of Ocean Wave Energy;* Springer: Berlin/Heidelberg, Germany, 2017; Volume 7. [CrossRef]

9. Tan, J.; Polinder, H.; Laguna, A.J.; Wellens, P.; Miedema, S. The Influence of Sizing of Wave Energy Converters on the Techno-Economic Performance. *J. Mar. Sci. Eng.* **2021**, *9*, 52. [CrossRef]
10. Tan, J.; Polinder, H.; Wellens, P.; Miedema, S. A feasibility study on downsizing of power take off system of wave energy converters. *Dev. Renew. Energies Offshore* **2020**, 140–148. [CrossRef]
11. Tan, J.; Polinder, H.; Miedema, S. The Fair Evaluation of Wave Energy Converters. In Proceedings of the 39th International Conference on Ocean, Offshore and Arctic Engineering, Online, 3–7 August 2020; Volume 84416, p. V009T09A016. [CrossRef]
12. Tokat, P. *Performance Evaluation and Life Cycle Cost Analysis of the Electrical Generation Unit of a Wave Energy Converter*; Chalmers Tekniska Hogskola: Göteborg, Sweden, 2018.
13. Cappelli, L.; Marignetti, F.; Mattiazzo, G.; Giorcelli, E.; Bracco, G.; Carbone, S.; Attaianese, C. Linear tubular permanent-magnet generators for the inertial sea wave energy converter. *IEEE Trans. Ind. Appl.* **2014**, *50*, 1817–1828. [CrossRef]
14. de Andres, A.; Maillet, J.; Todalshaug, J.H.; Möller, P.; Bould, D.; Jeffrey, H. Techno-economic related metrics for a wave energy converters feasibility assessment. *Sustainability* **2016**, *8*, 1109. [CrossRef]
15. Henderson, R. Design, simulation, and testing of a novel hydraulic power take-off system for the Pelamis wave energy converter. *Renew. Energy* **2006**, *31*, 271–283. [CrossRef]
16. Wang, L.; Isberg, J.; Tedeschi, E. Review of control strategies for wave energy conversion systems and their validation: The wave-to-wire approach. *Renew. Sustain. Energy Rev.* **2018**, *81*, 366–379. [CrossRef]
17. He, X.; Xiao, G.; Hu, B.; Tan, L.; Tang, H.; He, S.; He, Z. The applications of energy regeneration and conversion technologies based on hydraulic transmission systems: A review. *Energy Convers. Manag.* **2020**, *205*, 112413. [CrossRef]
18. Golbaz, D.; Asadi, R.; Amini, E.; Mehdipour, H.; Nasiri, M.; Nezhad, M.M.; Naeeni, S.T.O.; Neshat, M. Ocean Wave Energy Converters Optimization: A Comprehensive Review on Research Directions. *arXiv* **2021**, arXiv:2105.07180 [CrossRef]
19. Kofoed, J.P.; Frigaard, P.; Friis-Madsen, E.; Sørensen, H.C. Prototype testing of the wave energy converter wave dragon. *Renew. Energy* **2006**, *31*, 181–189. [CrossRef]
20. Heath, T.V. A review of oscillating water columns. *Philos. Trans. R. Soc. A* **2012**, *370*, 235–245. [CrossRef]
21. Binh, P.C.; Tri, N.M.; Dung, D.T.; Ahn, K.K.; Kim, S.J.; Koo, W. Analysis, design and experiment investigation of a novel wave energy converter. *IET Gener. Transm. Distrib.* **2016**, *10*, 460–469. [CrossRef]
22. Wu, J.; Qin, L.; Chen, N.; Qian, C.; Zheng, S. Investigation on a spring-integrated mechanical power take-off system for wave energy conversion purpose. *Energy* **2022**, *245*, 123318. [CrossRef]
23. Stansby, P.; Moreno, E.C.; Draycott, S.; Stallard, T. Total wave power absorption by a multi-float wave energy converter and a semi-submersible wind platform with a fast far field model for arrays. *J. Ocean. Eng. Mar. Energy* **2022**, *8*, 43–63. [CrossRef]
24. Falcão, A.F.O. Wave energy utilization: A review of the technologies. *Renew. Sustain. Energy Rev.* **2010**, *14*, 899–918. [CrossRef]
25. Prado, M.; Polinder, H. *Direct Drive Wave Energy Conversion Systems: An Introduction*; Woodhead Publishing Limited: Oxford, UK, 2013; pp. 175–194. [CrossRef]
26. Eriksson, M. *Modelling and Experimental Verification of Direct Drive Wave Energy Conversion*; Chalmers Tekniska Hogskola: Göteborg, Sweden, 2007.
27. Polinder, H.; Damen, M.E.C.; Gardner, F. Linear PM generator system for wave energy conversion in the AWS. *IEEE Trans. Energy Convers.* **2004**, *19*, 583–589. [CrossRef]
28. Hong, Y.; Temiz, I.; Pan, J.; Eriksson, M.; Boström, C. Damping Studies on PMLG-Based Wave Energy Converter under Oceanic Wave Climates. *Energies* **2021**, *14*, 920. [CrossRef]
29. Elwood, D.; Yim, S.C.; Prudell, J.; Stillinger, C.; von Jouanne, A.; Brekken, T.; Brown, A.; Paasch, R. Design, construction, and ocean testing of a taut-moored dual-body wave energy converter with a linear generator power take-off. *Renew. Energy* **2010**, *35*, 348–354. [CrossRef]
30. Faiz, J.; Nematsaberi, A. Linear permanent magnet generator concepts for direct-drive wave energy converters: A comprehensive review. In Proceedings of the 2017 12th IEEE Conference on Industrial Electronics and Applications, Siem Reap, Cambodia, 18–20 June 2017; pp. 618–623. [CrossRef]
31. Shuheng, Q.; Haifeng, W. Simulation of Tubular Transverse Flux Permanent Magnet Linear Generator. In Proceedings of the 2018 21st International Conference on Electrical Machines and Systems (ICEMS), Jeju, Korea, 7–10 October 2018; pp. 1766–1770. [CrossRef]
32. Chen, M.; Huang, L.; Hu, M.; Hu, B.; Ahmad, G. A Spiral Translator Permanent Magnet Transverse Flux Linear Generator Used in Direct Drive Wave Energy Converter. *IEEE Trans. Magn.* **2021**, *57*, 1–5. [CrossRef]
33. Chen, M.; Huang, L.; Li, Y.; Tan, P.; Ahmad, G.; Liu, Y.; Hu, M. Analysis of Magnetic Gearing Effect in Field-modulated Transverse Flux Linear Generator for Direct Drive Wave Energy Conversion. *IEEE Trans. Magn.* **2021**, *58*, 1–5. [CrossRef]
34. Liu, Z.; Wang, X.; Al Shami, E.; Baker, N.J.; Ji, X. A study of a speed amplified linear generator for low-frequency wave energy conversion. *Mech. Syst. Signal Process.* **2021**, *149*, 107226. [CrossRef]
35. Jing, H.; Maki, N.; Ida, T.; Izumi, M. Electromechanical design of an MW class wave energy converter with an HTS tubular linear generator. *IEEE Trans. Appl. Supercond.* **2018**, *28*, 1–4. [CrossRef]
36. Keysan, O.; Mueller, M.A. A linear superconducting generator for wave energy converters. In Proceedings of the 6th IET International Conference on Power Electronics, Machines and Drives (PEMD 2012), Bristol, UK, 27–29 March 2012; pp. 1–6. [CrossRef]

37. Jing, H.; Maki, N.; Ida, T.; Izumi, M. Design study of large-scale HTS linear generators for wave energy conversion. *IEEE Trans. Appl. Supercond.* **2017**, *27*, 1–5. [CrossRef]
38. Polinder, H.; Damen, M.E.; Gardner, F. Design, modelling and test results of the AWS PM linear generator. *Eur. Trans. Electr. Power* **2005**, *15*, 245–256. [CrossRef]
39. Prado, M.; Polinder, H. *Case Study of the Archimedes Wave Swing (AWS) Direct Drive Wave Energy Pilot Plant*; Woodhead Publishing Limited: Oxford, MI, USA, 2013; pp. 195–218. [CrossRef]
40. Polinder, H. *Principles of Electrical Design of Permanent Magnet Generators for Direct Drive Renewable Energy Systems*; Woodhead Publishing Limited: Oxford, MI, USA, 2013; pp. 30–50. [CrossRef]
41. Falnes, J.; Kurniawan, A. *Ocean Waves and Oscillating Systems: Linear Interactions Including Wave-Energy Extraction*; Cambridge University Press: Cambridge, UK, 2020. [CrossRef]
42. Penalba, M.; Kelly, T.; Ringwood, J.V. Using NEMOH for Modelling Wave Energy Converters : A Comparative Study with WAMIT. In Proceedings of the 12th European Wave and Tidal Energy Conference, Cork, Ireland , 27 August–1 September 2017; p. 10.
43. Shek, J.K.; Macpherson, D.E.; Mueller, M.A. Experimental verification of linear generator control for direct drive wave energy conversion. *IET Renew. Power Gener.* **2010**, *4*, 395–403. [CrossRef]
44. Polinder, H.; Van Der Pijl, F.F.; De Vilder, G.J.; Tavner, P.J. Comparison of direct-drive and geared generator concepts for wind turbines. *IEEE Trans. Energy Convers.* **2006**, *21*, 725–733. [CrossRef]
45. Queipo, N.V.; Haftka, R.T.; Shyy, W.; Goel, T.; Vaidyanathan, R.; Tucker, P.K. Surrogate-based analysis and optimization. *Prog. Aerosp. Sci.* **2005**, *41*, 1–28. [CrossRef]
46. De Andres, A.; Medina-Lopez, E.; Crooks, D.; Roberts, O.; Jeffrey, H. On the reversed LCOE calculation: Design constraints for wave energy commercialization. *Int. J. Mar. Energy* **2017**, *18*, 88–108. [CrossRef]
47. Kramer, M.M.; Marquis, L.; Frigaard, P. Performance Evaluation of the Wavestar Prototype. In Proceedings of the 9th European Wave and Tidal Conference, Southampton, UK, 5–9 September 2011; pp. 5–9.
48. Tokat, P.; Thiringer, T. Sizing of IPM Generator for a Single Point Absorber Type Wave Energy Converter. *IEEE Trans. Energy Convers.* **2018**, *33*, 10–19. [CrossRef]
49. Tan, J.; Polinder, H.; Laguna, A.J.; Miedema, S. The application of the spectral domain modeling to the power take-off sizing of heaving wave energy converters. *Appl. Ocean Res.* **2022**, *122*, 103110. [CrossRef]
50. Rinaldi, G.; Portillo, J.; Khalid, F.; Henriques, J.; Thies, P.; Gato, L.; Johanning, L. Multivariate analysis of the reliability, availability, and maintainability characterizations of a Spar–Buoy wave energy converter farm. *J. Ocean Eng. Mar. Energy* **2018**, *4*, 199–215. [CrossRef]
51. López-Ruiz, A.; Bergillos, R.J.; Lira-Loarca, A.; Ortega-Sánchez, M. A methodology for the long-term simulation and uncertainty analysis of the operational lifetime performance of wave energy converter arrays. *Energy* **2018**, *153*, 126–135. [CrossRef]
52. Majidi, A.G.; Bingölbali, B.; Akpınar, A.; Iglesias, G.; Jafali, H. Downscaling wave energy converters for optimum performance in low-energy seas. *Renew. Energy* **2021**, *168*, 705–722. [CrossRef]
53. Babarit, A. A database of capture width ratio of wave energy converters. *Renew. Energy* **2015**, *80*, 610–628. [CrossRef]
54. Babarit, A.; Hals, J.; Muliawan, M.J.; Kurniawan, A.; Moan, T.; Krokstad, J. Numerical benchmarking study of a selection of wave energy converters. *Renew. Energy* **2012**, *41*, 44–63. [CrossRef]

Journal of
Marine Science and Engineering

MDPI

Article

Improving Computational Efficiency in WEC Design: Spectral-Domain Modelling in Techno-Economic Optimization

Mauro Bonfanti * and Giuseppe Giorgi

Marine Offshore Renewable Energy Lab (MOREnergyLab), Department of Mechanical and Aerospace Engineering (DIMEAS), Politecnico di Torino, Corso Duca degli Abruzzi 24, 10129 Turin, Italy
* Correspondence: mauro.bonfanti@polito.it; Tel.: +39-3388006163

Abstract: Wave energy converter (WEC) optimization often underlines incremental and iterative approaches that result in suboptimal solutions, since all the elements that concur with a techno-economical evaluation are optimized separately due to computation constraints. A design process should rely on precise WEC models to ensure high result accuracy while minimizing the computational demand. These conflicting objectives can be addressed with non-linear time-domain models, known to be numerically accurate, and frequency-domain models due to their high computational efficiency. This work pursues the development of an all-encompassing optimization tool for a gyroscopic-type WEC called ISWEC that applies a new modelling technique named spectral-domain technique as a substitution to the complex time-domain model previously employed. In particular, the spectral-domain technique provides accurate and fast performance predictions of the ISWEC system and offers the possibility to model a hydraulic power take-off, not representable in the frequency domain. The article illustrates techno-economic trends associated with an early-stage design of the ISWEC in high-energy sea-sites, where the low-speed and high-torque profiles call for the use of hydraulic transmissions as opposed to the old electro-mechanical transmissions. The design tool proposed could facilitate the development of WEC technologies via efficient and accurate power assessment and via the possibility of carrying out advanced techno-economic optimisation that goes beyond linear models.

Keywords: genetic algorithm; optimal design; spectral-domain model; hydraulic PTO; techno-economic optimization

Citation: Bonfanti, M.; Giorgi, G. Improving Computational Efficiency in WEC Design: Spectral-Domain Modelling in Techno-Economic Optimization. *J. Mar. Sci. Eng.* **2022**, *10*, 1468. https://doi.org/10.3390/jmse10101468

Academic Editor: Domenico Curto

Received: 16 September 2022
Accepted: 7 October 2022
Published: 10 October 2022

1. Introduction

In the renewable energy context, wave energy converters (WECs) are not yet economically competitive with other technologies, i.e., onshore and offshore wind turbines, due to the excessive cost of energy at the current stage of development. Moreover, despite the enormous energy potential of ocean waves, the extraction capabilities of WECs could be poor due to non-optimal designs. The main challenges faced in designing WECs and their PTO systems include [1,2] the high variability of the wave resource in the short-term period; the slow, bi-directional and irregular kinematic behaviour of bodies forced by waves; and the harsh and corrosive marine environment that impacts the life cycle of the WEC. The optimization of a WEC system is dominated by iterative methods in which design parameters are managed in a trial and error fashion [3]. The authors of this work already stressed the use of a multi-objective techno-economic perspective by proposing an all-embracing design tool [4–6] that implements better design practices in respect to the current state of the art considering representative wave-to-wire models of WECs, technological and control boundaries, and economic evaluations. Recently, the use of heuristic optimizations gained popularity in the wave energy context due to their ability to deal with multi-objective targets even in the presence of numerous free variables. Two of the first application of a genetic optimization can be found in [7,8], where the SEAREV WEC was designed through a GA in a multi-objective framework with techno-economic considerations.

Both constraints regarding the design and control optimization were implemented. In [4–6], several applications of genetic algorithms (GA) has been reviewed, and many recent works deal with the use of GA for WEC optimization. In [9], an extensive review of the layout optimization strategies of offshore parks was collected, focusing, among other things, on evolutionary computation techniques, which are the most widespread. In [10], a GA method was applied to optimize the orifice diameter and draft depth of an oscillating water column. In [11–13], various GAs were proposed to optimize the layouts of a WEC array; in [14], a differential evolution algorithm was used to optimize an oscillating buoy-type WEC relying on submerged buoy volume to power the capture ratio. In [15], which is an extension of [16], both GAs and particle swarm optimization were applied to the geometric optimization of different WEC types including manufacturability considerations. In [17], a holistic optimization method that relates both energy production maximization (for operational conditions) and structural assessment (for severe conditions) were proposed at an early stage of WEC development. Despite the fact that valuable insights into WEC design have been reported in previous works, they mainly promote the use of frequency-domain or simplified time-domain models in order to bound the computational time required to optimize the WEC architecture, and on simplified key performances that lack of techno-economic considerations. In particular, to the best of the authors' knowledge, no unified guidelines or closed-form procedures are reported in the literature that effectively provide a design process with particular regard to the power take-off (PTO) system. Regarding the case of a hydraulic PTO (HPTO), Pedersen [18] assessed that *"in references where the hydraulic PTO system is mentioned, this is generally a simple non-optimized fluid power transmissions with very limited efficiency"*. In [19], a model-based design of a multiple cylinder HPTO for the Wavestar WEC [20] was proposed. The design relies on the transmission requirements, taking into account the efficiencies of the transmission components. In [21], which was an extension and a continuation of the research of [22], the authors investigated how the magnitude of the HPTO force influences the power capture ability of a two-raft-type WEC. According to Zheng, these would provide a valuable guidance for the optimal design of a PTO unit by altering the parameters of the hydraulic components. In [23], a parametric evaluation of the hydraulic transmission was carried out to highlight the influence of each variable on the capture width ratio of the device. Holistic optimization approaches were proposed in [24], where a GA is applied to the control parameters and the attachment point of the hydraulic cylinders to the absorber arms. The optimization was conducted on 8 different HPTO configurations. In [25], an HPTO system was optimised using a special GA optimisation tool in the Simcenter Amesim software [26]. The simulation results showed that GA was effective at determining the optimal configuration parameters of an HPTO system, such as the diameter of the piston, the diameter of the rod, accumulator capacity, accumulator pre-charge pressure, and hydraulic motor displacement. The objective of the work was to maximise the energy production of the system. In the work of Calvario [27], the GA was formulated as a constrained optimization problem with a death-penalty method, assigning a high cost to penalize infeasible solutions. The optimization aimed to maximize the average absorbed power. The decision variables were the distance of the mounting point of the cylinder and PTO control parameters. These variables were optimized with a heuristic approach based on the evolutionary computation already applied in [28,29].

Due to the proven effectiveness of holistic methods, this work applies a GA to design a well-known ISWEC (inertial sea wave energy converter) device [30] introducing a determinant innovation: the use of the spectral-domain modelling technique in a multi-objective design process. The proposed design logic was first introduced by Sirigu in [4] and applied to another device developed at the Politecnico di Torino, named PeWEC [31]. Afterwards, Carapellese [5] and Giorcelli [6] proposed a multi-objective optimization algorithm, know as non-dominated sorting genetic algorithm II (NSGA-II) [32], and compared it with the GA used in [4]. This work follows the optimization algorithm proposed in [4] introducing a thorough description of a HPTO, whereas previous optimizations was based on simpler electro-mechanical PTO. The spectral-domain model (SDM) applied to the ISWEC has been demonstrated to be more accurate than the frequency-domain model used in [4] and faster than its time-domain counterpart [33]. This allows to manage the time-consuming optimization process carried out by the GA in a reasonable time due to a fast and accurate SDM[1]. Moreover, the use of a frequency-domain model for the HPTO is not possible since no linear representation exists for a hydraulic transmission [35,36]. The work presents some techno-economic trends related to the ISWEC, focusing on a multi-objective evaluation of its performances in terms of: (I) the maximization of the annual energy production and (II) the minimization of the associated cost of energy. The ISWEC system will be optimized in respect to its main components: the hull, the gyroscope, and the PTO; their parametrization is presented and discussed along with the resultant techno-economic trends. The influence of the hull dimensions, gyroscope size, PTO components and control is reported against the objectives pursued by the GA optimization. Overall, the main outcome will be providing a detailed practice of optimising a WEC during its early-stage development through the use of SDMs, taking into account techno-economical aspects. This paper also shows how the SDM approach can be applied to various components, even those that are highly nonlinear, and still be implemented in a computationally demanding holistic optimization.

This work is organised in five sections: Section 2 presents the working principle of the ISWEC and its numerical model used into the design tool; Section 3 illustrates the architecture of the design tool and the parametrisation of the ISWEC, explaining how a single solution is evaluated on the sea-site of interest; Section 4 discusses the optimization results. Some final remarks are given in Section 5.

2. The ISWEC Concept

An ISWEC device is a well-known gyroscope based WEC developed by the Marine Offshore Renewable Energy Lab (MOREnergyLab). It is composed of a sealed hull carrying two or more gyroscopic units inside, as shown in Figure 1.

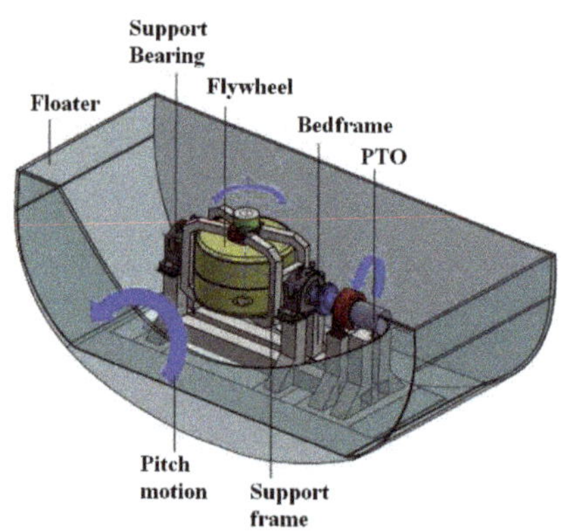

Figure 1. ISWEC architecture and main internal components.

Each unit is constituted of a flywheel driven by an electric motor, which is mounted on a support frame that is directly connected to an electrical generator. The wave energy is captured by damping the rocking motion of the floater by controlling the oscillation of the gyroscopes. Notwithstanding the energy required to keep the flywheel in rotation, the possibility to regulate its speed allows the adaptation of the natural resonant frequency of the system to the incoming wave [37,38]. Moreover, the ISWEC resonance could be enhanced through the so-called U-tank, a U-shaped water tank installed inside the floater, used to tune the pitch resonance frequency as a function of the prevalent incoming wave period [39]. However, the use of U-tank technology requires a different design of the ISWEC, accounting for its encumbrances and morphology, which affect both its fabrication and its dynamic behaviour. Its technological and control aspects are still under study.

Therefore, the U-tank system is neither reported in Figure 1 nor considered in this paper. The gyroscope mechanical energy can be converted into electrical energy through different PTO technologies. The gyroscope and PTO components are enclosed within the main structure, ensuring low maintenance and risk of failure. A detailed description of the ISWEC device components, along with its technological concept, can be found in [40–42].

2.1. Time-Domain Model

A time-domain model (TDM) represents an appealing solution to provide the final steps toward cost and performance estimates of a WEC system due to its ability to model non-linear hydrodynamic effects [43], to test complex control algorithms [44], and to simulate complex WEC subsystems [45] or structural components [46]. For a TDM, a generic non-linear state-space equation reads:

$$\mathbf{M}\ddot{\mathbf{X}} + \mathbf{B}\dot{\mathbf{X}} + \mathbf{K}\mathbf{X} + \boldsymbol{\Theta}(\mathbf{X}, \dot{\mathbf{X}}, \ddot{\mathbf{X}}) = \mathbf{f}(t) \tag{1}$$

System (1) includes the state variable of interest $\mathbf{X}$, the mass matrix $\mathbf{M}$, the damping matrix $\mathbf{B}$, the stiffness matrix $\mathbf{K}$, the non-linear function $\boldsymbol{\Theta}(\mathbf{X}, \dot{\mathbf{X}}, \ddot{\mathbf{X}})$ and the external forces $\mathbf{f}(t)$. The non-linear term $\boldsymbol{\Theta}(\mathbf{X}, \dot{\mathbf{X}}, \ddot{\mathbf{X}})$ includes several non-linearities involving the gyroscope and, depending on the ISWEC architecture, the PTO. The TDM presented hereafter is reported with the purpose to make explicit the equations used to derive its spectral-domain counterpart. In fact, the SDM is derived from the non-linear equations that describe the dynamic behaviour of the system in the time-domain.

2.1.1. Floater and Gyroscope System

For the hydrodynamic model, *Cummins' equation* [47] is used here with the assumption of linear potential flow theory [48] and mono-directional waves aligned with the longitudinal axis of the hull, as shown in Figure 2a. Concerning the gyroscope equation, a single degree of freedom (DoF) model of the gyroscope is considered coherently with the hypothesis of a mono-directional wave and the planar motion of the floater. The gyroscopic system is excited only along its precession axis, indicated with ε (see Figure 2b). Moreover, an eccentric mass, named the pendulum in Figure 2b, is fixed at the support frame of the gyroscope, avoiding the stalling position of the gyroscope around its 90 deg orientation due to the cancellation of the gyroscopic effect [42].

The floater-gyroscope equations read:

$$\left[\mathbf{M}_f + \mathbf{A}(\infty)\right]\ddot{\mathbf{X}}_f + \mathbf{C}_R\boldsymbol{\zeta} + \mathbf{D}_R\dot{\mathbf{X}}_f + \mathbf{K}_f\mathbf{X}_f = \mathbf{f}_E(t) - \mathbf{f}_{gf} \tag{2}$$

$$\dot{\boldsymbol{\zeta}} = \mathbf{A}_R\boldsymbol{\zeta} + \mathbf{B}_R\dot{\mathbf{X}}_f \tag{3}$$

$$I_g\ddot{\varepsilon} + m_p d_p g\sin(\varepsilon) = J\dot{\varphi}\dot{\delta}\cos(\varepsilon) - T_{HPTO} \tag{4}$$

where $\mathbf{X}_f = \begin{bmatrix} x & z & \delta \end{bmatrix}^T$ are the floater planar DoFs, $\mathbf{M}_f$ is the inertia matrix, $\mathbf{A}(\infty)$ is the added mass contribution evaluated for an infinite oscillation frequency, $\mathbf{K}_f$ is the hydrostatic stiffness matrix, $\mathbf{f}_E(t)$ is the wave excitation force, and $\mathbf{f}_{gf} = \begin{bmatrix} 0 & 0 & -J\dot{\varphi}\dot{\varepsilon}\cos(\varepsilon) \end{bmatrix}^T$ is the gyroscopic reaction acting on the hull, where J is the rotational inertia of the flywheel, $\dot{\varphi}$ is the angular speed of the flywheel, and ε and $\dot{\varepsilon}$ are the angular displacement and velocity of the gyroscope, respectively. The vector $\boldsymbol{\zeta}$ represents the state vector that approximates the radiation force contributions through the state space matrices $\mathbf{A}_R$, $\mathbf{B}_R$, $\mathbf{C}_R$ and $\mathbf{D}_R$ [49]. I_g is the inertia of the gyroscope around the precession axis, m_p is the pendulum mass, and d_p is the distance between the centre of mass of the pendulum and the ε DoF. T_{HPTO} indicates the HPTO reaction on the gyroscope.

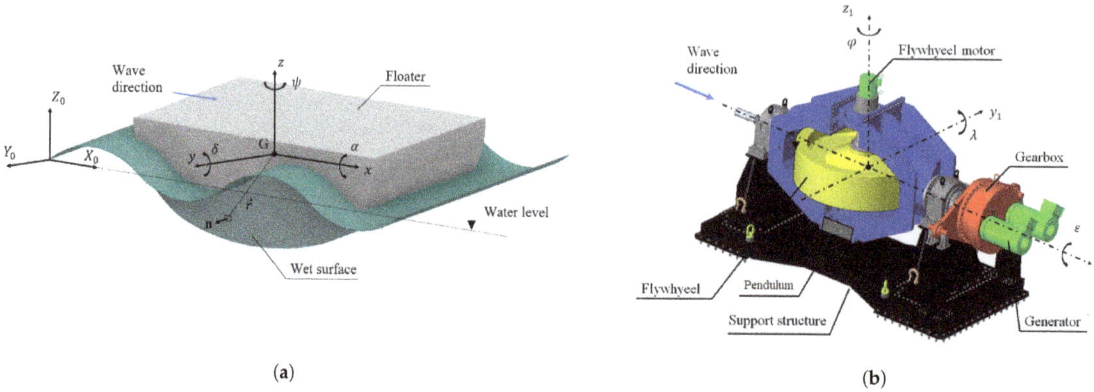

Figure 2. (**a**) Geometrical and inertial reference system of the ISWEC device. (**b**) Detailed design of the gyroscope system and its reference frame.

2.1.2. Hydraulic PTO

For our system, the HPTO architecture is shown in Figure 3. The prime mover (the gyroscope) is coupled to a hydraulic pump which is in a closed-circuit connection with a hydraulic motor. At the other end of the transmission, pressurised fluid is used to move a rotary hydraulic motor coupled with an electrical generator. A clutch-declutch valve is installed to perform the well-known declutch control [50,51]. Downstream of the clutch-declutch valve, a Graetz bridge, composed of four check-valves, is employed to rectify the bi-directional flow generated by the oscillating motion of the pump. Two cylinder-type accumulators are used to smooth the flow fluctuations and to provide a minimum charge pressure for the low-pressure ram in order to avoid cavitation. A simplified model of the HPTO is considered here to derive its spectral-domain version [35]. As proposed in [52,53], under the assumption of large high-pressure and low-pressure accumulators, the downstream of the hydraulic transmission is decoupled from the upstream part, and the pressure drop over the hydraulic pump and motor is considered as a constant. Thus, the magnitude of the T_{HPTO} depends only on such a pressure drop Δp and the pump displacement D_p as follows:

$$T_{HPTO} = u_c \operatorname{sign}(\dot{\varepsilon}) \Delta p D_p \tag{5}$$

Under this simplification, the system is truncated to the ε variable since the hydraulic PTO states, such circuit pressures, accumulator volumes and motor-generator speed, are not considered. The pressure drop Δp is considered as a control parameter, and u_c is the clutch-declutch variable, equal to 0 or 1 depending on the angular speed of the hydraulic pump [34,35,50]. The accuracy of this simple formulation is questionable in the presence of small accumulators since their dynamic behaviour could influence the power extraction of the whole system. In this regard, a TDM model that accounts for the accumulation system, the relief valve, the hydraulic motor and generator, should be used to accurate predict the transmission power performances. On the other hand, this formulation is justified by the need of derive a spectral-domain framework for the HPTO, with the main aim to improve the computational efficiency of the ISWEC design at an early stage. Despite the fact that the ISWEC system has been already deployed in full-scale, the research effort behind its development is not over, driven by the need to make the resultant cost of energy economically competitive. Since no previous work analysed the ISWEC with a HPTO, representing a promising solution to reduce the device cost, the design stage proposed here has to be considered an early stage, in which a novel mathematical representation (spectral-domain) has been implemented to speed up the optimization process. Further works will deal with the TDM applied to design the whole hydraulic transmission.

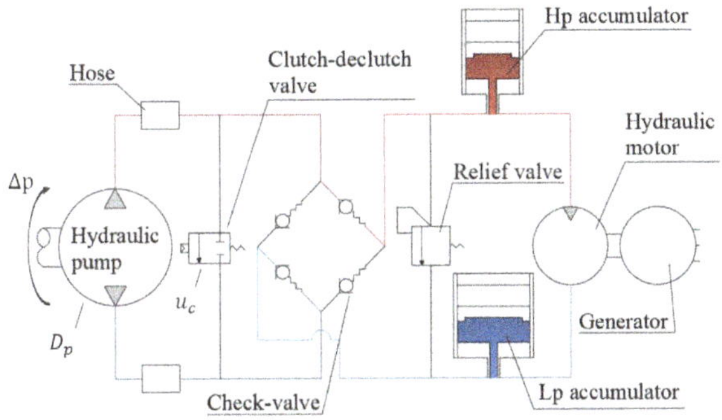

Figure 3. Hydraulic PTO components. Adapted from [34], with permission from IEEE, 2021.

2.2. Spectral-Domain Model

The spectral domain is a modelling technique based on probabilistic models of a system with the assumption that its input may be represented as a stochastic ergodic process [54,55]. In order to obtain an approximate solution of the generic time-domain Equation (1), it is useful to introduce an equivalent linear system, as follows:

$$\mathbf{M}_S\ddot{\mathbf{X}} + \mathbf{B}_S\dot{\mathbf{X}} + \mathbf{K}_S\mathbf{X} = \mathbf{f}(t) \tag{6}$$

where $\mathbf{X}$ is the augmented state vector comprising the hull states ($\mathbf{X}_f$), the radiation states (ζ) and the precession angle of the gyroscope (ε). The matrices $\mathbf{M}_S$, $\mathbf{B}_S$ and $\mathbf{K}_S$ are defined as:

$$\mathbf{M}_S = \mathbf{M} + \mathbf{M}_{eq} \tag{7a}$$

$$\mathbf{B}_S = \mathbf{B} + \mathbf{B}_{eq} \tag{7b}$$

$$\mathbf{K}_S = \mathbf{K} + \mathbf{K}_{eq} \tag{7c}$$

The matrices (7) are composed of a linear part, concerning the linear terms of the time-domain Equation (1), and an equivalent part indicated with the subscript *"eq"* [36,56]. Under the assumption of a Gaussian distribution of the system response, in order to minimize the mathematical expectation of the difference between the non-linear system and the linearised system, the equivalent matrices can be calculated as follows:

$$\mathbf{M}_{eq} = \langle \nabla_{\ddot{\mathbf{X}}} \boldsymbol{\Theta} \rangle \tag{8a}$$

$$\mathbf{B}_{eq} = \langle \nabla_{\dot{\mathbf{X}}} \boldsymbol{\Theta} \rangle \tag{8b}$$

$$\mathbf{K}_{eq} = \langle \nabla_{\mathbf{X}} \boldsymbol{\Theta} \rangle \tag{8c}$$

where the operator $\nabla_{\mathbf{X}}$ indicates the gradient with respect to the vector $\mathbf{X}$, and $\langle \cdot \rangle$ is the expected value operator. The equivalent matrices could be interpreted as a statistical representation of the time-varying behaviour of the system in time-domain. In fact, as reported in Equation (1), the non-linear time-domain function can be seen as time-varying inertial, damping and stiffness coefficients that are statistically linearised. Then, the linearised matrices can be used to compute the transfer function between the wave forces and the ISWEC states as:

$$\mathbf{H}_{\mathbf{fX}}(\omega) = \frac{\mathbf{I}}{-\omega^2\mathbf{M}_S + i\omega\mathbf{B}_S + \mathbf{K}_S} \tag{9}$$

where ω is the wave frequency, and $\mathbf{I}$ is an identity matrix of appropriate dimension. Similarly to a frequency-domain model, the WEC output statistics are obtained through the power spectral density (PSD) of the input wave force and the linear transfer function of the WEC, as follows:

$$\mathbf{S_{XX}}(\omega) = \mathbf{H_{fX}}(\omega) S_{\mathbf{ff}}(\omega) \mathbf{H}^{*}_{\mathbf{fX}}(\omega) \tag{10}$$

where $\mathbf{H}_{\mathrm{f}X}$ represents the WEC numerical model in the spectral domain, and $S_{\mathrm{ff}}(\omega)$ is the PSD of the input wave force. The operator $*$ stands for the complex-conjugate transpose. An SDM represents an intermediate solution between the time and frequency domains that exploits a statistical linearisation method to include non-linear effects inside the simulation. No analytical solution exists to this problem, and an iterative procedure is employed [36,54]. This modelling technique founds its main application in other engineering fields, e.g., wind loading and earthquake response calculations [55]. In the WEC industry, the first application dates back to 1983, when Gudmestad [57] applied an SDM to solve the dynamics of a floating offshore structure. More recent applications of the spectral-domain technique applied to WECs are reported in [36,56,58,59]. The full analytical description of the SDM for the ISWEC is derived in [34,35], demonstrating the high potential associated with this modelling framework: an almost perfect overlap between the SDM and TDM of the HPTO is reported for most of the control conditions of the ISWEC, while the reduction in computational time is demonstrated to be, at least, one order of magnitude between the two models. Hereafter, the resulting SDM equations are shown, which will feed into the GA presented in this paper.

The linear terms **M**, **B** and **K** can be defined as:

$$\mathbf{M} = \begin{bmatrix} \mathbf{M}_f + \mathbf{A}(\infty) & \mathbf{0} & \mathbf{0} \\ \mathbf{0} & \mathbf{0} & \mathbf{0} \\ \mathbf{0} & \mathbf{0} & I_g \end{bmatrix} \tag{11a}$$

$$\mathbf{B} = \begin{bmatrix} \mathbf{D}_R & \mathbf{0} & \mathbf{0} \\ -\mathbf{B}_R & \mathbf{I} & \mathbf{0} \\ \mathbf{0} & \mathbf{0} & 0 \end{bmatrix} \tag{11b}$$

$$\mathbf{K} = \begin{bmatrix} \mathbf{K}_f & \mathbf{C}_R & \mathbf{0} \\ \mathbf{0} & -\mathbf{A}_R & \mathbf{0} \\ \mathbf{0} & \mathbf{0} & 0 \end{bmatrix} \tag{11c}$$

The terms **0** and **I** are zeros and identity matrices of appropriate dimensions. The ISWEC shows non-linearities in term of gyroscope actions, pendulum elastic recall, and a constant-amplitude coulomb-like PTO torque acting on the gyroscope axis. The equivalent matrices are obtained as follows:

$$\mathbf{M}_{eq} = \mathbf{0} \tag{12a}$$

$$\mathbf{B}_{eq} = \begin{bmatrix} 0 & 0 & 0 & \mathbf{0} & 0 \\ 0 & 0 & 0 & \mathbf{0} & 0 \\ 0 & 0 & 0 & \mathbf{0} & -J\dot{\varphi}\langle\cos(\varepsilon)\rangle \\ \mathbf{0} & \mathbf{0} & \mathbf{0} & \mathbf{0} & \mathbf{0} \\ 0 & 0 & J\dot{\varphi}\langle\cos(\varepsilon)\rangle & \mathbf{0} & \langle\frac{\partial T_{HPTO}}{\partial\dot{\varepsilon}}\rangle \end{bmatrix} \tag{12b}$$

$$\mathbf{K}_{eq} = \begin{bmatrix} 0 & 0 & 0 & \mathbf{0} & 0 \\ 0 & 0 & 0 & \mathbf{0} & 0 \\ 0 & 0 & 0 & \mathbf{0} & J\dot{\varphi}\langle\dot{\varepsilon}\sin(\varepsilon)\rangle \\ \mathbf{0} & \mathbf{0} & \mathbf{0} & \mathbf{0} & \mathbf{0} \\ 0 & 0 & 0 & \mathbf{0} & m_p d_p g\langle\cos(\varepsilon)\rangle - J\dot{\varphi}\langle\dot{\delta}\sin(\varepsilon)\rangle \end{bmatrix} \tag{12c}$$

2.2.1. Floater and Gyroscope Equations

For what concerns the stiffness matrix $\mathbf{K}_{eq}$, the term $J\dot{\varphi}\langle\dot{\varepsilon}\sin(\varepsilon)\rangle$ is zero since $\dot{\varepsilon}$ and ε are uncorrelated [60] with zero-mean, and the $\sin(\cdot)$ function is odd. Second, the pendulum contribution and the gyroscopic effect are linearised as follows [35]:

$$m_p d_p g\langle\cos(\varepsilon)\rangle = m_p d_p g e^{-\frac{\sigma_\varepsilon^2}{2}} \tag{13}$$

$$-J\dot{\varphi}\langle\dot{\delta}\sin(\varepsilon)\rangle = -\frac{J\dot{\varphi}}{2\pi\sigma_{\dot{\delta}}\sigma_\varepsilon\sqrt{1-\rho_{\dot{\delta}\varepsilon}^2}}\int_{-\infty}^{\infty}\int_{-\infty}^{\infty}\dot{\delta}\sin(\varepsilon)e^{-\frac{1}{2\left(1-\rho_{\dot{\delta}\varepsilon}^2\right)}\left(\frac{\dot{\delta}^2}{\sigma_{\dot{\delta}}^2}-2\rho_{\dot{\delta}\varepsilon}\frac{\dot{\delta}\varepsilon}{\sigma_{\dot{\delta}}\sigma_\varepsilon}+\frac{\varepsilon^2}{\sigma_\varepsilon^2}\right)}d\dot{\delta}\,d\varepsilon \tag{14}$$

where σ_ε are $\sigma_{\dot{\delta}}$ are the standard deviations of the gyroscope precession motion and the pitching rate, respectively, $\rho_{\dot{\delta}\varepsilon} = \frac{\sigma_{\dot{\delta}\varepsilon}}{\sigma_{\dot{\delta}}\sigma_\varepsilon}$ is the correlation coefficient of $\dot{\delta}$ and ε, and $\sigma_{\dot{\delta}\varepsilon}$ is the covariance between $\dot{\delta}$ and ε. Considering the equivalent damping matrix $\mathbf{B}_{eq}$, the term $J\dot{\varphi}\langle\cos(\varepsilon)\rangle$ can be derived as:

$$J\dot{\varphi}\langle\cos(\varepsilon)\rangle = J\dot{\varphi}e^{-\frac{\sigma_\varepsilon^2}{2}} \tag{15}$$

2.2.2. Hydraulic PTO Equation

The HPTO torque defined in Equation (5) can be redefined as a piecewise constant function of the declutch and clutch speed limits $\dot{\varepsilon}_c$ and $\dot{\varepsilon}_d$ with the following formulation [35]:

$$T_{HPTO} = \begin{cases} 0 & \text{if } -\dot{\varepsilon}_d < \dot{\varepsilon} < \dot{\varepsilon}_d \\ \frac{\Delta p D_p}{\dot{\varepsilon}_c - \dot{\varepsilon}_d}(\dot{\varepsilon} - \dot{\varepsilon}_d) & \text{if } \dot{\varepsilon}_d \leq \dot{\varepsilon} < \dot{\varepsilon}_c \\ \frac{\Delta p D_p}{\dot{\varepsilon}_c - \dot{\varepsilon}_d}(\dot{\varepsilon} + \dot{\varepsilon}_d) & \text{if } -\dot{\varepsilon}_c < \dot{\varepsilon} \leq -\dot{\varepsilon}_d \\ \Delta p D_p & \text{if } \dot{\varepsilon} \geq \dot{\varepsilon}_c \\ -\Delta p D_p & \text{if } \dot{\varepsilon} \leq -\dot{\varepsilon}_c \end{cases} \tag{16}$$

Then, its spectral domain representation can be derived as:

$$\left\langle\frac{\partial T_{HPTO}}{\partial\dot{\varepsilon}}\right\rangle = \frac{\Delta p D_p}{\dot{\varepsilon}_c - \dot{\varepsilon}_d}\left(\text{erf}\left(\frac{\dot{\varepsilon}_c}{\sqrt{2}\sigma_{\dot{\varepsilon}}}\right) - \text{erf}\left(\frac{\dot{\varepsilon}_d}{\sqrt{2}\sigma_{\dot{\varepsilon}}}\right)\right) \tag{17}$$

From the hypotheses of a Gaussian sea as an input and Gaussian variables as an output, the expected value of the gross power extracted from the hydraulic pump follows [36,56]:

$$\begin{aligned} \overline{P}_g &= \left\langle\frac{\Delta p D_p}{\dot{\varepsilon}_c - \dot{\varepsilon}_d}\left(\text{erf}\left(\frac{\dot{\varepsilon}_c}{\sqrt{2}\sigma_{\dot{\varepsilon}}}\right) - \text{erf}\left(\frac{\dot{\varepsilon}_d}{\sqrt{2}\sigma_{\dot{\varepsilon}}}\right)\right)\dot{\varepsilon}^2\right\rangle = \\ &= \frac{\Delta p D_p}{\dot{\varepsilon}_c - \dot{\varepsilon}_d}\left(\text{erf}\left(\frac{\dot{\varepsilon}_c}{\sqrt{2}\sigma_{\dot{\varepsilon}}}\right) - \text{erf}\left(\frac{\dot{\varepsilon}_d}{\sqrt{2}\sigma_{\dot{\varepsilon}}}\right)\right)\sigma_{\dot{\varepsilon}}^2 \end{aligned} \tag{18}$$

Concerning the net power, the hydraulic circuit efficiency has to be considered for a correct quantification of its power extraction capabilities. Through the expected values of the mean absolute pump torque and angular speed, the expected pump efficiency is computed via the lookup table reported into the supplier catalogues (Bosch Rexroth in our case [61]). The motor-generator efficiencies along with the circuit power losses are considered constant and derived from the supplier catalogues and from the analyses reported in [62]. The circuit efficiency is fixed to 95% to obtain an average global efficiency ranging from 70% to 80% [62,63]. Then, the net power expression follows:

$$\overline{P}_n = \overline{P}_g\overline{\eta}_p\eta_c\eta_{mg} \tag{19}$$

where $\overline{\eta}_p$ is the expected value of the pump efficiency, η_c is the hydraulic circuit efficiency, and η_{mg} the motor-generator efficiency.

3. Design Tool Architecture

The optimization tool proposed here implements three consecutive steps, explained in detail hereafter and reported in Figure 4:

- Single device definition;
- Single device optimization;
- Key performance calculation.

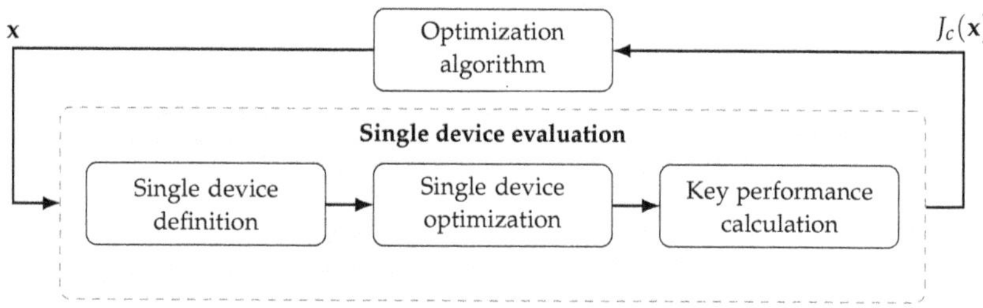

Figure 4. Architecture of the optimization tool. $\mathbf{x}$ is the vector of free parameters that define a unique ISWEC device, and $J_c(\mathbf{x})$ is the fitness function used as a performance indicator of the device $\mathbf{x}$.

Due to its numerical formulation and architecture, the hydraulic system should be evaluated both with an SDM and TDM. For the purpose of this work, the ISWEC device is designed accounting for the hull, gyroscope and the upstream part of the HPTO, considering only the hydraulic pump and its control. Further optimizations will be required to optimize the remaining components of the HPTO (accumulators, hydraulic motor and electrical generator) since only a time-domain framework is suitable to simulate all the hydraulic transmission [34,35]. An early optimization stage based on the SDM is suggested due to the high computational effort required by the TDM that is impossible to be used to design the whole system.

3.1. Device Parametrization and Assumptions

The number of parameters that describes the ISWEC device can be numerous and not handy to manage, especially in a preliminary design phase. Therefore, some simplifications and assumptions are needed to relate some system parameters in order to uniquely define a device. Here, 12 free parameters are chosen to describe the ISWEC, summarised in Table 1 and hereafter discussed in detail: six for the hull, four for the gyroscope and two for the HPTO. For what regards the hull and the gyroscope, the lower and upper bounds (LB and UB, respectively) of the optimization parameters have been chosen according to technological and manufacturing constraints, derived both from experience gained on two full-scale ISWEC prototypes deployed in the Mediterranean Sea and from limits imposed by the shipyards considered for system fabrication. Concerning the bearings and the hydraulic pump, they depend on commercial components available on the market, and the GA considers them as a "single solution on catalogue", as explained hereafter. Then, the hydraulic control refers to four different available logics to be implemented, and it is still considered as an integer "id".

Table 1. Design parameters used by the genetic algorithm to define a single ISWEC device. Lower (LB) and upper (UB) bounds are used to constraint the exploration domain.

		Hull		
Design Parameter	**Symbol**	**Unit**	**LB**	**UB**
Hull length	L	m	10	45
Hull width factor	W_f	–	1	4
Bow/stern circ. ratio	h_f	–	0.5	1
Height ratio	k_f	–	0.5	1
Maximum pitch angle	δ_0	deg	10	20
Ballast filling ratio	BFR	–	0.35	1
		Gyroscope		
Design Parameter	**Symbol**	**Unit**	**LB**	**UB**
Flywheel inertia	J	kgm^2	10,000	45,000
Gyroscope units	n_{gyros}	–	2	4
Pendulum mass	m_p	kg	500	8000
Bearings id	ID_b	–	1	15
		Hydraulic PTO		
Design Parameter	**Symbol**	**Unit**	**LB**	**UB**
Pump id	ID_p	–	1	36
Hydraulic control id	ID_c	–	1	4

3.1.1. Floater Geometry and Parameters

The floater shape and its inertial properties play an important role in the dynamics and power absorption of the device. The hull parametrization is based on a previous work of the authors [4] and relies on past experience gained in developing the full-scale prototypes of the ISWEC. For both technological effectiveness [64] and hydrodynamic performance (minimal viscous damping along the pitch degree of motion [65]), the hull profile is assumed to be cradle-shaped; then, it can be composed of a bottom circumference tangential to two circumferences in the bow/stern sections, while the transversal section is extruded for all the hull width, resulting in a floater symmetrical with respect to the $y-z$ and $x-z$ planes, as shown in Figure 5. A subset of six independent geometrical and inertial parameters is defined, which are used in the GA:

- $L = 2R$: length of the floater;
- $W_f = W/(W_g \times n_{gyros})$: hull width factor. It defines the width of the hull W as a function of the gyroscope unit dimension W_g and the number of gyroscopic units n_{gyros};
- $h = x_A/R$: bow/stern circumference ratio;
- $k = D/H$: height ratio;
- $\delta_0 = \tan^{-1}\left(\frac{H-D}{R}\right)$: maximum pitching angle, defined as the maximum pitch rotation to avoid the deck to be submerged. It defines the hull draft D function of the floater semi-length R and the floater height H;
- BFR: ballast filling ratio, which is defined as the ratio of ballast located in aft/fore ballast tanks over the total ballast ($BFR = 1$: all the ballast is stored in aft/fore ballast tanks; $BFR = 0$: all the ballast is stored in the bottom ballast tank).

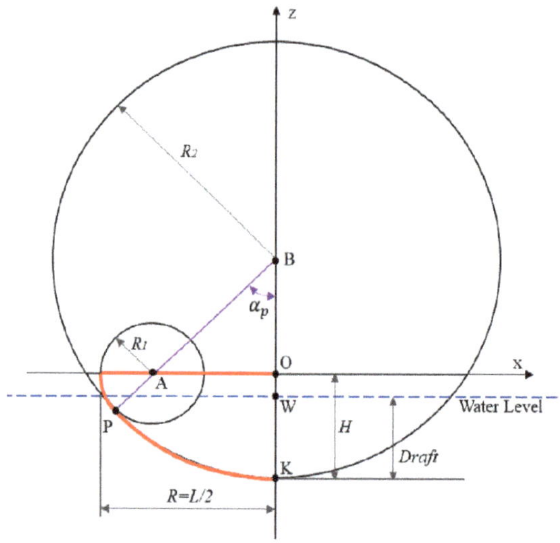

Figure 5. Parametric definition of the cross-sectional of the floater. Adapted from [4], with permission from MDPI, 2020.

Therefore, the floater profile can be expressed uniquely as a function of these design parameters:

$$R_1 = R - x_A = R(1-h) \tag{20a}$$

$$z_B = R\frac{k^2+1+2kh-2h-2k}{2-2h} \tag{20b}$$

$$R_2 = z_B + H = z_B + \frac{R}{1-k}tan(\delta_0) \tag{20c}$$

$$\alpha_p = \sin^{-1}\left(\frac{z_B}{x_A}\right) \tag{20d}$$

where x_A is derived from h and R. Then, the floater profile function $z(x)$ can be defined as:

$$z(x) = \begin{cases} \sqrt{R_1^2 - (x-x_A)^2} & \text{if } -R < x < R_2\sin(\alpha_p) \\ z_B - \sqrt{R_2^2 - (x-x_B)^2} & \text{if } R_2\sin(\alpha_p) < x \le 0 \end{cases} \tag{21}$$

Figure 6 shows how the geometrical ratios k and h influence the hull profile for a fixed hull length.

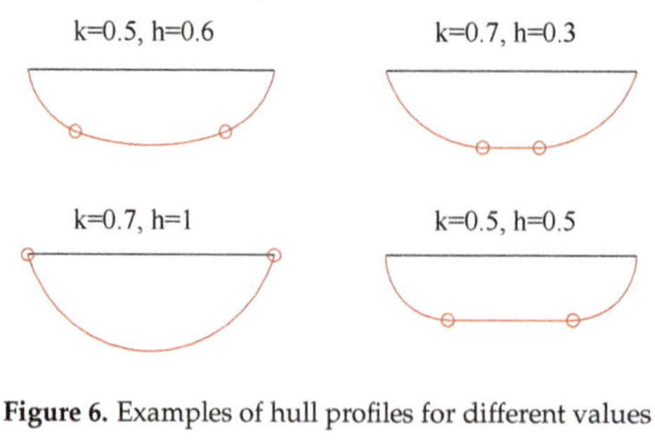

Figure 6. Examples of hull profiles for different values of k and h.

For what concern the ballast filling ration BRF, a schematic example of different filling ratios BFR is shown in Figure 7. The hull material is considered to be concrete to reduce the device costs with respect to a steel-made hull. Then, assuming the walls of the floater as thin plates, the equivalent structural thickness can be calculated from the lateral surface of the floater and the hull mass, making the computation of inertial properties of the hull possible.

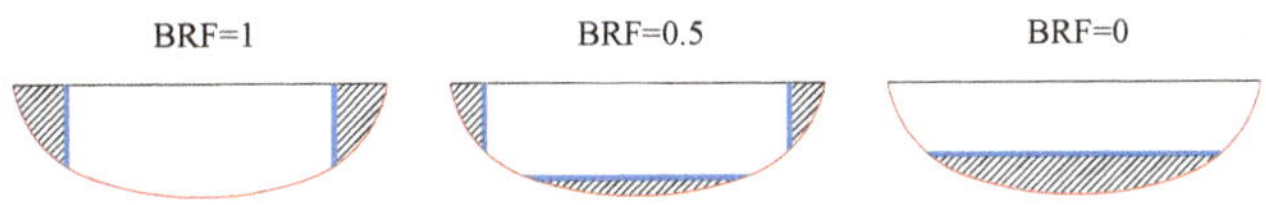

Figure 7. Examples of ballast mass distribution between the fore/aft and bottom compartments.

3.1.2. Gyroscope Parameters

The gyroscope represents the core of the energy conversion, and its geometrical optimization influences the maximization of power absorption, its costs and construction feasibility. Through the experience gained from the full-scale devices deployed in the Pantelleria and Adriatic Seas, the technological solution shown in Figure 2b is considered the most effective in terms of construction feasibility and costs. Two design parameters are used for the gyroscopic system:

- J: the flywheel inertia about its vertical rotational axis. J determines the angular momentum of the gyroscope, and it is used to derive its dimensions;
- n_{gyros}: the number of gyroscopic units inside the hull. The realization of one single gyroscope with a high inertia J can result in high loads and costs. Moreover, the device will be not balanced on the roll axis since the precession motion of one single gyroscope affects the equilibrium around the α-axis of the floater.

With reference to Figure 8a, the flywheel structure can be simplified as a hollow cylinder for which the geometrical properties are defined by an external diameter, $D_{e_{fw}}$, an internal diameter $D_{i_{fw}}$, and a height H_{fw}, expressed as follows:

$$D_{e_{fw}} = \left(\frac{2J}{0.5\pi\rho_{fw}(1-0.75^4)} \right)^{0.2} \tag{22a}$$

$$D_{i_{fw}} = 0.75 D_{e_{fw}} \tag{22b}$$

$$H_{fw} = 0.5 D_{e_{fw}} \tag{22c}$$

This parametrization is derived from the full-scale prototype deployed in 2015 in Patelleria (Italy) [66]. Then, the derivation of the principal moments of inertia for a hollow cylinder about its rotational axis is straightforward. Similar to the flywheel, the inertial ratios are derived from the ISWEC full-scale device design experience. Finally, it is assumed that the overall dimensions of the gyroscope unit are a linear function of the flywheel external diameter.

The verification of the internal encumbrances represents one of the feasibility checks performed. The number of gyro units n_{gyros} influence this feasibility check. It is assumed that the gyroscopic units are an even number to avoid roll torques on the hull. The algorithm tries to locate the units along the y-axis of the floater, and, if the total width of the gyroscope units exceeds the hull width, it tries to arrange the units in two parallel and adjacent lines. The aim is to fit all the gyros into the floater in relation to their dimensions. Examples of two feasible and one unfeasible arrangement are depicted in Figure 9.

The parameter used to define the pendulum subsystem is its mass m_p, which, combined with the distance d_p, determines the magnitude of the elastic recall provided on the gyroscope around its precession axis. Figure 8b shows the scheme of the pendulum installed at the base of the flywheel support frame in the $x - z$ reference plane. It is assumed that the distance d_p between the precession of the gyroscope and the centre of gravity of the eccentric mass is equal to the flywheel height:

$$d_p = H_{fw} \tag{23}$$

Then, given the mass and the distance of the eccentric mass, it is possible to compute the inertia tensor of the pendulum.

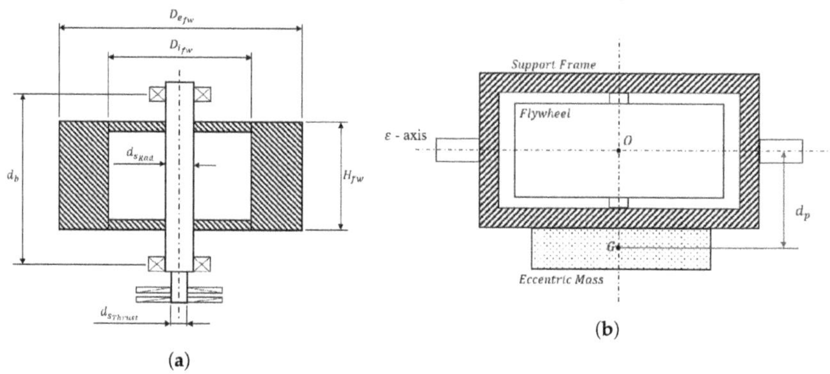

Figure 8. (**a**) Scheme of the flywheel geometrical parametrization. (**b**) Scheme of the pendulum fixed at the base of the flywheel support frame.

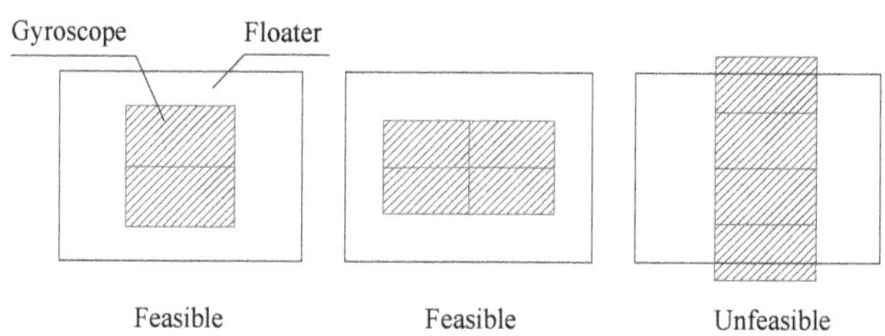

Figure 9. Different gyroscopic arrangement solutions: two feasible and one unfeasible disposals.

3.1.3. Flywheel Bearings

The flywheel bearings affect both the power losses of the system and its lifespan, determining part of the power consumed by the flywheel motor and the maximum gyroscopic load tolerable. The maximum static load is given in the bearing SKF catalogue [67]; during the optimization of the system control parameters, a load verification is performed. A total of 15 representative bearings couples are defined with the features shown in Figure 10, and the related optimization parameter is named ID_b. Such diagrams relate the inner diameter of the radial and axial bearings with their admissible static load. The optimization algorithm chooses from these bearing combinations to determine the optimal solution to fit the requirement of admissible loads (large bearings) and low power losses (small bearings). The distance between the radial bearings d_b is supposed to be related to the flywheel height H_{fw} as follows:

$$d_b = 1.75H_{fw} \tag{24}$$

Each bearing considered has been modelled according to the SKF catalogue [67] to carry out the power losses associated with it.

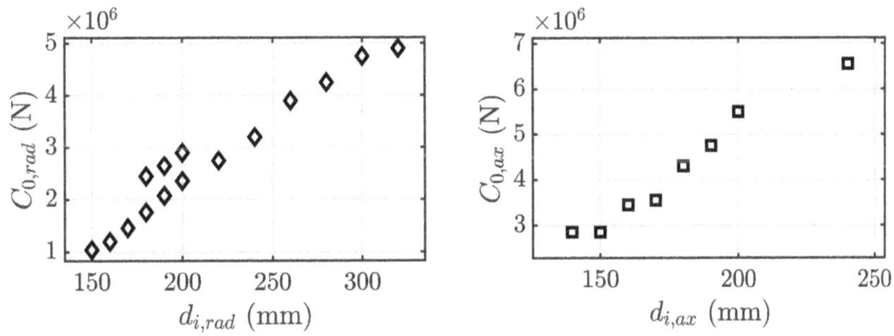

(**a**) (**b**)

Figure 10. Set of selected (**a**) radial bearings and (**b**) axial bearings with their inner diameter ($d_{i,rad}$ and $d_{i,ax}$) and the static load supported ($C_{0,rad}$ and $C_{0,ax}$).

3.1.4. Hydraulic PTO Parameters

The hydraulic pump is chosen from the Hägglunds CB and CBp radial piston pump catalogues offered by Bosch Rexroth [61]. The CBp series replicate the sizes of the CB ones, offering a powerful torque and speed range. Moreover, some tandem configurations are considered. The range of speed (ω_{pump}) and torque (T_{pump}) offered by the Hägglunds catalogue is reported in Figure 11, where 36 different solutions are considered. The pump model ID_p refers to one of the pump units summarized in Figure 11. Each series, identified with the prefix CB or CBp, can offer more than one solution (e.g., the series CBp 560 offers four different solutions: CBp 560-440, CBp 560-480, CBp 560-520 and CBp 560-560). The tandem configurations are composed of CBp units, for example CBp 280 + CBp 140. More details about the displacement, peak torque, peak speed, rated power, efficiency and main dimensions of each unit can be found in the online catalogue [61].

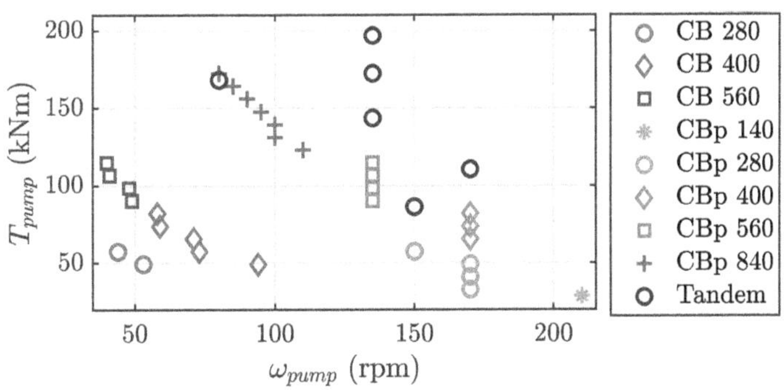

Figure 11. Hydraulic pump solutions offered by Bosch Rexroth in respect to their rated speed ω_{pump} and torque T_{pump}.

The control mode of the primary pump ID_c is an optimization parameter since different control architectures can be considered with different benefits and drawbacks. First, it is possible to implement or not implement the declutching control logic. Since the hydraulic pumps have a fixed displacement, the only way to regulate the torque acting on the gyroscope is to short-circuit the pump for some period of the wave cycle. This control could significantly improve the power production of the ISWEC system with respect to a solution without the clutch-declutch valve. However, the clutch-declutch valve can be prone to failures due to the number of open-close cycles demanded and thus can affect the reliability of the system, increasing the down-time, which can turn out to be costly in term of maintenance required (not accounted for in this work). A further degree of controllability is added to the system if tandem units are used, allowing to switch from one, two or three different total displacements and introducing regulation with respect to the PTO torque.

Additionally, in this case, the optimization algorithm can choose from considering or not considering the switching of the pump units. The four control possibilities are summarized in Table 2.

Table 2. Pump control combinations considered by the optimization algorithm. *: the pump switching can be enabled only with tandem configurations; if a single pump is used, only the first two control logics are implementable.

ID_c	Declutching Control	Pump Switching *
1	no	no
2	yes	no
3	no	yes
4	yes	yes

3.1.5. Device Cost Estimation

The ISWEC optimization includes a preliminary evaluation of the total cost of the device in order to evaluate the investment behind a specific configuration. Similar to [4], the overall cost is composed of: the floater expenditures (materials and constructions), the gyroscope units along with its main and auxiliary components (flywheel motor, cooling system, power electronics of the flywheel motor, etc.) and the PTO system with its power and electronic systems. The mooring system is not considered here since its parametrization is under study and no clear automatic definition could be tackled in such an early design stage. No operational and maintenance expenditure are considered in this work due to their low magnitude with respect to the capital cost [4].

Under these assumptions, the overall cost of the device C_{TOT} is given by the sum of the costs of each subsystem:

$$C_{TOT} = C_{HULL} + C_{GYRO} + C_{HPTO} \tag{25}$$

The cost of the floater C_{HULL} is proportional to its overall volume. As studied in [46], an interesting method to reduce the costs of ISWEC manufacturing and its environmental footprint could be found in the use of concrete structures, as opposed to traditional steel ones. In the naval carpentry sector, the cost per unit of overall volume can be reasonably assumed to be 275 €/m^3 if concrete is used as a construction material.

The gyroscope cost function includes the material and manufacturing costs of the flywheel, support frame structures, shafts and bearings. The cost of the gyroscope C_{GYRO} can be set to be proportional to its structural mass m_g. In mechanical manufacturing, the cost per unit mass produced can be reasonably assumed to be 6000 €/ton if steel is considered as a fabrication material.

The cost magnitudes of the hydraulic pumps are summarized in Figure 12 with respect to their rated torque T_{pump}.

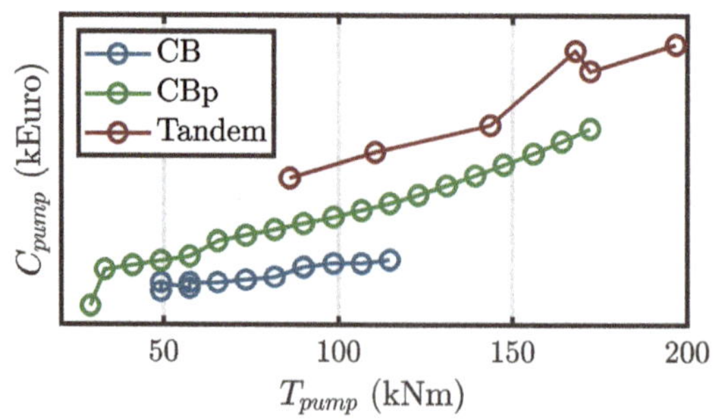

Figure 12. Cost magnitude of the pump units with respect to the rated torque and the pump type.

Such costs have been collected through close collaboration with Bosch Rexroth, which allowed us to obtain the costs of 36 of its hydraulic pump solutions. As specified in Section 3.1.4, the hydraulic pumps have been chosen from the Hägglunds CB and CBp radial piston pump catalogues, available at [61], and costs have been omitted for confidentiality. As shown in Figure 12, the cost of the CB units are considerably lower than the CBp ones, since a lower specific power is deliverable. On the other hand, the tandem units are the most expensive solutions since two hydraulic pumps are installed for each tandem unit. Despite the higher investment required, the tandem solution allows to switch between each other, giving different values of the total displacement available.

The rest of the hydraulic transmission, e.g., the accumulators, manifold, valves, hydraulic motor and electrical generator, is not modelled at this optimization stage. However, the cost of such components has to be taken into account. The cost of the accumulator can be expressed as a function of the accumulator's total volume, fixed to 600 L or 1000 L if two or four gyroscopic units are considered due to the experience gained designing hydraulic transmission for the ISWEC with the size of interest. An accurate market investigation allowed us to determine the cost per litre as equal to EUR 22/L when associated with the standard accumulator volume of 100 L. The cost can be determined considering multiple 100 L modules plus the cost of junctions and tubes needed for the connection of the accumulators with the manifold, which are fixed to EUR 250 per accumulator unit. Concerning the manifold, the cost has been fixed to EUR 17,500 or EUR 20,000 if two or four gyroscopic units are considered. As the number of gyroscopes, and thus the number of the associated hydraulic pump units connected to one single manifold increases, the dimensions and the auxiliary parts grow, increasing the cost per manifold. Concerning the auxiliary components, the cost associated with a single manifold unit is equal to EUR 5000, accounting for the clutch-declutch valve, check-valves, relief valves and sensors. The oil cost is given as equal to EUR 4.5/L.

The cost magnitudes of the motor-generator stage, consisting of the hydraulic motor, electric generator, and associated electrical and electronic components, are summarized in Figure 13 with respect to the rated power of the electric generator P_{gen}. The solutions offered by Moog [68], composed of a hydraulic motor coupled with a permanent magnet synchronous generator (PMSG), are completely comparable with the custom solutions obtained from the combination of the Bosch Rexroth hydraulic motors [69] and the SIEMENS electric generators [70] in terms of cost per kW of installed power. Since the motor-generator stage is not modelled here, its cost is selected depending on the number of gyroscopes considered (n_{gyros}) and on the mean cost of the units: the cost has been fixed to EUR 180,000 or EUR 300,000 if two or four gyroscopic units are considered, respectively. This procedure has to be considered as a first approximation to account for the motor-generator stage cost, and further analyses will be conducted. Similarly to the pump costs, the motor-generator quotations have been omitted for confidentiality.

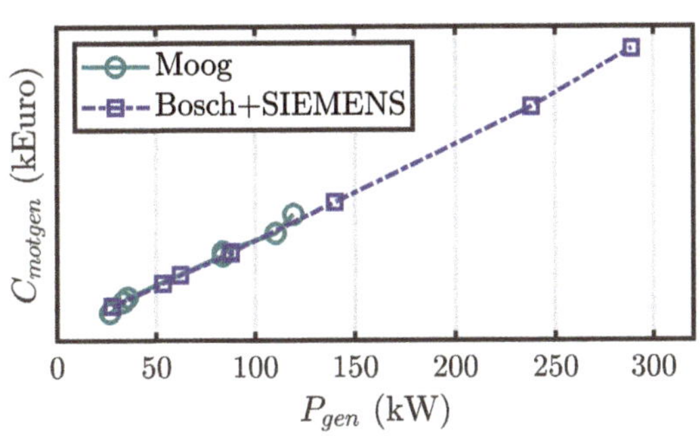

Figure 13. Cost magnitude of the motor-generator units, considering the cost of the associated power electronics and auxiliary components with respect to the rated power of the electric generator.

These assumptions have to be considered as preliminary, and further analysis will be performed to accurately design the power generation stage associated with an HPTO unit with the TDM of the transmission.

3.2. Single Device Optimization

The second step toward the evaluation of a single ISWEC individual consists of simulating and optimizing the annual productivity of the system. Once all the device parameters are defined and the hydrodynamic properties and forces are computed, spectral-domain simulations are performed under some hypotheses and restrictions. The aim is to maximize the performance for all the sea states considered.

3.2.1. Sea States Definition

The simulation waves were defined according to the scatter diagram of Figure 14, concerning the annual occurrences and wave energy carried in the Balder sea site, an oceanic site located in the North Sea. The data presented refer to hindcast data from the period 1958–2014 (56 years) and with a sea state duration of 3 h. As shown in Figure 14, the proper number of waves as well as their synthetic parameters (T_e and H_s) were chosen (highlighted with red dots) with the aim to cover at least 99% of the annual energy of the sea site. All the waves were simulated for a single device considering a JONSWAP (Joint North Sea Wave Project [71]) spectrum, since it is the most representative of the sea site of interest. At the end of the simulation process, the ISWEC performances are computed.

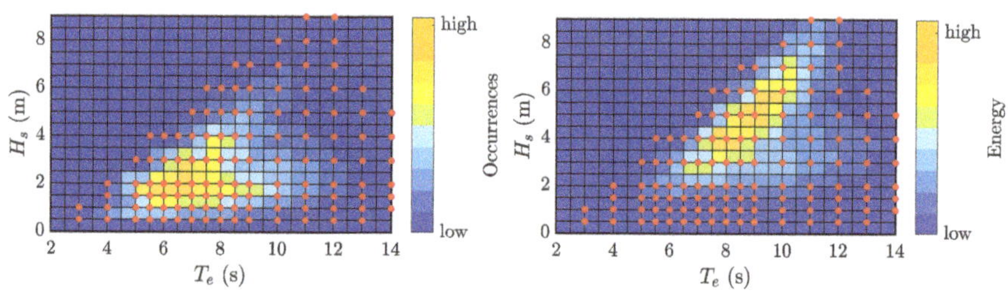

(a) (b)

Figure 14. Simulation waves (red dots) superposed to the (**a**) wave occurrences and (**b**) wave energy for Balder sea site.

3.2.2. Control Tuning

The single individual was simulated varying its control parameters ($\dot{\varphi}$, Δp, and $\dot{\varepsilon}_c$) with the aim of maximizing the power extracted for each wave considered. Note that, at this stage, it is essential to include a control-informed optimization, since it has the ability to significantly increase power production performance over the operational sea [72–74]. For each sea state, the optimization of the control parameters was performed with a Nelder–Mead simplex method [75] implemented in MATLAB with the native function `fminsearch`, considering the following cost function J_{opt}:

$$J_{opt} = -\overline{P}_n + \sum_{c}^{I_c} p_c \tag{26}$$

Equation (26) includes two terms: $\overline{P}_n$ is the mean net power extracted from the system for a single sea state, and $\sum_c^{I_c} p_c$ are the sum of the "penalty costs" associated to the violation of the physical constraints I_c. As discussed in [76], when an optimization algorithm is applied to a dynamical model, there is the probability that some state variables overcome the limit conditions, making the results not reliable or representative of the true system performance. Therefore, due to the impossibility of including explicit constraints into a

SDM, soft constraints were considered in the cost function by adding a penalty term to the optimization problem defined as:

$$p_c = \frac{1 + \tanh(k_c(|x_c| - x_{c,lim}))}{2}|x_c| \tag{27}$$

The penalty term p_c associated with the system variable x_c and constrained to the value $x_{c,lim}$ is computed with Function (27). k_c is the steepness coefficient to make the tanh function similar to a step function. When the variable x_c is below its limit $x_{c,lim}$, the value of the penalty function is equal to zero. On the other hand, as x_c exceeds the limit $x_{c,lim}$, the penalty function increases, affecting the value of the cost function J_{opt}. As specified in [77], larger loads on the gyroscope and on the hull structure will increase the investment cost for a given working principle or, in general, lead to larger failure rates for a given design and, consequently, higher maintenance and inspection costs. Then, the most critical variables are accounted for in the penalty function: hull and gyroscope angular displacements and torques, PTO angular speed and torque, and bearing loads.

3.3. Key Performance Indicators

Techno-economic optimization underlines the maximization of the economic efficiency of the system, giving different solutions that have to be selected by balancing the performance of the system and the investment required [78,79]. Two conflicting objectives are considered here:

- **AEP: annual energy production** (MWh/y). With AEP, we refer to the ISWEC production, computed through the net power produced by the system for each wave considered and with the annual occurence distribution associated to the sea site of interest:

$$AEP = \frac{8760}{10^6} \sum_{w=1}^{n_W} \overline{P}_{n,w} o_w \tag{28}$$

  where n_W is the total number of waves, $\overline{P}_{n,w}$ is the average net power extracted with the $w-th$ wave, and o_w is the annual occurrence associated with the $n-th$ wave. The coefficient $\frac{8760}{10^6}$ converts the annual productivity into MWh/y.
- **CoE: cost of energy** (Euro/MWh). In this work, the CoE is defined as the ratio between the device cost and the AEP multiplied by the lifetime of the device, fixed to 25 years:

$$CoE = \frac{C_{TOT}}{25 \times AEP} \tag{29}$$

The objective of the genetic algorithm is finally explicitly formulated: a multi-objective optimization regarding the AEP and the CoE. The multi-objective problem can be stated as follows:

$$\begin{aligned} &\min_{\mathbf{x}\in\Omega} \quad -AEP(\mathbf{x}) \quad , \quad \min_{\mathbf{x}\in\Omega} \quad CoE(\mathbf{x}) \\ &\text{s.t.} \quad \mathbf{x}_L \leq \mathbf{x} \leq \mathbf{x}_U \end{aligned} \tag{30}$$

where the objective functions $-AEP(\mathbf{x}) : \mathbb{R}^{n_x} \to \mathbb{R}$ and $CoE(\mathbf{x}) : \mathbb{R}^{n_x} \to \mathbb{R}$ are functions of a vector $\mathbf{x} \in \mathbb{R}^{n_x}$ subject to lower and upper bounds, $\mathbf{x}_L \in \mathbb{R}^{n_x}$ and $\mathbf{x}_U \in \mathbb{R}^{n_x}$. n_x is the dimension of the problem.

3.4. Optimization Algorithm

The optimization problem of a WEC falls within the field of multi-objective optimizations (known also as Pareto optimizations), where more than one objective concur. Two different multi-objective approaches have been used by the authors so far: (I) in [4], several single-objective optimisations have been performed, and the performance index has been computed as a weighted average of multiple objectives through the native MATLAB function `ga`; (II) in [5,6], the multi-objective NSGA-II method [32] has been used through the native MATLAB function `gamultiobj`. Here, approach (I) is employed to highlight different techno-economical trends obtained by pursuing contrasting objectives: maximization of the annual productivity and minimization of the cost of energy. In this regard, the Pareto front is built by aggregating the results of five single-objective optimisations with the pairs of weights reported in Table 3:

Table 3. Multi objective weights and performance metrices.

Set Label	Description	AEP Weight (w_1)	CoE Weight (w_2)
1	AEP-driven	1	0
2	AEP-weighed-CoE-weighed	0.75	0.25
3	AEP-weighed-CoE-weighed	0.5	0.5
4	AEP-weighed-CoE-weighed	0.25	0.75
5	CoE-driven	0	1

The cost function, concerning the mininimization problem, can finally be presented as follows:

$$J_c(\mathbf{x}) = \begin{cases} -AEP(\mathbf{x})w_1 + CoE(\mathbf{x})w_2 & \text{if feasible device} \\ C + \Delta_c & \text{if unfeasible device} \end{cases} \tag{31}$$

In order to deal with unfeasible solutions, constraints are enforced by applying a death penalty that worsens the fitness values by a constant C increased by a measure Δ_c of how much that constraint is violated [80,81]. The most common reasons for solution unfeasibility are the geometrical intersection of the internal subsystems, the gyroscope and the ballast encumbrances with respect to the available space, and the hydrodynamic instability. They are recognized by the algorithm and labelled as unfeasible solutions.

The tuning factors of the algorithm are summarized in Table 4. It is worthwhile to specify that, although beyond of the scope of this work, a sensitivity analysis on the ga parameters can potentially improve the convergence rate of the algorithm. The interested reader should refer to [82–84] for a detailed description of evolutionary algorithms.

Table 4. Parameters of the genetic algorithm.

Name	Symbol	Value
Population size	P_s	75
Maximum generations	G_m	150
Maximum stall generations	G_s	50
Convergence threshold	Δ	10^{-6}
Elitism percentage	e_p	5%

4. Optimization Results

In this chapter, the optimization results are discussed. Attention is given to the annual energy production, the device cost, its manufacturability and feasibility. The reader should note that all the AEP, CoE and cost values have been hidden for confidentiality, and only qualitative results and relative percentage differences are reported here. The purpose is to point out the techno-economic trends of a HPTO-based ISWEC device preliminarily designed with the spectral domain technique. However, the order of magnitude of AEP, CoE and cost are reported in the axes' legend to give insight into the power capabilities and cost of the ISWEC.

4.1. Convergence Analysis

The ISWEC parameters cannot be combined at will, and some unfeasible solutions appear during the optimization process. The so-called mortality rate, defined as the ratio between the number of unfeasible individuals and the total individuals, is shown through generations in Figure 15.

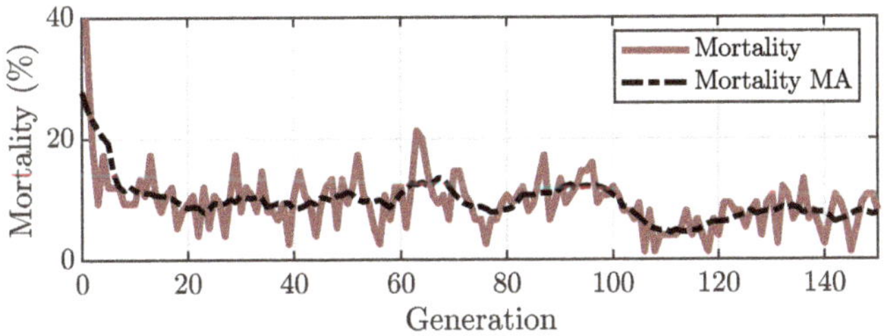

Figure 15. Mortality rate in respect to the generation number.

The dashed line is the moving average (MA) of the mortality rate computed by grouping 10 generations at once. The mortality drastically decreases in the first 20 generations and slighty increases again between generation 80 and 100; then, it moves below 10% at the end of the optimization process. Overall, the method proposed to handle unfeasible devices is demonstrated to be effective since the mortality rate decreases as generations progress. However, the mortality does not go to zero due to the exploratory nature of the GA, which explores the solution domain in a stochastic way during the crossover and mutation operations.

The development of the algorithm is shown in Figure 16 for the example of the AEP and CoE performances.

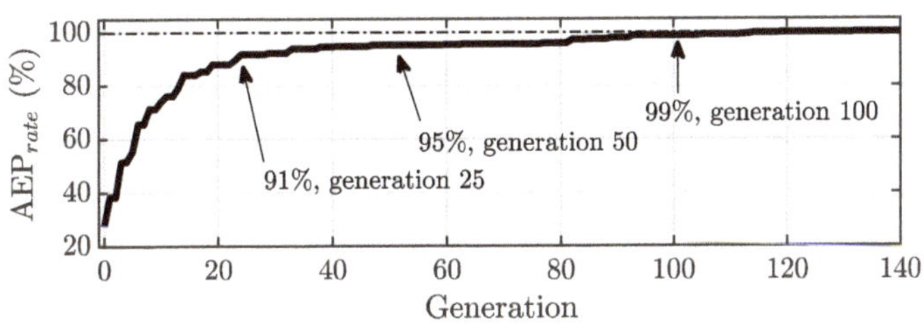

(**a**)

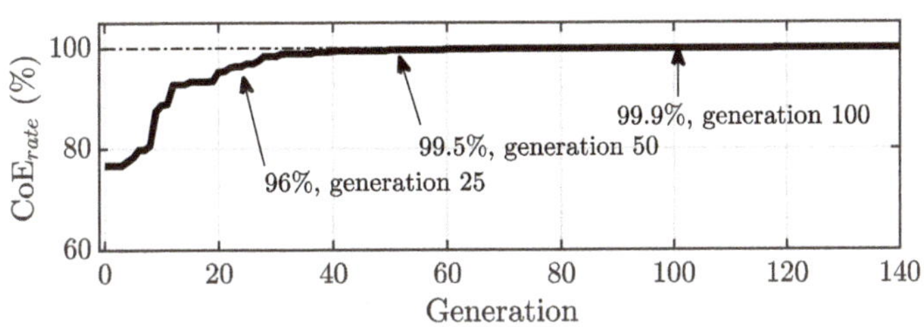

(**b**)

Figure 16. (**a**) AEP and (**b**) CoE rate of convergence to the best-performing individual with respect to the generation number.

Figure 16 has the purpose of showing the rate of the fittest individuals over the whole population as the generation number increases with respect to the best individual for each performance index; the improvement is abrupt in early generations (until generation 25–50) and slows down afterwards (until generation 100). It is believed that the current setup provides appropriate convergence and satisfactory results; thus, the maximum number of generations allowed, G_m, could be reduced to 100 or 120 to reduce the computational time.

This result is confirmed in Figure 16a,b, where the AEP and CoE rate is reported in respect to the generation number, defined as follows:

$$AEP_{rate} = \left(1 - \frac{AEP_{max} - AEP_g}{AEP_g}\right)100\% \quad (32a)$$

$$CoE_{rate} = \left(1 + \frac{CoE_{min} - CoE_g}{CoE_g}\right)100\% \quad (32b)$$

where AEP_g is the AEP of the fittest value at generation g, and AEP_{max} is the best AEP value among all individuals. The same nomenclature is valid for the CoE_{rate} definition. It is shown that at generation 100, the best individual performs 99% equal to the absolute best, and for the subsequent 50 generations no relevant improvement of the fitness function are found. The GA is able to find the optimal solution near generation 100 by evaluating approximately 7500 individuals (75 individuals in 100 generations) despite the billions of feasible combinations.

4.2. Optimization Output and Techno-Economic Trends

Figure 17a shows the ensemble of all the individuals generated from the optimisation. Each point represents a feasible individual with its related AEP and CoE. The colour code is proportional to the device cost magnitude. Figure 17b highlights the Pareto front (red line) concerning the best individuals according to the two evaluation metrics considered. Each individual on the Pareto front can be interpreted as a non-dominated solution of the solution set, and here stand the optimal devices. As expected, the optimal device with the best AEP differs from the one with lowest CoE. In fact, designing a system with high productivity leads to an increase in the system dimensions and costs; on the other hand, since there is not a linear relationship between the system performances and its investment, the economic efficiency of the ISWEC has to be maximized with a CoE-driven optimization. The CoE-driven area is attracted towards the AEP axis, while the AEP-driven optimisation extends the furthest along the CoE axis. In between, the multi-objective optimisation will reach a good compromise of the two metrics. The optimal devices stand in the north-west corner of the diagram, where the AEP is maximized and the CoE is minimized. As expected, the most expensive devices appear in the top part of the device point cloud where the AEP is maximized.

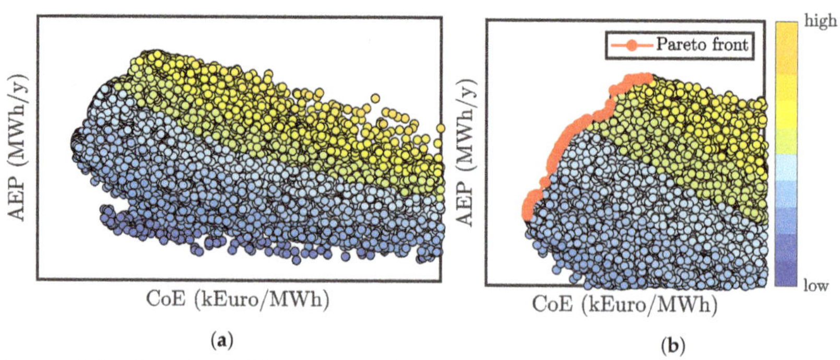

Figure 17. Productivity (AEP) versus cost of energy (CoE) of the ensemble of the optimisation; each marker (individual) is coloured proportional to the device cost (C_{TOT}). (**a**) Best feasible individuals and (**b**) a zoom on the Pareto front (red line) concerning the optimal individuals according to the evaluation metrics.

Figure 18 aims to highlight the convergence of the hull length L, the gyroscope moment of inertia J, the pump displacement D_p and the hydraulic control ID_c driven by the maximization of AEP.

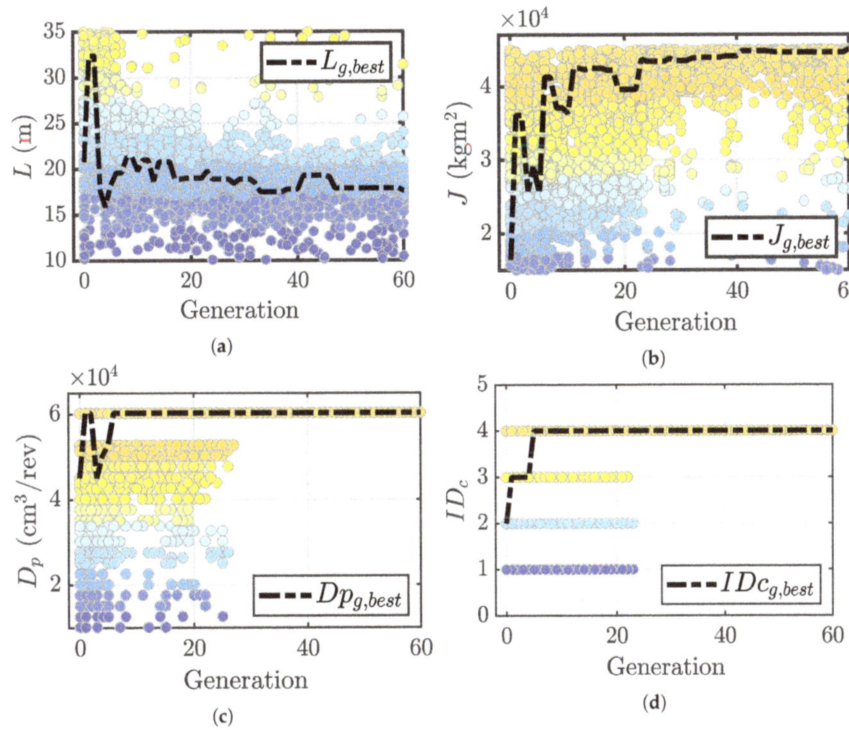

Figure 18. (**a**) Hull length L, (**b**) gyroscope moment of inertia J, (**c**) pump total displacement D_p and (**d**) hydraulic control mode ID_c with respect to the generation number for an AEP-driven optimization. The colour code represents the magnitude of the variable of interest. The dashed line refers to the best individual of each generation.

First, in Figure 18a, the behaviour of the hull length L through generations shows that the hull length rapidly converges to the optimal value (represented with a black dashed line) that refers to the best individual of the generation g. An optimal value around 17 m is found for the most well-performing device; small and large hull length are not optimal. This result is consistent with the hydrodynamic theory and, more in specific, the hydrodynamic of floating bodies: short devices compared with the incident wavelength are expected to behave like a point-absorber-WEC, inducing a predominant heave motion that, in our case, does not contribute to the power generation; on the other hand, long devices acts like a low-pass filter due to the peak-to-trough distance of the wave crest, which is not suitable to induce their oscillation. Concerning the gyroscope's moment of inertia, J, Figure 18b demonstrates that, as expected, in order to maximize the system productivity, large gyroscopic units are required, and thus the $J_{g,best}$ value converges to the upper bound imposed on J, equal to 45,000 kgm^2 (see Table 1). Concerning the hydraulic parameters, the pump displacement D_p and the hydraulic control type ID_c are reported in Figure 18c and 18d, respectively. It is worthwhile to remember that the free parameter of the pump is an ID that uniquely corresponds to a pump unit in the Bosch Rexroth catalogue. However, in order to give an engineering meaning to the results presented, the pump displacement is used to represent the trend obtained during the optimization. The pump displacement D_p is directly related to its maximum torque capabilities and gives insight into the size of the pump. The results of Figure 18c,d are also consistent with intuition: the AEP is maximized through the largest pump unit available in the catalogue, which is able to exert the highest torque on the gyroscope unit, and, thus, no theoretical limits exist in extracting the available power generated by the gyroscope motion. In this regard, the pump unit can be interpreted as an electrical generator: the larger the torque available, the larger its saturation limits, and all the available power transferred to the gyroscope can be converted into electricity (unless the HPTO efficiency). Figure 18d determines that the most well-performing hydraulic control logic is the $ID_c = 4$ regarding the switching of the pump displacements and the clutching speed threshold (see Table 2). The control mode $ID_c = 4$ is the most versatile and ensures the best controllability of the hydraulic system: the clutch-declutch valve is able to regulate the pressure acting on the primary pump, and, if a tandem configuration is installed, the possibility to switch between three different pump displacement allows a further degree of regulation of the PTO torque. Similarly to L and J, the algorithm finds the best D_p and ID_c before the 50th generation.

Concerning the CoE, the resultant optimal parameters are shown in Figure 19.

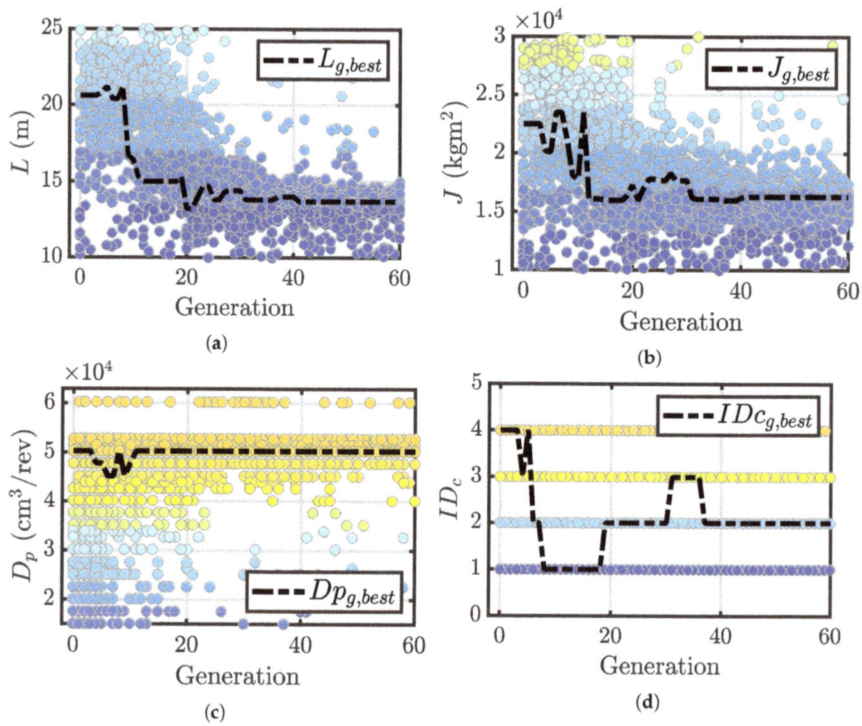

Figure 19. (**a**) Hull length L, (**b**) gyroscope moment of inertia J, (**c**) pump total displacement D_p and (**d**) hydraulic control mode ID_c with respect to the generation number for a CoE-driven optimization. The colour code represents the magnitude of the variable of interest. The dashed line refers to the best individual of each generation.

In Figure 19a, it is shown that the hull length converges to lower values compared to a AEP-driven optimization, almost equal to 14 m. However, such a result confirms that the optimal hull longitudinal dimension stands in the same range regardless of the optimization target in order to maximize the hydrodynamic performance of the system. Conversely, the gyroscope moment of inertia J is almost three times lower for a CoE-optimum device, standing around 16,300 kgm^2 (see Figure 19b). This represents the main difference emerging from the two optimizations: as shown hereafter, the gyroscope cost represents almost 40 ÷ 45% of the total device cost, and, in a CoE-driven design, the economic efficiency is favoured over the system performances, reducing the size of the gyroscopic units. The pump displacement optimal value, reported in Figure 19c, is slightly lower than the AEP-optimum one, and it refers to a single displacement unit. Despite the economical efficiency of the system being preferred to its annual productivity, the selected pump is again one of the larger units in the catalogue. Balder is a high-energy sea site, and the algorithm typically explores the most powerful PTO solution available to maximize system productivity. Moreover, the hydraulic pump requires a low investment compared to the gyroscope unit, and no relevant impact on the overall cost is obtained. As shown in Figure 19d, the most performing hydraulic control logic is $ID_c = 2$, concerning the clutch-declutch logic. The pump switch is not possible here because the optimal pump is a single displacement unit. Overall, the best value of L, J, D_p and ID_c are found in generation 50, again demonstrating the high rate of convergence of the GA.

In order to investigate significant and more informative trends, Figure 20 shows relevant results and design parameters for a selection of ten devices standing on the Pareto Front, chosen in order to be representative of the fittest individuals in the whole population.

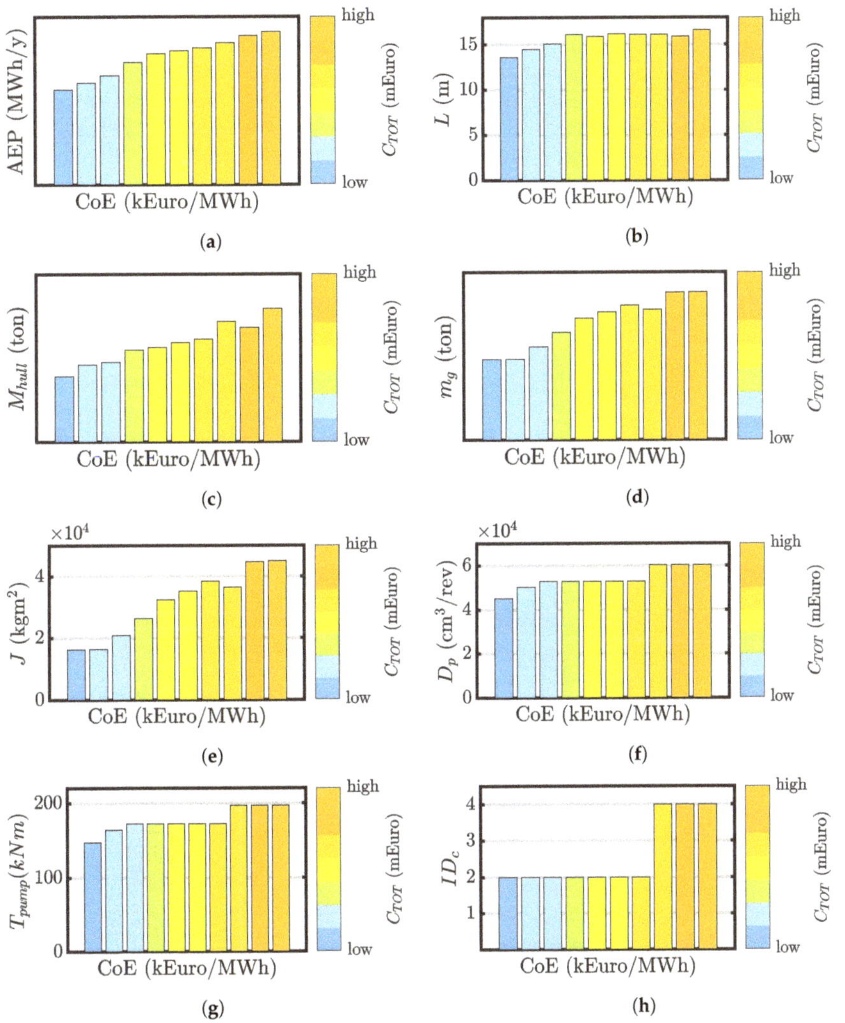

Figure 20. Ten individuals on the Pareto front with respect to their CoE and the associated performance and system parameters. The results refer to a Stage 1a optimization. Bars coloured proportionally to the device cost C_{TOT}. (**a**) AEP, (**b**) hull length L, (**c**) hull mass M_{hull}, (**d**) gyroscope mass m_g, (**e**) gyroscope inertia J, (**f**) pump displacement D_p, (**g**) pump-rated torque T_{pump} and (**h**) hydraulic control ID ID_c. AEP, CoE, cost, and mass data have been hidden for confidentiality.

In order to highlight the most convenient individual, according to each metric, bars colour proportional to the device cost are used, so that cheapest devices are represented by sky-blue bars, while most expensive ones in orange. The highest AEP and the lowest CoE are achieved with significantly different individuals. In particular, it is evident that the increase in AEP is slower than the increase in cost, so the lowest CoE is found in the low cost region in spite of a lower AEP. Contrary, achieving high AEP generally requires such an increase in device cost that the overall economical effort becomes unfavourable. Specifically, Figure 20b,c explores variations in the hull length and mass, which are the main drivers of hydrodynamic performance and hull cost. While CoE consistently increases with the hull length, longer hulls generate high AEP, despite the hull longitudinal dimension remaining bounded between 14 m and 17 m. The same behaviour is shown in the hull mass, which reports a slight upward trend as the AEP increases. Similarly, Figure 20e,d depicts that productive devices call for heavy and large gyroscopes. Despite the relevant increase in the gyroscope unit inertia, from J = 16,500 kgm^2 to J = 45,000 kgm^2, the gain in productivity does not follow the same rate of growth, confirming that the incremental investment required to produce more energy is not cost-effective. Concerning the pump unit, mainly as a consequence of the high AEP demand, the optimal pump displacement and rated torque grow from left to right, resulting in more powerful hydraulic units installed (see Figure 20f,g). Finally an optimal hydraulic control requires the use of the clutch-declutch logic and (if a tandem unit is installed) the pump switching.

The device cost and the relative impact of different system parts on the total device cost is reported in Figure 21. The costs are divided into hull, gyroscope unit and HPTO. Figure 21a demonstrates that the absolute cost of each subcomponent tends to increase more rapidly than the resultant increase in AEP, leading to higher CoE for high investments. The rates of increase seem to be different: the hull cost tends to become less relevant in favour of the gyroscope cost, although the mass of the hull increases, as shown in Figure 20c; therefore, the proportional increase in gyroscope unit size has a higher impact on the overall cost than the increase in hull dimensions and mass. In conclusion, a cost-effective solution seeks compact gyroscopic units rather than small hulls. Overall, hull and gyroscope almost equally split up 80 $\div$ 85% of the device cost, while the hydraulic transmission is just 15 $\div$ 20% of the investment.

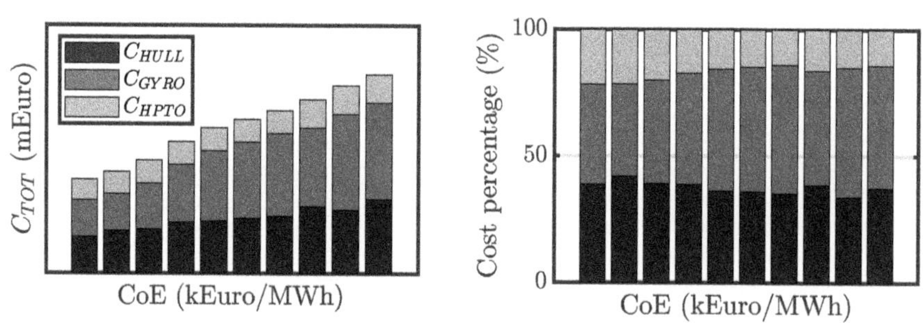

(a) (b)

Figure 21. Ten individuals on the Pareto front with their CoE and the associated cost: hull, gyroscope and HPTO; (**a**) absolute cost; and (**b**) percentage cost.

4.3. Optimal Devices

Finally, the results of the two optimum devices for Balder according to different objective functions are summarized in Table 5. The AEP optimum is wider than the CoE optimum, intercepting more wavecrests, hence enhancing the capture of incoming energy. The same rationale applies to the flywheel inertia, pump displacement and rated torque, which are driven to the upper limit available to extract as much energy as possible, regardless of the device cost. Conversely, a lower CoE requires a smaller floater and gyroscope units, along with a less powerful pump unit. The optimal control mode ID_c

is, as expected, the one that gives the higher controllability to the system: $ID_c = 4$, for the AEP-optimum, means that both the clutch-declutch logic and the switch of the tandem units is enabled; $ID_c = 2$, for the CoE-optimum, consists of enabling clutch-declutch logic (here, the pump switch is not possible since a single-unit pump is considered). Overall, dimensions and mass of the AEP-driven optimum are higher compared with the CoE-driven one. This suggests that increasing the device size requires a higher increment in cost than the consequent increase in AEP, hence increasing the resulting CoE. Regarding the performances, the CoE optimum device underperforms the AEP optimum by up to 37.4% considering the annual energy extraction. However, its cost of energy is 14.9% lower due to a reduction in the total device cost almost equal to 52%.

Table 5. Key performance differences from the AEP optimum and CoE optimum.

Performance	Symbol	Delta AEP-CoE-Optimum
Delta Annual energy production	ΔAEP	−37.4%
Delta Cost of energy	ΔCoE	−14.9%
Delta Device cost	ΔC_{TOT}	−52.3%

5. Conclusions

In this work, the results of a techno-economic design tool implementing the SDM of the ISWEC equipped with an HPTO are discussed. The main novelty of this new design procedure concerns the introduction of the spectral-domain technique to design a HPTO in the early stage, limited until now to the use of complicated and computationally inefficient time-domain frameworks. Several aspects have to be considered during the design process: the power generation capacity, named here AEP, and the economic efficiency, identified as CoE. Additionally, the initial investment play an important role in the decision process. Originally, the ISWEC device was designed for the Mediterranean Sea or, more generally, for low-medium energetic seas. The main aim of this work is to explore the performances of the ISWEC device in high energetic seas toward the implementation of an HPTO. In this regard, the ISWEC device is designed for the Balder sea site, an oceanic site located in the North Sea.

First, the ISWEC system equipped with HPTO was modelled in the spectral-domain, deriving its equation following the approach proposed in the previous works of the first author. Second, the optimization tool was presented defining the device parametrization along with its key performance indicators. The optimization relies on the SDM targeting a multi-objective function to maximize the annual energy production, named AEP, and to minimize the associated cost of energy, labelled CoE. The two most well-performing devices, one designed to maximize the AEP and one to minimize the CoE, were compared, showing the influence of the design parameters on the selected performance metrics. In particular, the algorithm converges on costly devices if the AEP is to be maximized: large hulls, massive gyroscopic units and powerful hydraulic pumps are preferred to enhance the annual productivity, regardless of the economic efficiency of the system. On the other hand, the CoE minimization is obtained through smaller and lighter devices with medium-size hydraulic units. Concerning the hydraulic control mode, the GA converges to the declutching control, and, if a tandem pump unit is considered, to the implementation of the switching logic, as expected. Synthetic results demonstrate that the AEP optimum outperforms the CoE optimum by almost 37% in the annual energy extracted. In contrast, the performance indicators show a reduction of 14.9% of the CoE passing from the AEP optimum to the CoE optimum, supported by a cost reduction of almost 52%, making the ISWEC more attractive for investors.

The results discussed so far highlights the limitations of a single-objective design or, even worse, of a sequential design process, considering different parts of the power conversion chain independently. In particular, it was demonstrated that high overall cost consistently results in high AEP/high CoE, and vice versa, remarking that the economic ability and the conversion efficiency can be contrasting objectives. The design tool presented aims to offer an all-encompassing optimization tool to cover both technical and economical aspects in order to help the design process of an WEC. Such a design method could be extended to other WEC concepts with the main aim of accelerating the development of the wave energy industry. Further work will be done on the basis of these preliminary design results. The SDM of the HPTO will be expanded to also account for the dynamic behaviour of the accumulator and the hydraulic motor in order to improve its accuracy, reducing the performance gap with the TDM. A further optimization tool will be presented and used to design the whole hydraulic transmission in the time-domain; the final aim will be to compare different PTO technologies designed for the ISWEC to assess the best PTO configuration from a techno-economic point of view. Moreover, a formal uncertainty analysis will be added based on a previous work of the first author [6], applied to the PeWEC device, in order to account for performance prediction variation based on parameter uncertainty in designing the ISWEC system with HPTO.

Author Contributions: Conceptualization, M.B. and G.G.; Data curation, M.B.; Formal analysis, M.B.; Investigation, M.B.; Methodology, M.B.; Project administration, M.B. and G.G.; Resources, M.B.; Software, M.B.; Validation, M.B. and G.G.; Visualization, M.B.; Writing—original draft, M.B. and G.G.; Writing—review and editing, G.G. All authors will be informed about each step of manuscript processing including submission, revision, revision reminder, etc. via emails from our system or assigned Assistant Editor. All authors have read and agreed to the published version of the manuscript.

Funding: This research received no external funding.

Institutional Review Board Statement: Not applicable.

Informed Consent Statement: Not applicable.

Data Availability Statement: Not applicable.

Conflicts of Interest: The authors declare no conflict of interest.

Abbreviations

The following abbreviations are used in this manuscript:

MDPI	Multidisciplinary Digital Publishing Institute
DOAJ	Directory of open access journals
WEC	Wave Energy Converter
PTO	Power Take-Off
GA	Genetic Algorithm
NSGA-II	Non-dominated Sorting Genetic Algorihm II
HPTO	Hydraulic PTO
TDM	Time-Domain Model
SDM	Spectral-Domain Model
AEP	Annual Energy Production
CoE	Cost of Energy
MA	Moving Average

Note

1 the accuracy of the spectral-domain modelling technique has been demonstrated against the time-domain technique by the author in [33–35] and in several recent studies. The most relevant ones are cited in this paper.

References

1. Salter, S.H.; Taylor, J.R.; Caldwell, N.J. Power conversion mechanisms for wave energy. *Proc. Inst. Mech. Eng. Part J. Eng. Marit. Environ.* **2002**, *216*, 1–27. doi:10.1243/147509002320382103. [CrossRef]
2. Cruz, J. *Ocean Wave Energy*; Springer: Berlin/Heidelberg, Germany, 2008. doi:10.1007/978-3-540-74895-3. [CrossRef]
3. Trueworthy, A.; Dupont, B. The wave energy converter design process: Methods applied in industry and shortcomings of current practices. *J. Mar. Sci. Eng.* **2020**, *8*, 932. doi:10.3390/jmse8110932. [CrossRef]
4. Sirigu, S.A.; Foglietta, L.; Giorgi, G.; Bonfanti, M.; Cervelli, G.; Bracco, G.; Mattiazzo, G. Techno-Economic optimisation for a wave energy converter via genetic algorithm. *J. Mar. Sci. Eng.* **2020**, *8*, 482. doi:10.3390/JMSE8070482. [CrossRef]

5. Carapellese, F.; Sirigu, S.A.; Giorgi, G.; Bonfanti, M.; Mattiazzo, G. Multiobjective optimisation approaches applied to a wave energy converter design. In Proceedings of the European Wave and Tidal Energy Conference, European Wave and Tidal Energy Conference Series, Plymouth, UK, 5–9 September 2021; pp. 2114-1–2114-8.
6. Giorcelli, F.; Sirigu, S.A.; Pasta, E.; Gioia, D.G.; Bonfanti, M.; Mattiazzo, G. Wave Energy Converter Optimal Design Under Parameter Uncertainty. In Proceedings of the ASME 2022 41th International Conference on Ocean, Offshore and Artic Engineering, Hamburg, Germany, 5–10 June 2022; *in press*.
7. Babarit, A.; Clément, A.H.; Gilloteaux, J.C. Optimization and Time-Domain Simulation of the SEAREV Wave Energy Converter, In Proceedings of the ASME 2005 24th International Conference on Offshore Mechanics and Arctic Engineering, Halkidiki, Greece, 12–17 June 2005. doi:10.1115/OMAE2005-67286. [CrossRef]
8. Babarit, A.; Clement, A.H. Shape optimisation of the SEAREV wave energy converter. In Proceedings of the World Renewable Energy Congress IX, Florence, Italy, 19–25 August 2006.
9. Teixeira-Duarte, F.; Clemente, D.; Giannini, G.; Rosa-Santos, P.; Taveira-Pinto, F. Review on layout optimization strategies of offshore parks for wave energy converters. *Renew. Sustain. Energy Rev.* **2022**, *163*, 112513. doi:10.1016/j.rser.2022.112513. [CrossRef]
10. Li, L.; Gao, Z.; Yuan, Z.M. On the sensitivity and uncertainty of wave energy conversion with an artificial neural-network-based controller. *Ocean. Eng.* **2019**, *183*, 282–293. doi:10.1016/j.oceaneng.2019.05.003. [CrossRef]
11. Neshat, M.; Mirjalili, S.; Sergiienko, N.Y.; Esmaeilzadeh, S.; Amini, E.; Heydari, A.; Garcia, D.A. Layout optimisation of offshore wave energy converters using a novel multi-swarm cooperative algorithm with backtracking strategy: A case study from coasts of Australia. *Energy* **2022**, *239*, 122463. doi:10.1016/j.energy.2021.122463. [CrossRef]
12. Zhu, K.; Shi, H.; Han, M.; Cao, F. Layout study of wave energy converter arrays by an artificial neural network and adaptive genetic algorithm. *Ocean Eng.* **2022**, *260*, 112072. doi:10.1016/j.oceaneng.2022.112072. [CrossRef]
13. Zeng, X.; Wang, Q.; Kang, Y.; Yu, F. Hydrodynamic interactions among wave energy converter array and a hierarchical genetic algorithm for layout optimization. *Ocean Eng.* **2022**, *256*, 111521. doi:10.1016/j.oceaneng.2022.111521. [CrossRef]
14. He, Z.; Ning, D.; Gou, Y.; Zhou, Z. Wave energy converter optimization based on differential evolution algorithm. *Energy* **2022**, *246*, 123433. doi:10.1016/j.energy.2022.123433. [CrossRef]
15. Garcia-Teruel, A.; Forehand, D.I. Manufacturability considerations in design optimisation of wave energy converters. *Renew. Energy* **2022**, *187*, 857–873. doi:10.1016/j.renene.2021.12.145. [CrossRef]
16. Garcia-Teruel, A.; DuPont, B.; Forehand, D.I. Hull geometry optimisation of wave energy converters: On the choice of the optimisation algorithm and the geometry definition. *Appl. Energy* **2020**, *280*, 115952. doi:10.1016/j.apenergy.2020.115952. [CrossRef]
17. Giannini, G.; Rosa-Santos, P.; Ramos, V.; Taveira-Pinto, F. Wave energy converters design combining hydrodynamic performance and structural assessment. *Energy* **2022**, *249*, 123641. doi:10.1016/j.energy.2022.123641. [CrossRef]
18. Pedersen, H.C.; Hansen, R.H.; Hansen, A.H.; Andersen, T.O.; Bech, M.M. Design of full scale wave simulator for testing Power Take off systems for wave energy converters. *Int. J. Mar. Energy* **2016**, *13*, 130–156. doi:10.1016/j.ijome.2016.01.005. [CrossRef]
19. Hansen, R.H.; Andersen, T.O.; Pedersen, H.C. Model Based Design of Efficient Power Take-Off Systems for Wave Energy Converters. In Proceedings of the 12th Scandinavian International Conference on Fluis Power, Tampere, Finland, 18–20 May 2011; pp. 1–15.
20. Wave Star. Unlimited Clean Energy with the Wavestar Machine. Available online: http://wavestarenergy.com/ (accessed on 30 November 2020).
21. Liu, C.H.; Yang, Q.J.; Bao, G. Performance investigation of a two-raft-type wave energy converter with hydraulic power take-off unit. *Appl. Ocean. Res.* **2017**, *62*, 139–155. doi:10.1016/j.apor.2016.12.002. [CrossRef]
22. Zheng, S.M.; Zhang, Y.H.; Zhang, Y.L.; Sheng, W.A. Numerical study on the dynamics of a two-raft wave energy conversion device. *J. Fluids Struct.* **2015**, *58*, 271–290. doi:10.1016/j.jfluidstructs.2015.07.008. [CrossRef]
23. Liu, C.; Yang, Q.; Bao, G. Influence of hydraulic power take-off unit parameters on power capture ability of a two-raft-type wave energy converter. *Ocean. Eng.* **2018**, *150*, 69–80. doi:10.1016/j.oceaneng.2017.12.063. [CrossRef]
24. Gaspar, J.F.; Calvário, M.; Kamarlouei, M.; Soares, C.G. Design tradeoffs of an oil-hydraulic power take-off for wave energy converters. *Renew. Energy* **2018**, *129*, 245–259. doi:10.1016/j.renene.2018.05.092. [CrossRef]
25. Jusoh, M.A.; Ibrahim, M.Z.; Daud, M.Z.; Yusop, Z.M.; Albani, A.; Rahman, S.J.; Mohad, S. Parameters estimation of hydraulic power take-off system for wave energy conversion system using genetic algorithm. *IOP Conf. Ser. Earth Environ. Sci.* **2020**, *463*, 012129. doi:10.1088/1755-1315/463/1/012129. [CrossRef]
26. Siemens Digital Industries. Siemens Simcenter Amesim. 2020. Available online: https://www.plm.automation.siemens.com/global/en/products.html (accessed on 5 November 2020).
27. Calvário, M.; Gaspar, J.F.; Kamarlouei, M.; Hallak, T.S.; Guedes Soares, C. Oil-hydraulic power take-off concept for an oscillating wave surge converter. *Renew. Energy* **2020**, *159*, 1297–1309. doi:10.1016/j.renene.2020.06.002. [CrossRef]
28. Sarkar, D.; Doherty, K.; Dias, F. The modular concept of the Oscillating Wave Surge Converter. *Renew. Energy* **2016**, *85*, 484–497. doi:10.1016/j.renene.2015.06.012. [CrossRef]
29. Sarkar, D.; Contal, E.; Vayatis, N.; Dias, F. Prediction and optimization of wave energy converter arrays using a machine learning approach. *Renew. Energy* **2016**, *97*, 504–517. doi:10.1016/j.renene.2016.05.083. [CrossRef]
30. MOREnergy Lab. ISWEC, 2022. Available online: http://www.morenergylab.polito.it/iswec/ (accessed on 22 August 2022).

31. Pozzi, N.; Bonetto, A.; Bonfanti, M.; Bracco, G.; Dafnakis, P.; Giorcelli, E.; Passione, B.; Sirigu, S.; Mattiazzo, G. PeWEC: Preliminary design of a full-scale plant for the mediterranean sea. In Proceedings of the NAV International Conference on Ship and Shipping Research, Trieste, Italy, 20–22 June 2018; Number 221499. doi:10.3233/978-1-61499-870-9-504. [CrossRef]
32. Deb, K.; Pratap, A.; Agarwal, S.; Meyarivan, T. A fast and elitist multiobjective genetic algorithm: NSGA-II. *IEEE Trans. Evol. Comput.* **2002**, *6*, 182–197. doi:10.1109/4235.996017. [CrossRef]
33. Sirigu, S.; Bonfanti, M.; Carapellese, F.; Bracco, G. Stochastic modelling via statistical linearization of a wave energy converter. In Proceedings of the European Wave and Tidal Energy Conference, Plymouth, UK, 5–9 September 2021; pp. 2132-1–2132-9.
34. Bonfanti, M.; Sirigu, S.; Mattiazzo, G. ISWEC approaching the spectral-domain: Modelling and numerical experiments. In Proceedings of the International Conference on Electrical, Computer, Communications and Mechatronics Engineering (ICECCME 2021), Flic En Flac, Mauritius, 7–8 October 2021. doi:10.1109/ICECCME52200.2021.9590910. [CrossRef]
35. Bonfanti, M.; Bracco, G. Non-linear Frequency Domain Modelling of a Wave Energy Harvester. In *The International Conference of IFToMM ITALY*; Springer: Cham, Switzerland, 2022; Volume 122 MMS, pp. 874–881. doi:10.1007/978-3-031-10776-4_100. [CrossRef]
36. Silva, L.S.; Sergiienko, N.Y.; Pesce, C.P.; Ding, B.; Cazzolato, B.; Morishita, H.M. Stochastic analysis of nonlinear wave energy converters via statistical linearization. *Appl. Ocean. Res.* **2020**, *95*, 102023. doi:10.1016/j.apor.2019.102023. [CrossRef]
37. Pozzi, N.; Bonfanti, M.; Mattiazzo, G. Mathematical Modeling and Scaling of the Friction Losses of a Mechanical Gyroscope. *Int. J. Appl. Mech.* **2018**, *10*, 1850024. doi:10.1142/S1758825118500242. [CrossRef]
38. Sirigu, S.A.; Gallizio, F.; Giorgi, G.; Bonfanti, M.; Giovanni, B.; Mattiazzo, G. Numerical and experimental identification of the aerodynamic power losses of the ISWEC. *J. Mar. Sci. Eng.* **2020**, *8*, 1–25. doi:10.3390/jmse8010049. [CrossRef]
39. Sirigu, S.; Bonfanti, M.; Dafnakis, P.; Bracco, G.; Mattiazzo, G.; Brizzolara, S. Pitch Resonance Tuning Tanks: A novel technology for more efficient wave energy harvesting. In Proceedings of the OCEANS 2018 MTS/IEEE Charleston, Charleston, SC, USA, 22–25 October 2018; doi:10.1109/OCEANS.2018.8604591. [CrossRef]
40. Sirigu, S.A.; Bracco, G.; Bonfanti, M.; Dafnakis, P.; Mattiazzo, G. On-board sea state estimation method validation based on measured floater motion. *IFAC-PapersOnLine* **2018**, *51*, 68–73. doi:10.1016/j.ifacol.2018.09.471. [CrossRef]
41. Bonfanti, M.; Hillis, A.; Sirigu, S.A.; Dafnakis, P.; Bracco, G.; Mattiazzo, G.; Plummer, A. Real-time wave excitation forces estimation: An application on the ISWEC device. *J. Mar. Sci. Eng.* **2020**, *8*, 1–30. doi:10.3390/jmse8100825. [CrossRef]
42. Bonfanti, M.; Sirigu, S.A.; Giorgi, G.; Dafnakis, P.; Bracco, G.; Mattiazzo, G. A passive control strategy applied to the iswec device: Numerical modelling and experimental tests. *Int. J. Mech. Control.* **2020**, *21*, 143–154.
43. Giorgi, G.; Sirigu, S.; Bonfanti, M.; Bracco, G.; Mattiazzo, G. Fast nonlinear Froude–Krylov force calculation for prismatic floating platforms: A wave energy conversion application case. *J. Ocean. Eng. Mar. Energy* **2021**, *7*, 439–457. doi:10.1007/s40722-021-00212-z. [CrossRef]
44. Genuardi, L.; Bracco, G.; Sirigu, S.A.; Bonfanti, M.; Paduano, B.; Dafnakis, P.; Mattiazzo, G. An application of model predictive control logic to inertial sea wave energy converter. In *IFToMM World Congress on Mechanism and Machine Science*; Springer: Cham, Switzerland, 2019; Volume 73, pp. 3561–3571. doi:10.1007/978-3-030-20131-9_351. [CrossRef]
45. Niosi, F.; Parrinello, L.; Paduano, B.; Pasta, E.; Carapellese, F.; Bracco, G. On the Influence of Mooring in Wave Energy Converters Productivity: the PeWEC case. In Proceedings of the 2021 International Conference on Electrical, Computer, Communications and Mechatronics Engineering (ICECCME), Flic En Flac, Mauritius, 7–8 October 2021; pp. 1–6. doi:10.1109/ICECCME52200.2021.9590867. [CrossRef]
46. Anerdi, C.; Paduano, B.; Casalone, P.; Mattiazzo, G.; Giordano, L. Design of a Reinforced Concrete Wave Energy Converter in Extreme Wave Conditions. In Proceedings of the I4SDG Workshop 2021, Online, 25–26 November 2021; Quaglia, G., Gasparetto, A., Petuya, V., Carbone, G., Eds.; Springer International Publishing: Cham, Switzerland, 2022; pp. 70–77.
47. Hatecke, H. The impulse response fitting and ship motions. *Ship Technol. Res.* **2015**, *62*, 97–106. doi:10.1179/2056711115Y.0000000001. [CrossRef]
48. Temarel, P. Loads on ships and offshore structures. In Proceedings of the 26th International Conference on Offshore Mechanics and Arctic Engineering (OMAE 2007), San Diego, CA, USA. 10–15 Jun 2007; p. 10. doi:10.1115/OMAE2007-29768. [CrossRef]
49. Peña-Sanchez, Y.; Faedo, N.; Penalba, M.; Giorgi, G.; Merigaud, A.; Windt, C.; Garc, D.; Wang, L.; Ringwood, J.V. Finite-Order hydrodynamic Approximation by Moment-Matching (FOAMM) toolbox for wave energy applications. In Proceedings of the 13th European Wave and Tidal Energy Conference, Napoli, Italy, 1–6 September 2019.
50. Babarit, A.; Guglielmi, M.; Clément, A.H. Declutching control of a wave energy converter. *Ocean. Eng.* **2009**, *36*, 1015–1024. doi:10.1016/j.oceaneng.2009.05.006. [CrossRef]
51. Wu, J.; Yao, Y.; Zhou, L.; Göteman, M. Latching and declutching control of the solo duck wave-energy converter with different load types. *Energies* **2017**, *10*, 70. doi:10.3390/en10122070. [CrossRef]
52. Folley, M.; Whittaker, T. Preliminary cross-Validation of wave energy converter array interactions. In Proceedings of the International Conference on Offshore Mechanics and Arctic Engineering-OMAE. American Society of Mechanical Engineers Digital Collection, Nantes, France, 9–14 June 2013; Volume 8. doi:10.1115/OMAE2013-10837. [CrossRef]
53. Babarit, A.; Hals, J.; Kurniawan, A.; Muliawan, M.; Moan, T.; Krokstad, J. *The NumWEC Project: Numerical Esitmation of Energy Delivery from a Selection of Wave Energy Converters*; Technical Report December; Norwegian University of Science and Technology: Trondheim, Norway, 2011. doi:10.13140/RG.2.1.3807.8885. [CrossRef]

54. Folley, M. Chapter 4 - Spectral-Domain Models. In *Numerical Modelling of Wave Energy Converters*; Academic Press: Cambridge, MA, USA, 2016; pp. 67–80. ISBN 9780128032107, doi:10.1016/b978-0-12-803210-7.00004-9. [CrossRef]
55. Roberts, J.; Spanos, P. *Random Vibration and Statistical Linearization by Roberts JB, Spanos PD*; John Wiley & Sons: New York, NY, USA, 1990.
56. Da Silva, L.S.; Morishita, H.M.; Pesce, C.P.; Gonçalves, R.T. Nonlinear analysis of a heaving point absorber in frequency domain via statistical linearization. In Proceedings of the International Conference on Offshore Mechanics and Arctic Engineering—OMAE, Glasgow, Scotland, UK, 9–14 June 2019; Volume 9. doi:10.1115/OMAE2019-95785. [CrossRef]
57. Gudmestad, O.T.; Connor, J.J. Linearization methods and the influence of current on the nonlinear hydrodynamic drag force. *Appl. Ocean. Res.* **1983**, *5*, 184–194. doi:10.1016/0141-1187(83)90032-9. [CrossRef]
58. Spanos, P.D.; Arena, F.; Richichi, A.; Malara, G. Efficient dynamic analysis of a nonlinear wave energy harvester model. *J. Offshore Mech. Arct. Eng.* **2016**, *138*, 1–8. doi:10.1115/1.4032898. [CrossRef]
59. Spanos, P.D.; Strati, F.M.; Malara, G.; Arena, F. Stochastic dynamic analysis of U-OWC wave energy converters. In Proceedings of the International Conference on Offshore Mechanics and Arctic Engineering—OMAE, Trondheim, Norway, 25–30 June 2017; Volume 10. doi:10.1115/OMAE2017-61522. [CrossRef]
60. Bain, L.J.; Engelhardt, M. *Introduction to Probability and Mathematical Statistics*; Classic Series; Brooks/Cole Cengage Learning: Boston, MA, USA, 1993; Volume 49, p. 673. doi:10.2307/2532587. [CrossRef]
61. Rexroth, B. Hägglunds Drive Systems and Solutions, 2020. Available online: https://dc-us.resource.bosch.com/media/us (accessed on 22 January 2021).
62. Hansen, R.H.; Kramer, M.M.; Vidal, E. Discrete displacement hydraulic power take-off system for the wavestar wave energy converter. *Energies* **2013**, *6*, 4001–4044. doi:10.3390/en6084001. [CrossRef]
63. Hansen, R.H. Aalborg Universitet Design and Control of the PowerTake-Off System for a Wave Energy Converter with Multiple Absorbers Hansen, Rico Hjerm Publication Date. Ph.D. Thesis, Aalborg Universitet, Aalborg, Denmark, 2013.
64. Garcia-Teruel, A.; Forehand, D.I.; Jeffrey, H. Wave Energy Converter hull design for manufacturability and reduced LCOE. In Proceedings of the 7th International Conference on Ocean Energy (ICOE), Cherbourg, Normandy, France, 12–14 June 2018, pp. 1–9.
65. Fontana, M.; Casalone, P.; Sirigu, S.A.; Giorgi, G.; Bracco, G.; Mattiazzo, G. Viscous damping identification for a wave energy converter using CFD-URANS simulations. *J. Mar. Sci. Eng.* **2020**, *8*, 355. doi:10.3390/JMSE8050355. [CrossRef]
66. Bracco, G.; Giorcelli, E.; Giorgi, G.; Mattiazzo, G.; Passione, B.; Raffero, M.; Vissio, G. Performance assessment of the full scale ISWEC system. In Proceedings of the IEEE International Conference on Industrial Technology, Seville, Spain, 17–19 March 2015; Volume 2015-June, pp. 2499–2505. doi:10.1109/ICIT.2015.7125466. [CrossRef]
67. SKF. Cuscinetti Super-Precision, 2015. Available online: https://www.skf.com/binaries/pub44/Images/0901d196803e9f7d-13383_1-IT_tcm_44-129877.pdf (accessed on 26 January 2020).
68. Moog. Electro-Hydrostatic Actuation System, 2020. Available online: https://www.moog.com/content/sites/global/en/products.html (accessed on 5 February 2021).
69. Rexroth, B. Fixed Motor, 2020. Available online: https://www.boschrexroth.com/it/xc/products/product-groups/industrial-hydraulics/motors/axial-piston-motors/fixed-motors (accessed on 5 February 2021).
70. SIEMENS. 1PH8 Servomotors (Synchronous). Available online: https://mall.industry.siemens.com/mall/it/it/Catalog/Products/10045151 (accessed on 5 February 2021).
71. Hasselmann, K.; Barnett, T.P.; Bouws, E.; Carlson, H.; Cartwright, D.E.; Eake, K.; Euring, J.A.; Gicnapp, A.; Hasselmann, D.E.; Kruseman, P.; et al. Measurements of wind-wave growth and swell decay during the joint North Sea wave project (JONSWAP). *Deut. Hydrogr. Z.* **1973**, *8*, 1–95.
72. Ruellan, M.; Benahmed, H.; Multon, B.; Josset, C.; Babarit, A.; Clement, A. Design methodology for a SEAREV wave energy converter. *IEEE Trans. Energy Convers.* **2010**, *25*, 760–767. doi:10.1109/TEC.2010.2046808. [CrossRef]
73. Garcia-Rosa, P.B.; Bacelli, G.; Ringwood, J.V. Control-informed geometric optimization of wave energy converters: The impact of device motion and force constraints. *Energies* **2015**, *8*, 13672–13687. doi:10.3390/en81212386. [CrossRef]
74. Giorgi, G.; Ringwood, J.V. Parametric motion detection for an oscillating water column spar buoy. In Proceedings of the Advances in Renewable Energies Offshore—Proceedings of the 3rd International Conference on Renewable Energies Offshore, RENEW, Lisbon, Portugal, 8–10 October 2018; pp. 505–512.
75. Lagarias, J.C.; Reeds, J.A.; Wright, M.H.; Wright, P.E. Convergence properties of the Nelder-Mead simplex method in low dimensions. *SIAM J. Optim.* **1998**, *9*, 112–147. doi:10.1137/S1052623496303470. [CrossRef]
76. Sirigu, S.A.; Vissio, G.; Bracco, G.; Giorcelli, E.; Passione, B.; Raffero, M.; Mattiazzo, G. ISWEC design tool. *Int. J. Mar. Energy* **2016**, *15*, 201–213. doi:10.1016/j.ijome.2016.04.011. [CrossRef]
77. Ambühl, S.; Kramer, M.; Sørensen, J.D. Reliability-based structural optimization of wave energy converters. *Energies* **2014**, *7*, 8178–8200. doi:10.3390/en7128178. [CrossRef]
78. de Andres, A.; Maillet, J.; Todalshaug, J.H.; Möller, P.; Bould, D.; Jeffrey, H. Techno-economic related metrics for a wave energy converters feasibility assessment. *Sustainability* **2016**, *8*, 1109. doi:10.3390/su8111109. [CrossRef]
79. Driscoll, F.R.; Weber, J.W.; Jenne, D.S. *Methodology to Calculate the ACE and HPQ Metrics Used in the Wave Energy Prize*; Technical Report; NREL/TP-5000-70592; National Renewable Energy Lab.(NREL): Golden, CO, USA, 2018. doi:10.2172/1426063. [CrossRef]

80. Joines, J.A.; Houck, C.R. On the use of non-stationary penalty functions to solve nonlinear constrained optimization problems with GA's. In Proceedings of the IEEE Conference on Evolutionary Computation—Proceedings, IEEE, Walt Disney World Dolphin Hotel, Orlando, FL, USA, 27–29 June 1994; Number 2/-, pp. 579–584. doi:10.1109/icec.1994.349995. [CrossRef]
81. Homaifar, A.; Qi, C.X.; Lai, S.H. Constrained optimization via genetic algorithms. *Simulation* **1994**, *62*, 242–254. doi:10.1177/003754979406200405. [CrossRef]
82. Sivanandam, S.N.; Deepa, S.N. *Introduction to Genetic Algorithms*; Springer: Berlin/Heidelberg, Germany, 2008; pp. 1–442. doi:10.1007/978-3-540-73190-0. [CrossRef]
83. Horst, R.; Panos, P. *Handbook of Global Optimization Volume 2*; Springer: Berlin, Germany, 2002; Volume 53, pp. 1689–1699. doi:10.1007/978-1-4757-5362-2. [CrossRef]
84. Arora, J. *Introduction to Optimum Design*; Academic Press: Cambridge, MA, USA, 2012. doi:10.1016/C2009-0-61700-1. [CrossRef]

Journal of
Marine Science and Engineering

MDPI

Article

Design and Research of Slope-Pendulum Wave Energy Conversion Device

Zhanhong Wan [1], Ze Li [1,*], Dahai Zhang [1] and Honghao Zheng [2]

1 Ocean College, Zhejiang University, Zhoushan 316021, China
2 School of Aerospace Engineering and Applied Mechanics, Tongji University, Shanghai 200092, China
* Correspondence: ocean-lize@zju.edu.cn

Abstract: Wave energy is a kind of clean energy that is rich in reserves and has not been exploited on a large scale. The slope-pendulum wave energy conversion (S-PWEC) device has been optimized in structure and its capture efficiency has been increased. Taking the selection of the Zhejiang sea area as the research background, this paper performs numerical simulation and array WEC experimental testing of S-PWEC under 66 major sea conditions. The experimental results show that S-PWEC adds a slope structure to the bottom, which can effectively improve the motion response ability and resistance to extreme sea conditions. In the regular wave and irregular wave tests, the electron power output efficiency can be increased by 13.24% and 10.06%, respectively; in the array WEC experiment, the diffraction effect and radiation effect will affect the work of the array WEC, and the optimal arrangement distance can be selected to maximize the power output of the WEC system.

Keywords: WEC; Zhejiang sea area; numerical model; power output efficiency; array WEC

Citation: Wan, Z.; Li, Z.; Zhang, D.; Zheng, H. Design and Research of Slope-Pendulum Wave Energy Conversion Device. *J. Mar. Sci. Eng.* **2022**, *10*, 1572. https://doi.org/10.3390/jmse10111572

Academic Editors: Mauro Bonfanti and Giuseppe Giorgi

Received: 8 September 2022
Accepted: 19 October 2022
Published: 24 October 2022

1. Introduction

Wave energy converters (WECs) are devices used to absorb energy in waves and convert the output into electrical energy. Although wave energy conversion projects have not fully reached the commercial stage, it is an ancient discipline, and hundreds of different forms of wave energy conversion device concepts and models have been proposed and experimented with over the past few hundred years [1–4].

There are many ways to classify wave energy conversion devices, which can be divided into onshore, nearshore, and offshore according to the installation location. The offshore WEC can be in a higher-energy wave environment, which is conducive to absorbing more wave energy, but it will bring great challenges to the structural stability, cost, installation, and maintenance of the device [5], which is unfavorable for the commercialization of wave energy conversion devices. Therefore, the WEC built onshore or nearshore does not need to arrange long submarine cables, manual maintenance is relatively easy, the cost is greatly reduced, and the long-term operation of the WEC can be achieved, which may be more attractive for commercial investment. How to develop a stable and efficient WEC is a very challenging research project.

Companies or research teams have developed several different modes of wave energy conversion devices near the shore and on land. The Oyster device developed by Queen's University in the United Kingdom has been updated for two generations. It has performed well in various indicators and has successfully achieved long-term stable operation under real sea conditions [6]. Unlike Oyster, WaveRoller uses a fully immersed floating pendulum, which can form an array power generation module. All floating pendulum movements are completed under the water surface, and its power range is 350 kW-1000 kW [7]. Eco Wave Power combines a pendulum-type wave energy conversion device with a breakwater, and currently has more than 325.7 MW of important project pipelines worldwide [8]. Wavestar combines offshore wind energy capture devices to achieve multi-energy hybrid absorption and improve commercial competitiveness [9].

The main technologies for optimizing the WEC are physical experiments, numerical simulation, and real sea condition testing. With the development of modern computers, numerical simulation technology provides more space for optimization. Its advantage is that it is relatively convenient and fast, does not require the construction of physical objects, and the cost is low [10]. Poguluri et al. used CFD technology to test WEC-rotor, which can be well consistent with the results of physical experiments [11]. This article mainly focuses on a set of open-source programs developed by the WEC-Sim team to optimize the design of a new type of WEC. Ruehl et al. used WEC-Sim to make meaningful optimizations and designs of floating oscillating surge wave energy converters [12]. Wei et al. learned the feasibility of the multi-pump multi-piston power take-off system for the WEC [13], and Sricharan and Chandrasekaran used WEC-SIM and PTO-Sim to evaluate the output of a bean-shaped multi-body floating wave energy converter to maximize power [14].

Complex and changeable sea conditions, and the size and shape of the WEC will affect the work efficiency of the WEC. If the amplitude of WEC motion is increased while the original cost remains unchanged, wave energy capture and power output can be increased, which may be an effective method. The slope-pendulum wave energy conversion device designed in this paper is based on this concept. Through the reflection of waves under the free liquid level, the floating plate is subjected to greater wave excitation force, thereby increasing the incoming torque from the power take-off (PTO) to the hydraulic cylinder, and realizing the first-level wave energy capture efficiency improvement, and ultimately realizing the entire WEC device conversion efficiency improvement.

In short, the development and research of WECs are accelerating, and it is the goal of many researchers and countries interested in developing wave energy to design a WEC device that is suitable for real sea conditions and has a high yield. In this paper, the concept and model of a new wave energy conversion device are studied and developed under the conditions of Zhoushan, Zhejiang, China. In Section 2, this model concept is introduced and described in detail. Next, Section 3 defines and describes the steps and methods of this device experiment. Finally, meaningful results are obtained from these numerical simulation experiments, and discussions and research are carried out.

2. Mathematical

2.1. S-PWEC

To be able to maximize wave energy capture, Wan et al. [15] designed a slope-pendulum wave energy converter, as shown in Figure 1. The device consists of two main modules:

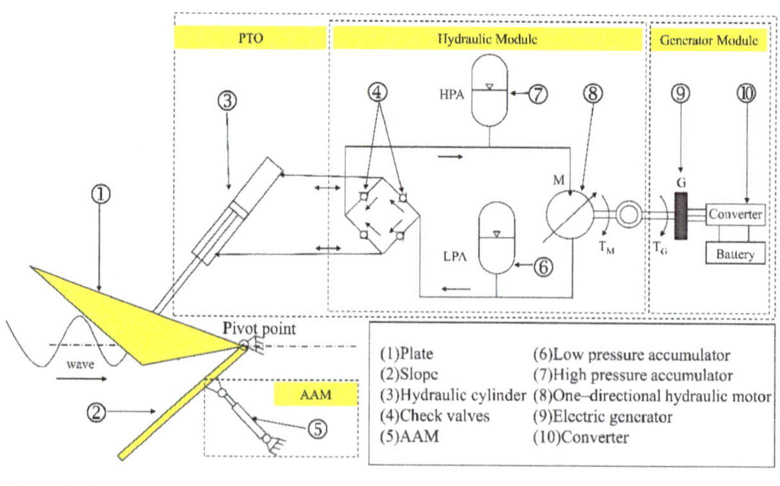

Figure 1. Structure schematic of S-PWEC.

PTO Module: The part is mainly composed of a hydraulic system and generator, a floating plate under the action of wave excitation, and rotating movement around the fixed hinge pivot point, pushing the hydraulic rod. The hydraulic rod movement will compress the hydraulic cylinder and convert the mechanical energy of the floating plate into hydraulic energy. The hydraulic system is a hydraulic circuit consisting of a high-pressure accumulator (HPA), a low-pressure accumulator (LPA), and a check valve. Afterward, the hydraulic energy will be converted into electrical energy using a generator and stored in a battery, or the electricity can also be directly integrated into the grid system.

AAM Module: Different angular positions of the slope in the water have different effects on the reflection and transmission of the incident waves [16]. Under different sea conditions, the AAM system is used to adjust the position of the slope to make the best reflection of the incident waves, which allows the floating plate to absorb as much wave energy as possible to maximize the power output.

When the S-PWEC model does not have a slope structure, it is defined as NS-PWEC in this article.

2.2. Motion Response

Based on the Cummins time domain equation [17], and considering the floating plate viscous effect [12], the motion response equation of the floating plate to the pivot point can be obtained:

$$(I + A_{55}(\infty))\ddot{\theta} = M_{ex} - \int_0^t K_R(t-\xi)\dot{\theta}(t)d\xi + M_b - M_G + M_{vis} + M_{PTO} \tag{1}$$

In the above equation:

- $\theta, \dot{\theta}$, and $\ddot{\theta}$ are the angular displacement, angular velocity, and angular acceleration of the device moving at the pivot point, respectively.
- I and $A_{55}(\infty)$ are the moment of inertia of the device and the additional moment of inertia at infinite frequency, respectively.
- $\int_0^t K_R(t-\xi)\dot{\theta}(t)d\xi$ is the convolution term of the radiation moment, which represents the memory effect of the fluid, where ξ is a small time step.

$$K_R = \frac{2}{\pi}\int_0^{+\infty} B(\omega)\cos(\omega t)dt \tag{2}$$

$B(w)$ is radiation damping.

- M_b and M_G are the buoyancy moment and the gravitational moment, respectively, which constitute the recovery moment of the floating plate, which is $K_S\theta$.
- M_{ex} is the excitation moment; the excitation moment of the regular wave can be given by the following equation:

$$M_{ex}(t) = Re\left\{R_f(t)\frac{H}{2}\widetilde{M}_{ex}(\omega,\theta)e^{i\omega t}\right\} \tag{3}$$

where Re denotes taking the real part, $\widetilde{M}_{ex}(\omega,\theta)$ denotes the moment per unit wave amplitude of the floating plate at frequency ω, and $R_f(t)$ is used to avoid strong transient flow, which is expressed as:

$$R_f(t) = \begin{cases} \frac{1}{2}\left(1+\cos\left(\pi + \frac{\pi t}{t_r}\right)\right) & \frac{t}{t_r} < 1 \\ 1 & \frac{t}{t_r} \geq 1 \end{cases} \tag{4}$$

t is the computation time and t_r is the continuous function computation time.

The excitation moment for the irregular wave case can be expressed as follows:

$$M_{ex}(t) = Re\left\{R_f(t)\sum_{j=1}^{N}\widetilde{M}_{ex}(\omega,\theta)e^{i(\omega_j t+\phi_j)}\sqrt{2S(\omega_j)\Delta\omega}\right\} \tag{5}$$

$S(\omega_j)$ is for calculating the random wave spectrum under the sea state.

- M_{vis} is the secondary viscous drag moment; in practice, the wave energy converter device in the process of action with the wave will induce vortex detachment, resulting in a viscous effect [18].

$$M_{vis} = (-C_D|V - V_0|(V - V_0)) \times L_{vis} \quad (6)$$

where $C_D = \frac{1}{2}\rho C_d A_d$ is the coefficient of the secondary damping term in the viscous torque. C_D can be obtained from the relevant literature or hydrodynamic experiments. V is the speed of the floating plate, V_0 is the speed of the surrounding fluid, and L_{vis} represents the moment of the viscous force on the pivot point.

- M_{PTO} is the damping torque of the hydraulic power take-off.

$$M_{PTO} = \Delta p_{piston} A_{piston} \times L_{PTO} \quad (7)$$

where Δp_{piston} is the pressure difference in the hydraulic rod, A_{piston} is the cross-sectional area of the hydraulic rod, and L_{piston} is the distance from the intersection of the hydraulic rod and the floating plate to the pivot point. Assuming that the hydraulic rod is always perpendicular to the floating plate, and the nonlinearity of the PTO is not considered, the influence of the nonlinear PTO on the wave energy capture efficiency will be studied in future work.

For measuring the merit of a wave energy conversion device, it is necessary to define the capture width ratio (CWR), the ratio of the average power absorbed by the device for the converted wave energy to the incident wave energy power per unit time [19]. In addition, this paper compares the electron power output efficiency (EPOE) of the device, which can be defined as the ratio of the electron power output of the WEC device to the incident wave power:

$$CWR = \frac{P_1}{P_{Wave} b} \quad (8)$$

$$EPOE = \frac{P_3}{P_{wave} b} \quad (9)$$

where b is the device width, P_1 and P_3 are the primary wave energy capture efficiency and tertiary electrical output efficiency of the WEC per unit time, respectively, and P_{wave} is the incoming wave power per unit width of the incident wave.

3. Results and Discussion

The geometry files and hydrodynamic coefficients are imported in WEC-Sim (using boundary element software such as WAMIT, NEMOH, and ANSYS AQWA) and the motion response of the imported objects under different sea conditions is determined with the help of control equations, wave loads, and the PTO-Sim module [20], as shown in Figure 2.

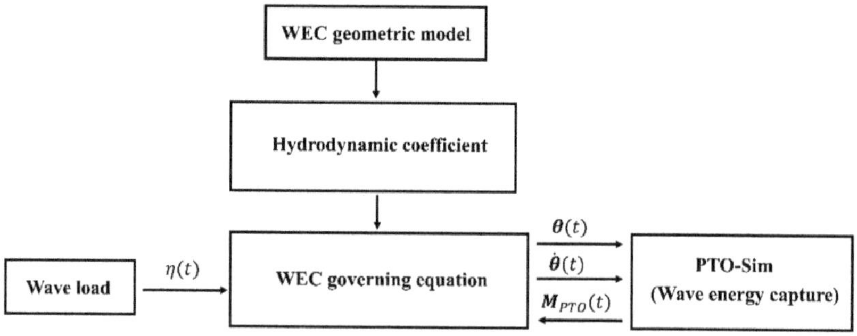

Figure 2. Workflow diagram.

3.1. Initial Condition Setting

3.1.1. Analysis of Sea State Conditions

The wave period and wave height data from 1 January 2021 to 1 January 2022 in the sea area of Zhoushan, Zhejiang Province, China are obtained on NOAA, as shown in Figure 3. The statistical results show that the wave energy generated by waves with a spectral peak period of 6–9 s and a wave height between 0.5 and 2 m in the Zhoushan sea area accounts for about 60% of the annual wave energy, which is close to the research results of Wang et al. [21].

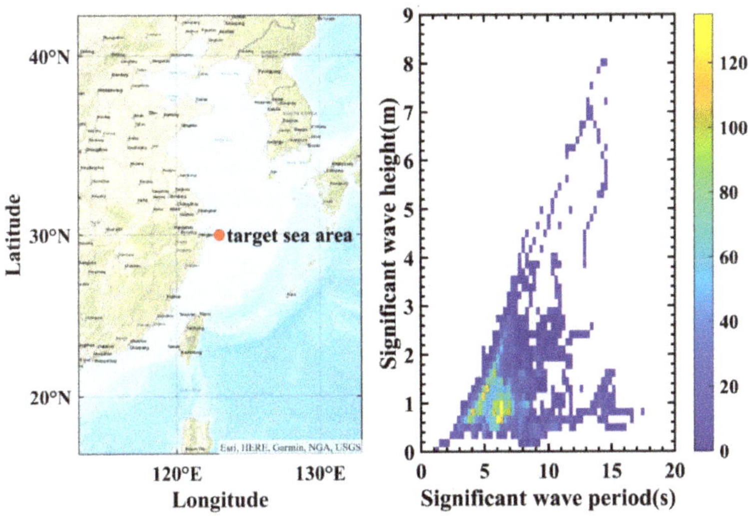

Figure 3. Statistical of sea state.

3.1.2. Single WEC Model

The floating plate and the slope structure are used to obtain BEM hydrodynamic data in ANSYS AQWA, some of which are shown in Table 1. Meanwhile, a slope-pendulum wave energy conversion device model is established in MATLAB Simulink [22], and a hydraulic power take-off (PTO) is used for power generation to complete the conversion from wave energy capture to mechanical energy and finally realize the output result of electrical energy.

Table 1. S-PWEC components parameters.

Property	Floating Plate	Slope
Mass (kg)	802.36	5065.41
$X_{CG}(\mathrm{m})$	−2.24	−2.651
$Y_{CG}(\mathrm{m})$	0	0
$Z_{CG}(\mathrm{m})$	0.22	−2.7
$I_{xx}\left(\mathrm{kgm}^2\right)$	1379.1	60,802
$I_{yy}\left(\mathrm{kgm}^2\right)$	4211.8	42,574
$I_{zz}\left(\mathrm{kgm}^2\right)$	5439.5	17,467

As shown in Figure 4, "plate" represents the main body of the floating plate and "Slope" represents the slope structure, both of which are constrained by "PTO" so that the floating plate can only rotate around the fixed axis. "PTO" transmits the wave energy captured by the floating plate to "PTO-Sim" as a moment, and calculates the power that can be obtained, which is the hydraulic force extractor chosen in this model. The device is fixed to the "global reference frame" through the "constraint", which acts as the seabed in the model and is used for various data input in the simulation. The "PTO-Sim" works by transferring the torque at the connection point between the PTO and the floating plate, and converting the mechanical energy into the hydraulic system's pressure energy to achieve a stable output of electricity. Some of the essential data for the WEC-Sim simulation are shown in Table 2.

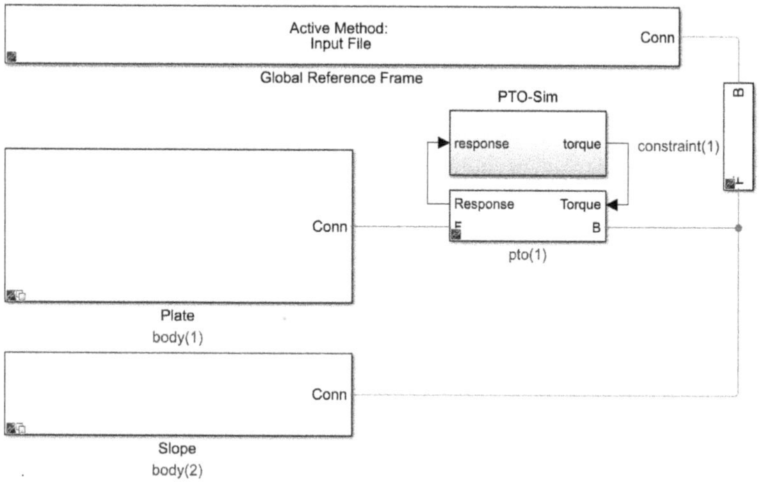

Figure 4. Single WEC model.

Table 2. WEC-Sim simulation parameters.

Description	Value
Wave height (m)	H
Wave period (s)	T
Simulink solver	ode45
Maximum time step (s)	T/200
Wave ramp time (s)	5 T
Convolution integral time (s)	25
Simulation end time (s)	25 T

3.2. Motion Response

The new slope-pendulum wave energy conversion device fixes an adjustable angle slope structure in the lower part of the floating plate, the presence of which can be used for wave reflection and limit the angular displacement of the floating plate immersed in seawater; when it is under severe sea conditions; its maximum angular displacement is related to the angle of the slope.

As shown in Figure 5, when the floating plate works under a regular wave condition with a wave height of 2 m, the floating plate will not touch the slope structure and the overall operation is relatively continuous. However, when the wave height increases to 6 m, the amplitude of the floating plate increases accordingly, and the downward amplitude will exceed the initial displacement between it and the bottom slope (the angle between the floating plate and the slope in hydrostatic water); then, it will touch the bottom slope structure, and a small oscillation will occur, which is used to reduce the speed of the floating plate while waiting for the next wave peak to come. This process, which can only be triggered under larger sea conditions, is an effective self-locking protection mechanism that uses its own oscillations to consume excess energy, which is useful for trying to apply captured wave energy under extreme sea conditions (typhoon weather) and for maintaining the stability of its own system.

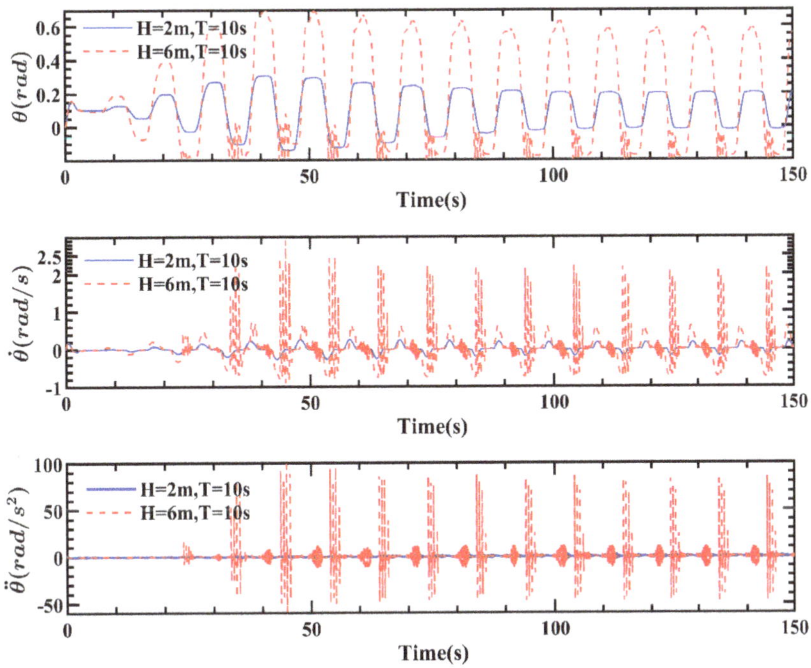

Figure 5. Self-locking response of S-PWEC.

3.3. Power Generation Efficiency of Single WEC

In order to better illustrate that the bottom-inclined plate structure can reflect the wave energy under the free liquid surface and facilitate the absorption of the floating plate, it is necessary to compare the hydrodynamic response results and each power output result of S-PWEC and NS-PWEC. The initial conditions of the incident wave are: regular wave; wave height H = 2 m; period T = 6 s. The steady working phase t = 100–150 s is selected, and the hydrodynamic response results are shown in Figure 6. It can be clearly seen that the angular displacement amplitude of S-PWEC is larger compared with NS-PWEC, which is nearly doubled, indicating that the wave excitation force on the floating plate of the S-PWEC device is larger than that of the N-SPWEC device. The addition of the slope structure can significantly increase the motion amplitude of the device and provide positive feedback to the wave energy capture results, which can also be greatly enhanced.

The peak speed of S-PWEC is increased by 58.3%, which again shows that the motion of the floating plate is more violent and subject to the strong impact of waves, which is very meaningful for the design of the wave energy conversion device (provided that it is within the load-bearing range of the device), and by adding a slope structure at the bottom, the wave field energy reflection is concentrated near the free liquid surface, which increases the wave intensity near the floating plate and makes the floating plate as high as possible. It is an essential part of the commercialization to capture wave energy and improve the efficiency of wave energy capture to obtain more profit with small cost.

The output power of S-PWEC and NS-PWEC in each stage is shown in Figure 7. "Abspower" represents the power transferred from the floating plate to the hydraulic rod, which is the first-stage wave energy capture power; "genpower" is the power transferred from the hydraulic system to the generator; "elepower" is the final output power of the generator. It is known from Figure 6 that the addition of the slope structure at the bottom of the floating plate can effectively increase the motion response of the floating plate, the first-level hydrodynamic capture efficiency and power (despite the loss in the conversion of mechanical and electrical energy, the final converted electrical power must be increased), the wave energy first-level instantaneous capture power by 73.1%, and the final electrical output power by two times. This result is exciting, showing that a slope structure can at least have a favorable impact on the wave energy capture, and this optimization on the WEC structure is certain.

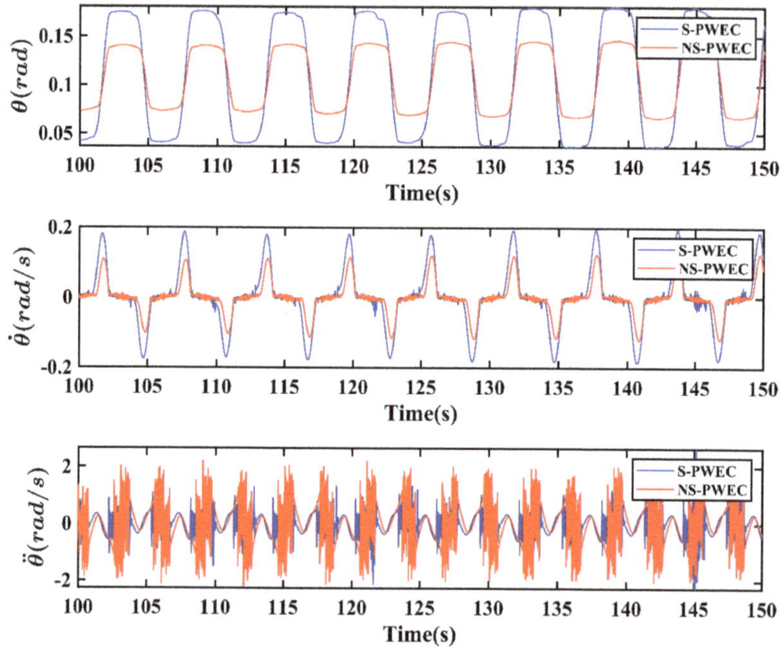

Figure 6. Motion Response of S-PWEC and NS-PWEC.

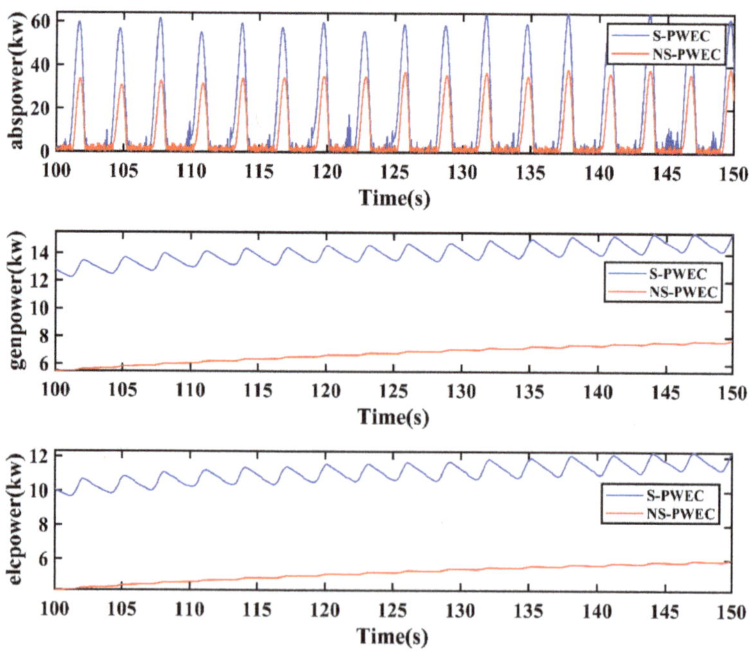

Figure 7. Power output of S-PWEC and NS-PWEC.

In Section 3.1.1, the collation of the sea state is carried out, so the main sea state of the Zhoushan sea area is the condition for simulation experiments, and the range of the test sea state is shown in Table 3. After the free arrangement and combination, there is a total of 66 groups of sea state experiments, the experimental objects are S-PWEC and NS-PWEC, and the following will obtain the electron power output efficiency (EPOE) of S-PWEC and NS-PWEC in the form of a matrix, as shown in Figures 8 and 9.

Table 3. Study of sea state.

Description	T(s)	H(m)
Value	2–12	0.5–3

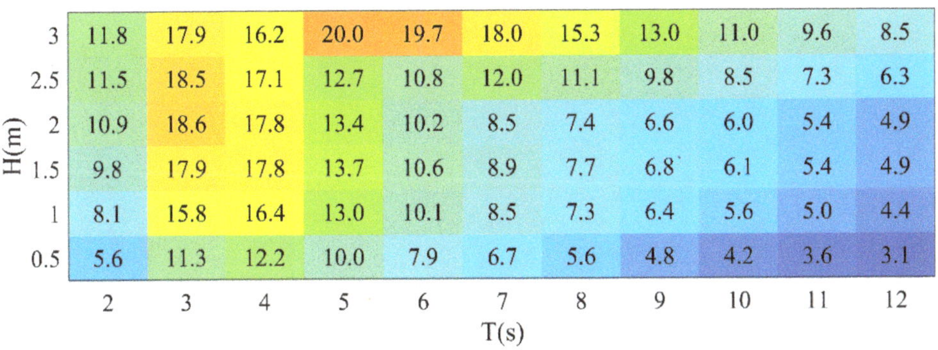

Figure 8. NS-PWEC power generation efficiency under regular wave sea conditions.

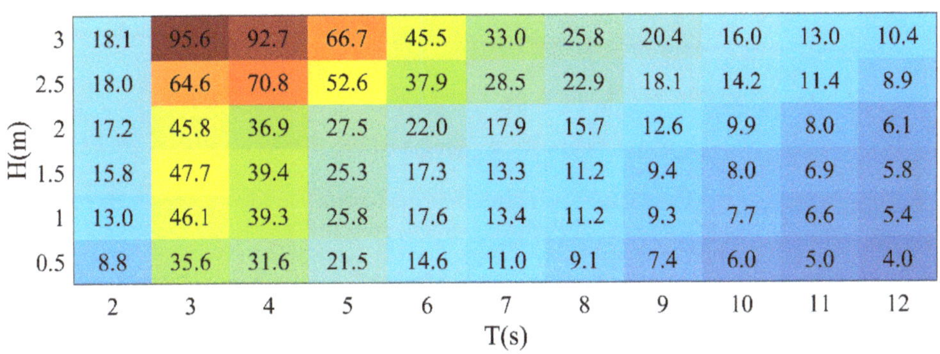

Figure 9. S-PWEC power generation efficiency under regular wave sea conditions.

As shown in Figure 9, among 66 kinds of sea state tests, the optimal sea state for the NS-PWEC device is T = 5 s, H = 3 m sea state, and the power generation efficiency is 20%, which shows that the better power generation sea state for this wave energy conversion device is roughly concentrated in such a sea state with small period and large wave height.

As can be seen from Figure 9, for the S-PWEC device, the power output efficiency is greater than the NS-PWEC power output efficiency results in all of these 66 sets of sea state tests, once again indicating that the S-PWEC has a significant improvement in wave energy capture and absorption compared to the conventional NS-PWEC device. Its optimal sea state is at T = 3 s and H = 3 m, and the power output EPOE reaches 95.58%, which is an incredible power generation efficiency. For the 3D device, wave energy beyond its width is absorbed, resulting in a CWR greater than 100%, as has been confirmed by many researchers [23], but it is also sufficient to show that this sea state is particularly favorable for the new wave energy conversion device for power output. Of course, the high output efficiency of the S-PWEC is also due to the resonance of the SOFWEC device in this sea state, which means that the efficiency of the WEC is several times higher than usual if the WEC resonates with the surrounding wave field when capturing wave energy.

Obviously, the optimal capture sea state of the S-PWEC device calculated in this paper is not the highest-frequency sea state in the Zhoushan sea area, which means that S-PWEC still has room for optimization. From the above calculation results, it can be seen that the inherent period of S-PWEC is close to three seconds, which is consistent with the results of the free decay test. As shown in Figure 10, the subsequent optimization of the S-PWEC device will start from changing the inherent period of the device and adjusting the shape, density, and cavity volume of the floating plate to make its inherent period close to the wave period of the Zhoushan sea, to achieve high efficiency in capturing wave energy.

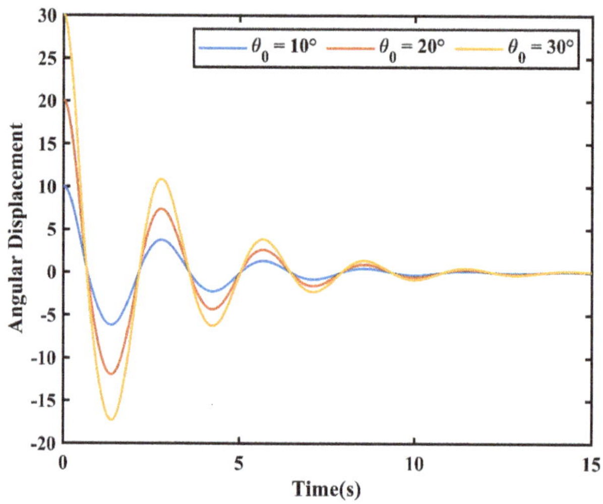

Figure 10. S-PWEC free decay test.

The above-mentioned multiple sets of regular incident wave experiments can analyze some of the working characteristics of S-PWEC, but in the actual ocean, it can be regarded as a random wave sea condition formed by the random superposition of multiple regular incident waves. Therefore, it is very necessary to study the working characteristics of S-PWEC under random waves. Figures 11 and 12 are the results of the power generation efficiency of the S-PWEC and NS-PWEC models under the random wave spectrum (Jonswap spectrum). Compared with regular wave sea conditions, the EPOE of both has been improved as a whole, but the power generation power has been significantly reduced. This is because the incident wave power under irregular wave sea conditions is also significantly reduced, and the captured wave energy is also reduced accordingly.

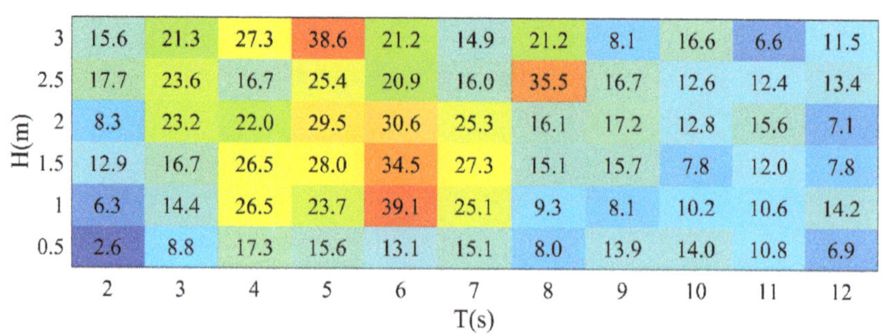

Figure 11. Power generation efficiency of NS-PWEC under irregular wave sea conditions.

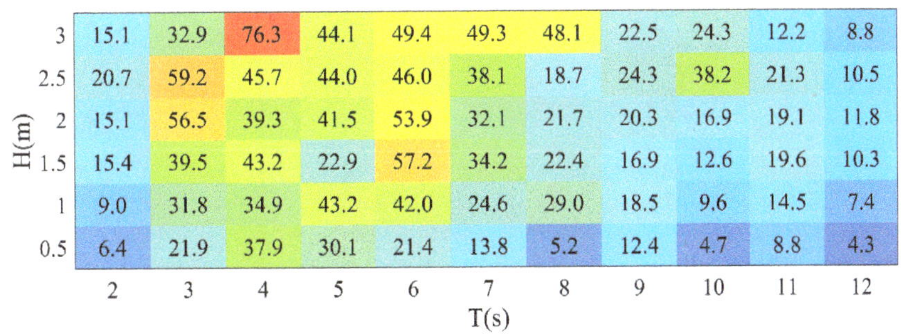

Figure 12. Power generation efficiency of S-PWEC under irregular wave sea conditions.

The average power generation efficiency of NOSWEC and S-PWEC under regular and irregular wave sea conditions is shown in Table 4. From the results, adding a slope structure at the bottom of the floating plate can effectively improve its wave energy capture and absorption efficiency, which is meaningful for the development of wave energy conversion devices, which can exchange a smaller cost for a higher yield by adding a slope structure.

Table 4. Average power generation efficiency.

Average Power GenerationEfficiency (%)	WEC	
	NS-PWEC	S-PWEC
Regular wave	10.32	23.56
Irregular wave	17.26	27.32

3.4. Power Generation Efficiency of Array WEC

A single wave energy conversion device can only absorb limited energy. To realize the large-scale utilization and development of wave energy, it is necessary to study the arrangement of multiple wave energy devices. As shown in Figure 13, the S-PWEC is arranged in the direction of wave propagation, and then on the basis of WEC-Sim, a set of arrays of S-PWEC numerical models arranged along the direction of the incident wave are developed, as shown in Figure 14. Each floating plate works independently and is interrelated. In a single floating plate model, the effect of waves on the structure is mainly considered. However, in the array model, the diffraction effects and radiation effects of adjacent floats also need to be considered. These will change the local wave field, thereby affecting the WEC's capture and absorption of wave energy. Finally, the output electrical energy of each independent WEC is incorporated into the power grid or stored in the battery.

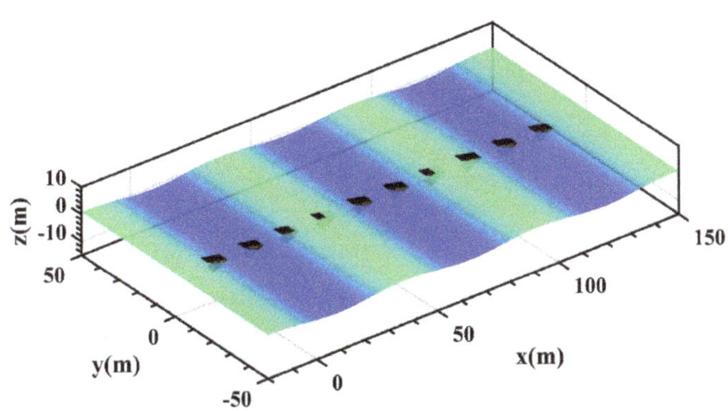

Figure 13. Array S-PWEC model layout diagram.

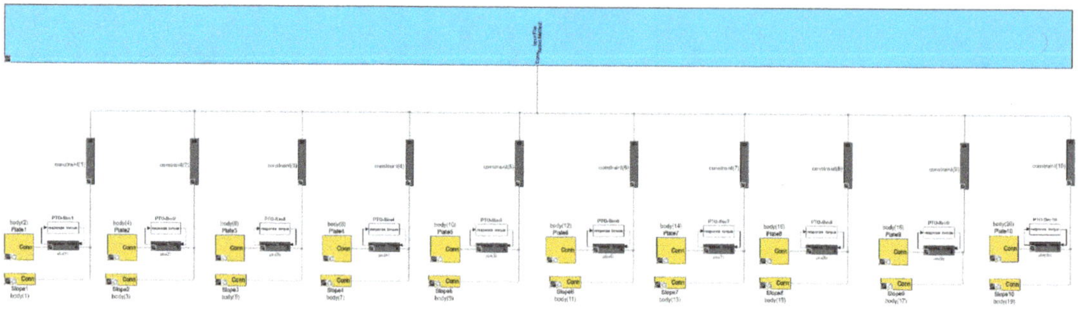

Figure 14. Array S-PWEC numerical model.

In the array WEC experiment, the incident waves pass through each floating plate in sequence, and ten identical wave energy conversion devices are arranged at the same distance. Starting from the pivot point of the first S-PWEC, the distance between the hinge points of two adjacent WECs is defined as S_d. The interaction between waves and floating plates can be treated by ANSYS AQWA, and many hydrodynamic coefficients can be obtained.

When the interval distance of the new wave energy conversion device S_d = 15 m, and the sea condition is set as a regular wave with T = 6 s and H = 2 m, the PTO-Sim module can calculate the power output of 10 different floating plates. As shown in Figure 15, the output power of the first floating plates is highest, which is 6.9 kW, and the power generation efficiency is 15.39%. The latter floating plates are reduced in different degrees. It is not difficult to understand that this is because every time an incident wave passes a floating plate, part of the energy carried by it will be absorbed. The energy absorbed by each S-PWEC is related to the change in local wave field, and the distance S_d between adjacent S-PWECs must be considered.

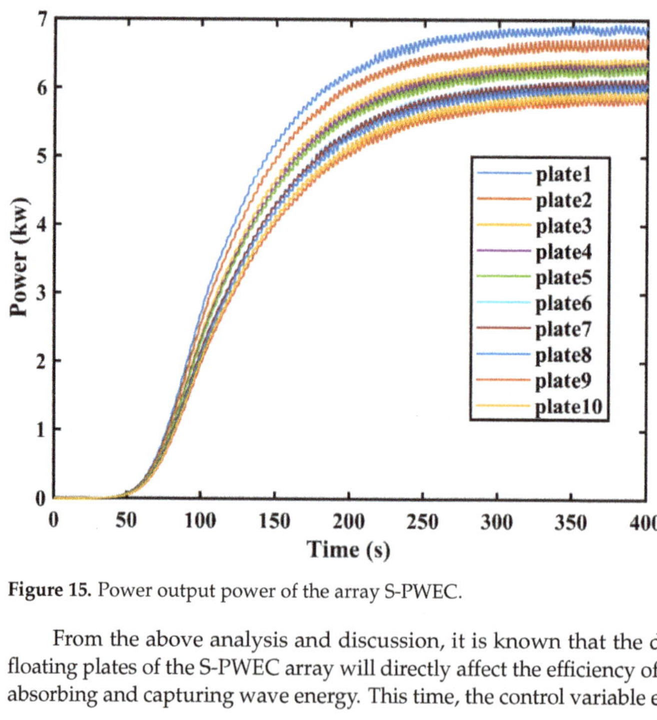

Figure 15. Power output power of the array S-PWEC.

From the above analysis and discussion, it is known that the distance between the floating plates of the S-PWEC array will directly affect the efficiency of the floating plates in absorbing and capturing wave energy. This time, the control variable experimental analysis of the floating plate distance S_d along the wave direction was carried out. Under the sea conditions with a period of T = 6 s and a wave height of H = 2 m, the numerical calculation was used to obtain the results shown in Figure 16.

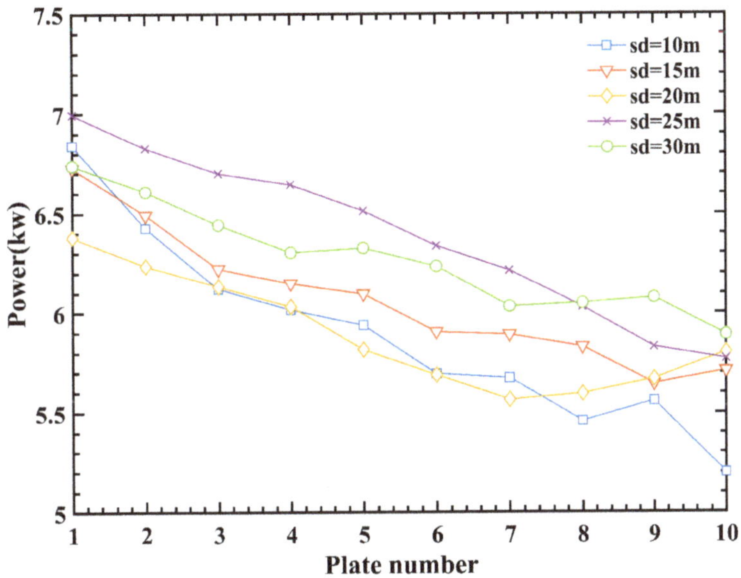

Figure 16. Effect of different spacing on array S-PWEC power.

As shown in Figure 16, the electrical output power of the floating plate at different spacing is different. In the WEC system with a spacing of 10 m, the electrical output power of the floating plate of different serial numbers changes significantly, showing that under small spacing, the hydrodynamic interaction of the adjacent WEC has a greater impact, and the overall absorption power will decrease. The spacing distance is too close to be conducive to wave energy capture, and the arrangement distance needs to be increased. When the spacing increases, the hydrodynamic interaction of the adjacent WEC weakens and the absorbed power rises. However, if the floating plate has a beneficial interaction with the adjacent floating plate in this process, the overall power will increase significantly. From the results, it can be found that the best power generation effect is the working condition of the array floating plate with a spacing of 25 m. The beneficial interaction of the adjacent floating plates increases the absorbed power. Under the prescribed sea conditions, it is a very good arrangement distance. The power generation power of the floating plate serial number 1–7 is significantly better than those of other working conditions, and the total power generation power of the ten floating plates reaches 63.87 kW. In addition, if the distance is too far apart, the number of wave-farm WECs in the same area will be reduced, and the utilization rate of the sea area will be low, which is not conducive to the absorption and utilization of wave energy.

4. Conclusions and Future Work

In this paper, the hydrodynamic performance and electric power output of a new wave energy conversion device are investigated using an open-source program to establish a numerical model, and the following conclusions can be drawn.

- The slope-pendulum wave energy conversion device adds a slope structure at the bottom to reflect through waves. The innovation of this structure can effectively enhance the ability of the floating plate to capture wave energy and improve the overall efficiency at a smaller cost.

- It can also cope with extreme sea conditions with large wave heights and has a certain "self-locking" function to keep the floating plate working normally, and the power generation efficiency of the wave energy conversion device is effectively improved under regular and irregular wave sea conditions.
- The array floating plate has a longitudinal optimal arrangement distance, the spacing is too large or too small, and it is not conducive to the array floating plate power generation. By experimental calculation, when the arrangement distance is 25 m, the whole wave energy conversion system electrical energy output power reaches the maximum.

In the current work, the wave energy capture efficiency has been improved through the optimization of the structure. In the future work, the angle of the bottom oblique plate structure will continue to be studied, the optimal angle under different sea conditions will be given, and the other arrangement of the array floating plate will be studied, and we will focus on optimizing the design and active control to improve the overall wave energy absorption capacity.

Author Contributions: Z.W., conceptualization, methodology, analysis; Z.L., data curation, analysis, writing—original draft, writing—review and editing; D.Z., conceptualization, methodology, writing—review and editing, supervision; H.Z., software, data curation, analysis, writing—original draft. All authors have read and agreed to the published version of the manuscript.

Funding: The support by the National Natural Science Foundation of China (Nos. 11572283; 11602179), the National Key Research & Development Plan of China (Nos. 2017YFC1403306; 2016YFC1401603); the Program for Zhejiang Leading Team of S & T Innovation (No. 2010R50036); the Public Science Technology Research Funds Projects of Ocean (No. 20110518-5) and HPC Center OF ZJU (ZHOUSHAN CAMPUS) are all gratefully acknowledged.

Institutional Review Board Statement: Not applicable.

Informed Consent Statement: Not applicable.

Data Availability Statement: Data available on request due to restrictions of privacy.

Conflicts of Interest: The authors declare no conflict of interest.

References

1. Ahamed, R.; McKee, K.; Howard, I. Advancements of Wave Energy Converters Based on Power Take off (PTO) Systems: A Review. *Ocean Eng.* **2020**, *204*, 107248. [CrossRef]
2. Guo, B.; Ringwood, J.V. Geometric Optimisation of Wave Energy Conversion Devices: A Survey. *Appl. Energy* **2021**, *297*, 117100. [CrossRef]
3. Rusu, E.; Onea, F. A Review of the Technologies for Wave Energy Extraction. *Clean Energy* **2018**, *2*, 10–19. [CrossRef]
4. Titah-Benbouzid, H.; Benbouzid, M. An Up-to-Date Technologies Review and Evaluation of Wave Energy Converters. *Int. Rev. Electr. Eng. IREE* **2015**, *10*, 52–61. [CrossRef]
5. López, I.; Andreu, J.; Ceballos, S.; Martínez de Alegría, I.; Kortabarria, I. Review of Wave Energy Technologies and the Necessary Power-Equipment. *Renew. Sustain. Energy Rev.* **2013**, *27*, 413–434. [CrossRef]
6. Whittaker, T.; Folley, M. Nearshore Oscillating Wave Surge Converters and the Development of Oyster. *Philos. Trans. R. Soc. Math. Phys. Eng. Sci.* **2012**, *370*, 345–364. [CrossRef] [PubMed]
7. Waveroller. Available online: https://aw-energy.com/waveroller/ (accessed on 7 September 2022).
8. Tul Huda Ahmad, N.H.; Zamri Ibrahim, M.; Rahman, S.J.A.; Albani, A.; Mohad, S. The Development of Wave Energy Converter System Using Hydraulic Power Take Off at Terengganu Shoreline. In Proceedings of the 2018 International Conference and Utility Exhibition on Green Energy for Sustainable Development (ICUE), Phuket, Thailand, 24–26 October 2018; IEEE: New York, NY, USA, 2018; pp. 1–7.
9. Ghafari, H.R.; Ghassemi, H.; Neisi, A. Power Matrix and Dynamic Response of the Hybrid Wavestar-DeepCwind Platform under Different Diameters and Regular Wave Conditions. *Ocean Eng.* **2022**, *247*, 110734. [CrossRef]
10. Ransley, E.J.; Greaves, D.M.; Raby, A.; Simmonds, D.; Jakobsen, M.M.; Kramer, M. RANS-VOF Modelling of the Wavestar Point Absorber. *Renew. Energy* **2017**, *109*, 49–65. [CrossRef]
11. Poguluri, S.K.; Kim, D.; Bae, Y.H. Performance Assessment of Pitch-Type Wave Energy Converter in Irregular Wave Conditions on the Basis of Numerical Investigation. *Ocean Syst. Eng.* **2022**, *12*, 23–38.
12. Ruehl, K.; Forbush, D.D.; Yu, Y.-H.; Tom, N. Experimental and Numerical Comparisons of a Dual-Flap Floating Oscillating Surge Wave Energy Converter in Regular Waves. *Ocean Eng.* **2020**, *196*, 106575. [CrossRef]

13. Wei, Y.; Barradas-Berglind, J.J.; van Rooij, M.; Prins, W.A.; Jayawardhana, B.; Vakis, A.I. Investigating the Adaptability of the Multi-Pump Multi-Piston Power Take-off System for a Novel Wave Energy Converter. *Renew. Energy* **2017**, *111*, 598–610. [CrossRef]
14. Sricharan, V.V.S.; Chandrasekaran, S. Time-Domain Analysis of a Bean-Shaped Multi-Body Floating Wave Energy Converter with a Hydraulic Power Take-off Using WEC-Sim. *Energy* **2021**, *223*, 119985. [CrossRef]
15. Wan, Z.; Zheng, H.; Sun, K.; Zhang, D.; Yao, Z.; Song, T. Simulation of Wave Energy Converter with Designed Pendulor-Slope Combination. *Energy Procedia* **2019**, *158*, 733–737. [CrossRef]
16. Midya, C.; Kanoria, M.; Mandal, B.N. Scattering of Water Waves by Inclined Thin Plate Submerged in Finite-Depth Water. *Arch. Appl. Mech.* **2001**, *71*, 827–840. [CrossRef]
17. Cummins, W.E. The Impulse Response Function and Ship Motions. *Schiffstechnik* **1962**, *47*, 101–109.
18. Giorgi, G.; Ringwood, J.V. Comparing Nonlinear Hydrodynamic Forces in Heaving Point Absorbers and Oscillating Wave Surge Converters. *J. Ocean Eng. Mar. Energy* **2018**, *4*, 25–35. [CrossRef]
19. Babarit, A. A Database of Capture Width Ratio of Wave Energy Converters. *Renew. Energy* **2015**, *80*, 610–628. [CrossRef]
20. Yu, Y.-H.; Jenne, D. Analysis of a Wave-Powered, Reverse-Osmosis System and Its Economic Availability in the United States. In Proceedings of the International Conference on Offshore Mechanics and Arctic Engineering, Trondheim, Norway, 25 June 2017; American Society of Mechanical Engineers: New York, NY, USA; p. V010T09A032.
21. Wang, L.; Tang, X.; Xia, Y.; Liu, H. Assessment of Wave Energy Resources in Zhoushan Islands. *Renew. Energy* **2014**, *32*, 1939–1944.
22. Simulink Documentation—MathWorks. Available online: https://ww2.mathworks.cn/help/simulink/index.html (accessed on 8 September 2022).
23. Retzler, C. Measurements of the Slow Drift Dynamics of a Model Pelamis Wave Energy Converter. *Renew. Energy* **2006**, *31*, 257–269. [CrossRef]

Journal of
Marine Science and Engineering

MDPI

Article

Comparison of Offline, Real-Time Models and Hardware-in-the-Loop Test Results of a Power Take-Off for Wave Energy Applications

Luca Castellini [1], Federico Gallorini [2], Giacomo Alessandri [2], Erick Fernando Alves [3], Dan Montoya [3], Bhavana Mudigonda [3,4,*] and Elisabetta Tedeschi [3,4,*]

[1] UMBRAGROUP s.p.a, 06034 Foligno, Italy
[2] VGA S.R.L., 06053 Deruta, Italy
[3] Department of Electric Power Engineering, Norwegian University of Science and Technology (NTNU), 7034 Trondheim, Norway
[4] Department of Industrial Engineering, University of Trento, 38123 Trento, Italy
* Correspondence: bhavana.mudigonda@ntnu.no (B.M.); elisabetta.tedeschi@ntnu.no (E.T.)

Abstract: The power take-off (PTO) of a wave energy converter (WEC) converts mechanical power extracted from the waves into electrical power. Increasing PTO performance under several operational conditions is therefore essential to reduce the levelized cost of energy of a given wave energy concept and to achieve higher levels of technology readiness. A key task in the WEC design will then be the holistic assessment of the PTO performance in combination with other subsystems. It is hence important that WEC designers are aware of the different modeling options. This paper addresses this need and presents two alternative wave-to-wire modeling approaches based on a 250 kW modular electromechanical PTO coupled to an oscillating wave surge converter (OWSC) device. The first is a detailed and accurate offline model. The second model is a simplified and faster version of the first, being adequate for rapid analyses and real-time (RT) simulation. The paper presents the benchmarking of the offline model against the RT model and the hardware-in-the-loop (HIL) tests of the PTO. The normalized root-mean-square error (NRMSE) is considered as a quantitative indicator for the measurement of real-time and HIL test results against the offline simulation. Results show that the dynamics of the offline model are well represented by the RT model with execution times up to 10 times faster. The offline model also depicts well the behavior observed in the HIL tests with the NRMSE values for the PTO position, velocity, and force above 0.90, which shows the HIL test results replicates with fidelity the dynamic behavior of the complete model. Meaningful differences are however present and highlighted in this paper. An understanding of the advantages and drawbacks of these three approaches is fundamental to properly design a WEC during its project cycle and validate PTO concepts with a certain level of simplification.

Keywords: wave energy conversion; systems modeling; power take-off; ballscrew; electric generator; test bench

Citation: Castellini, L.; Gallorini, F.; Alessandri, G.; Alves, E.F.; Montoya, D.; Mudigonda, B.; Tedeschi, E. Comparison of Offline, Real-Time Models and Hardware-in-the-Loop Test Results of a Power Take-Off for Wave Energy Applications. *J. Mar. Sci. Eng.* **2022**, *10*, 1744. https://doi.org/10.3390/jmse10111744

Academic Editors: Giuseppe Giorgi and Mauro Bonfanti

Received: 10 October 2022
Accepted: 10 November 2022
Published: 14 November 2022

1. Introduction

There is a renewed urge for innovative and efficient ways of converting, transmitting and storing energy as the world transitions to carbon neutral and sustainable energy systems. The European Union has set ambitious goals towards carbon neutrality by 2050 [1] and ocean energy can play a crucial role to reach those, having a declared target of 40 GW of installed power by 2050 [2].

Wave energy has, on one hand, a tremendous potential at a global level and is one of the most promising forms of ocean energy. Developing a WEC is, on the other hand, still a complex and evolving subject. Aspects that prevented wave energy to reach a technical and commercial maturity comparable to other renewable energy technologies include the abundance of substantially different design concepts, the variability of input loads, the harsh marine environment, and high installation and maintenance costs [3–5].

PTOs are of primary significance among the subsystems in a WEC. They are critical to define aspects such as reliability, survivability, controllability, and maintainability and have a direct impact on the energy yield of the overall device [3–5]. Moreover, the PTO may account for up to 22% of the costs in a typical wave energy project [6].

A modular PTO system has been recently developed by UMBRAGROUP based on a novel electrical machine that converts slow-speed, reciprocating linear motion into electric power with a high efficiency and reliability [7,8]. The modular nature of this PTO system was also explored during the selection and early design phases of a WEC [9,10] under the Innovative Method for Affordable Generation IN ocean Energy (IMAGINE) project within the European Union's Horizon 2020 research and innovation program.

Testing and integration with other subsystems should however be pursued from an early design phase considering how important the PTO is in a WEC. These tasks are typically achieved through a simulation of wave-to-wire models [11]. Wave-to-wire models that model the energy conversion chain from wave to grid help understanding the system dynamics and its overall response for various sea states. The accuracy [12] and computational speed [13] of the model is critical when developing control algorithms targeting the improved efficiency of a PTO. There is also a need for wave-to-wire models' fidelity and low computational requirements in the early stages of WEC development for component selection [14] and site assessment.

Moreover, the HIL testing scheme is considered as a state-of-the-art tool to validate models and bridge the gap between the laboratory and the real operational environment in several industries, such as automotive, aerospace, and power systems. HIL testing has already been used in the wave energy sector for different PTO concepts [15–20].

Two alternative wave-to-wire modeling approaches were presented in [21] to assess the expected performance of a WEC employing UMBRAGROUP's PTO system. The first was an offline model including an OWSC device, a modular PTO system based on a parallel electromechanical unit (EMU), power converters, and all controllers and functionalities required to operate a WEC safely, reliably, and efficiently. The second model was a simplified and faster version of the first, which was suitable for RT HIL testing. The performances of these two models were compared using the NRMSE as a quantitative metric.

This paper extends and further analyzes the two mentioned wave-to-wire approaches introduced in [21] and compares their performance against a HIL test setup of UMBRAGROUP's PTO system. The main goals of this work were to: (1) describe these two wave-to-wire models and the HIL test setup; (2) compare the offline model to the RT model and the HIL tests; and (3) emphasize the advantages and drawbacks of each model, highlighting for which purposes they are adequate during the WEC project cycle.

2. Background

The aim of the IMAGINE project was designing, manufacturing, and testing a novel linear generator to decrease the cost of current PTO technologies, while increasing their efficiency and reliability. A ballscrew in association with an electromechanical generator (EMG) was the core element of the PTO in the proposed concept. Ballscrews are load-carrying mechanical components where the balls convert the slow-speed linear motion of the input shaft into a high-speed rotary motion of the nut and vice-versa. They have typically high efficiencies due to the low rolling friction between balls and threads. Moreover, permanent magnets integrated in the nut produce a magnetic rotating field coupled to an externally mounted stator. The screw pitch (i.e., the axial distance between adjacent threads) and the number of poles of the rotor act, respectively, as mechanical and magnetic gears, increasing the frequency of the induced voltage.

In the context of the IMAGINE project, the target of designing a 250 kW modular EMG started by identifying the loading envelope that the generator would have been subject to. Three different WEC types and three deployment sites were evaluated in this task [9]. The selection process concluded with the identification of the OWSC concept as the preferred reference device for the estimation of the detailed EMG input data. The choice was primarily related to the capability of this architecture to constrain the stroke under different working conditions. A scaling exercise was conducted to have a WEC hosting a PTO system consisting of four 250 kW peak power EMGs in parallel given the available energy at the selected deployment sites. The deployment of modular units was then performed; each EMG would be composed of two 125 kW EMUs working in unison. This arrangement was selected to provide redundancy and to demonstrate the capability of the modular EMG to work in parallel. Figure 1 provides an overview of the EMG along the WEC width and of the kinematics linking the EMG and the OWSC flap.

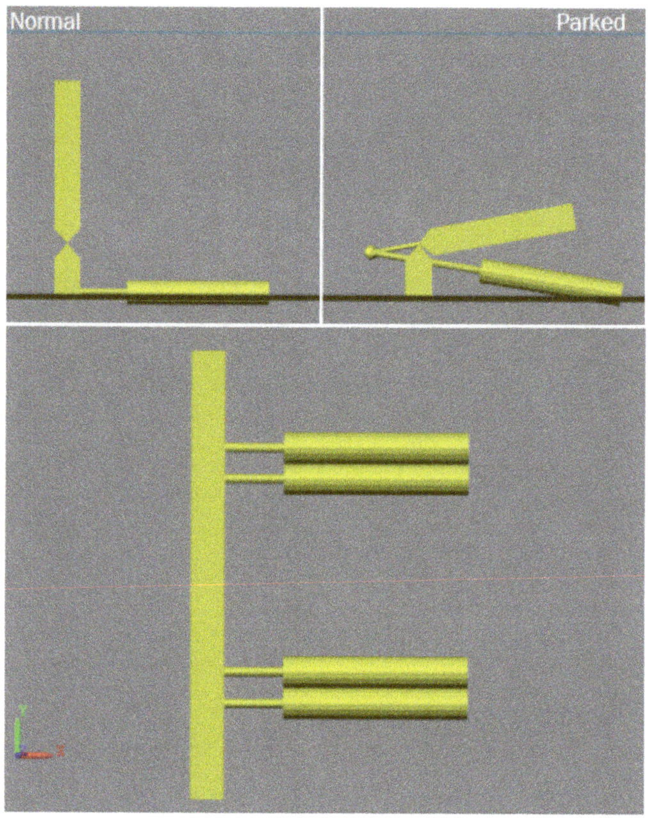

Figure 1. Views with rest and extended PTO conditions of the IMAGINE OWSC model (IMAGINE project, 2018).

Generic load cases were analyzed to identify the overall EMG's loading envelope. Those included power production, occurrence of faults, start-up, and parking in several relevant sea states. Force and power limits were then integrated to allow the EMG working within its operative conditions, 165 kN and 250 kW, respectively, with a maximum achievable EMG speed of 1500 rpm. The power limit was a constraint of the IMAGINE project, while the force limit was set according to a preliminary design of the ballscrew. The PTO damping was identified in an optimization procedure starting from a pure passive control law to maximize the mean annual energy production for each relevant sea state. From the optimal damping the optimization algorithm then tried to tune the PTO stiffness or inertia to achieve an optimal reactive control law aiming to increase the energy production even further. The maximum values of relevant variables were extracted from the numerical simulation results to define the overall loading envelope.

Taking these into account, the final EMG design contained two EMUs with moving ball nuts and rotating screws, connected to the same moving flange. The screw was connected to the rotor of a permanent magnet synchronous machine (PMSM) at one end, while a support with a roller bearing withstood the screw weight and allowed its rotation on the other end. Figure 2 provides an overview of the EMG's main components.

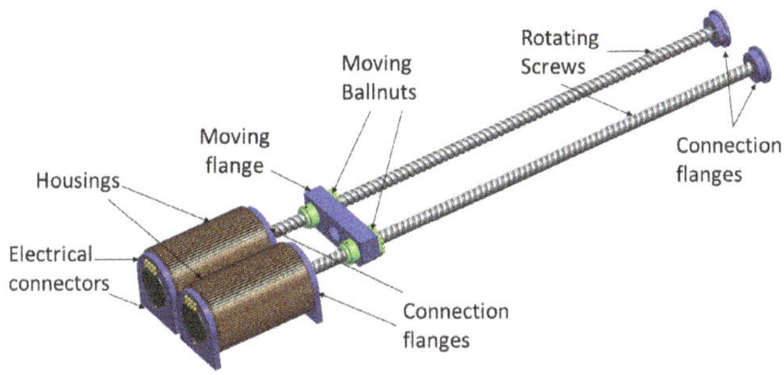

Figure 2. EMG design (IMAGINE project, 2019).

A test rig was designed by VGA according to the proposed PTO concept by connecting an EMG and an EMU to the same carriage. This approach allowed the testing of the EMG both as a generator and an actuator, thanks to the reversibility of the ballscrew and PMSM technologies. The EMU could exert a load when the EMG was used as a generator. Tests under this configuration allowed the demonstration of the EMG's modularity when employing EMUs in parallel. The EMU could be tested up to its maximum operating limits when the EMG worked as an actuator. This configuration was used to demonstrate the coherence with the input design data. Figure 3 provides an overview of the test-rig layout, while Figure 4 depicts the EMU installed in the HIL test rig.

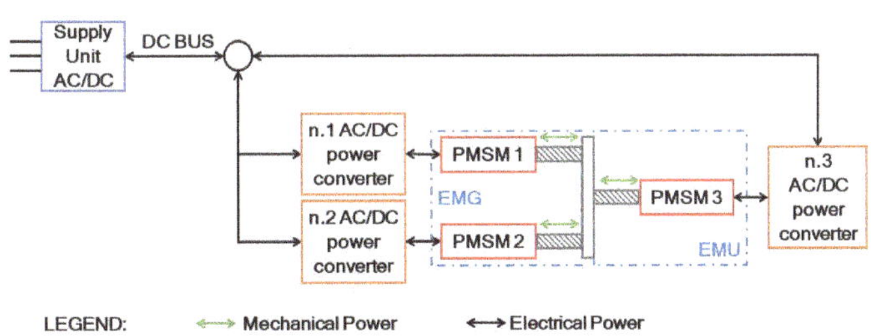

Figure 3. IMAGINE's test rig layout (IMAGINE project, 2020).

Figure 4. EMU installed on the HIL test rig at VGA facilities.

3. Offline Model

The PTO performance was evaluated by a detailed wave-to-wire model developed by NTNU and integrating the following subsystems: the OWSC device, the PTO system based on parallel EMUs, the power converters, the controllers, and electrical components. The simulation model, shown in Figure 5, was developed in Matlab/Simulink v2018a using the WEC-Sim [22] toolbox v3.0 and the Simscape Power Systems 6.9 library.

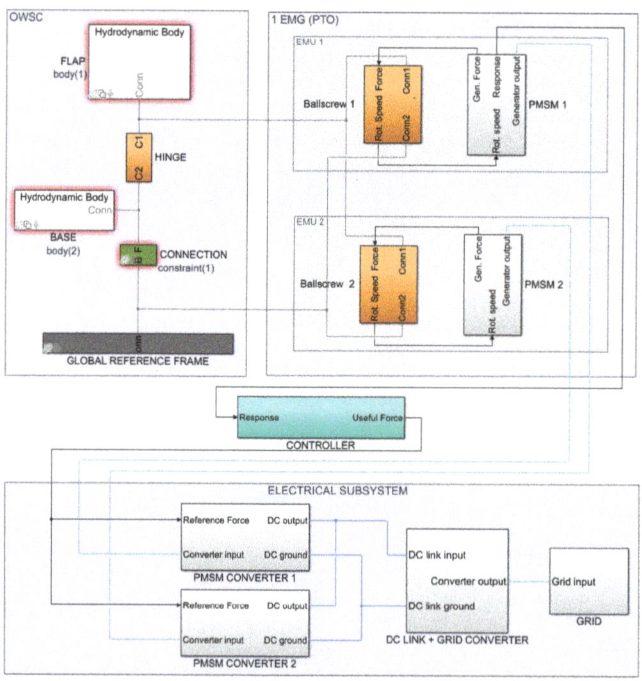

Figure 5. Offline Matlab/Simulink Model including the WEC, PTO, and control and electrical subsystems.

The hydrodynamic model of the OWSC's WEC is formed by the *FLAP*, *HINGE*, *BASE*, *fixed CONNECTION*, and *seabed* blocks. The EMG (Figure 2) is formed by two EMUs, each composed of a ballscrew and a PMSM. In the corresponding model, (Figure 5), the input to each *Ballscrew* block is the applied force originated in the PMSM, while its output is the angular speed of the PMSM.

The *PMSM* block models the electrical generator, its input being the angular speed from the ballscrew and its output the PMSM's force, further connected to the *Ballscrew* block. An additional port, containing the PMSM's stator variables, is connected to the *PMSM converter* block, composed of the average model of a 2-level, three-phase voltage source converter (VSC) using sinusoidal pulse width modulation (SPWM). The two *PMSM CONVERTER* blocks share the same dc link (input port), while their three-phase (output) port is linked to the *PMSM* blocks. The reference signal to control the converter comes from the *CONTROLLER* block, which is further explained later in this section. Finally, the *DC LINK + GRID-SIDE CONVERTER* block models the grid-side converter. As for the PMSM converter, this is also the average model of a 2-level, three-phase VSC applying SPWM. It has two input/output ports: the dc link port and the connection to the grid, respectively. The *GRID* block includes a transformer and an infinite bus.

The *CONTROLLER* block is a high-level controller that implements the previously described control optimization strategy, which selects between a passive and reactive control law to maximize power production according to the sea state. The controller sends each EMU a reference force signal, which is then converted into a PMSM's reference torque (current) on each *PMSM CONVERTER* block. In turn, in order to set the optimum EMU's reference force, the controller considers the signals provided by the EMG, namely, the PTO position, speed, and acceleration.

Table 1 presents a summary of the main parameters of the offline model. For the rest of this paper, the variables $\psi_{pm}, I_s^*, i_d^*, i_q^*, I_{max}, V_{max}, L_d, L_q, \omega_e,$ and p refer to the permanent magnet flux linkage, stator total current, d- and q-axis reference currents, stator maximum current and voltage, total inductance in the d- and q-axis, electrical angular speed, and number of pole pairs of the PMSM.

Table 1. Parameters of the offline model blocks.

	Flap	
Width 14.3 m	Thickness 1.1 m	Height 6.6 m
	Base	
Width 14.3 m	Thickness 1.1 m	Height 2.2 m
Seabed	**Hinge**	**Ballscrew**
Water depth 10.8 m	Height from seabed 2.2 m	Pitch 60 mm
	PMSM	
V_{max} 690 V_{rms}	p 20 pairs	ψ_{pm} 0.235 Wb
I_{max} 185.2 A_{rms}	L_d 1.003 mH	L_q 1.203 mH
	Generator Converter	
V_{ac} 690 V_{rms}	2-level VSC with SPWM	V_{dc} 1000 V
	Grid Converter	
V_{ac} 400 V_{rms}	2-level VSC with SPWM	V_{dc} 1000 V
DC Link	**Grid**	
Capacit. 100 mF	Series induct. 0.102 mH	Freq. 50 Hz

MTPA and FW

The PTO reference force is converted into a PMSM reference torque (T^*), which is then transformed into I_s^* inside each generator's power converter block. A low-level controller adjusts this reference to comply with the I_{max} and V_{max} constraints applying both maximum torque per ampere (MTPA) and field-weakening (FW) techniques [23,24]. The nonlinear equation system formed by Equations (1)–(3) is solved to set i_d^*, i_q^* using MTPA. The FW technique readjusts i_q^* to satisfy the V_{max} limit of the PMSM solving the fourth degree polynomial seen in Equation (4) [25].

$$T^* = \frac{3}{2}p(\psi_{pm}i_q^* + (L_d - L_q)i_q^*i_d^*) \tag{1}$$

$$i_d^* = \frac{\psi_{pm} - \sqrt{\psi_{pm}^2 + 8(L_q - L_d)^2 I_s^{*2}}}{4(L_q - L_d)} \tag{2}$$

$$i_q^* = sign(I_s^*)\sqrt{I_s^{*2} - i_d^{*2}} \tag{3}$$

$$A_4 i_q^{*4} + A_3 i_q^{*3} + A_2 i_q^{*2} + A_1 i_q^* + A_0 = 0 \tag{4}$$

where

$$A_4 = 9p^2(L_d - L_q)^2 L_q^2 \omega_e^2 \tag{5}$$

$$A_3 = 0 \tag{6}$$

$$A_2 = 9p^2\psi_{pm}^2 L_q^2 \omega_e^2 - 9p^2(L_d - L_q)^2 V_{max}^2 \tag{7}$$

$$A_1 = -12T^* p\psi_{pm} L_d L_q \omega_e^2 \tag{8}$$

$$A_0 = 4T^{*2} L_d^2 \psi_{pm}^2 \tag{9}$$

The readjusted value of i_d^* given by the FW algorithm is obtained using Equation (10).

$$i_d^* = -\frac{\psi_{pm}}{L_d} + \frac{1}{L_d}\sqrt{\left(\frac{V_{max}}{\omega_e}\right)^2 - L_q^2 i_q^{*2}} \tag{10}$$

4. Pretest Real-Time Model

The RT model was developed by NTNU and VGA from the offline version presented in Section 3 to: (1) check the feasibility of running the offline model in a RT HIL platform; (2) be integrated in the test rig described in Section 2 for the EMG's qualification tests under the IMAGINE project. The model included two subsystems: (1) *SC_GUI*, integrating all elements required for the supervision and control of the test rig; (2) *SM_PLANT* (Figure 6), including all the dynamic models required for the RT simulation.

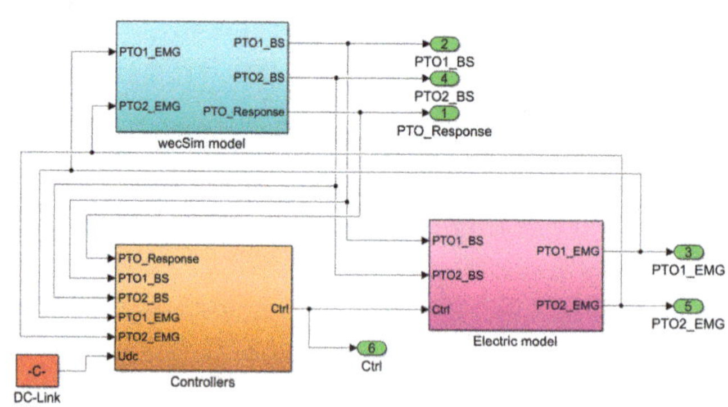

Figure 6. Components of the *SM_PLANT* subsystem.

The *wecSim model* block integrates the OWSC model and the kinematics of the PTO. It receives the torque applied by each EMU and calculates the equilibrium between PTOs and hydrodynamic loads defining the position, speed, and acceleration at the EMG's axis.

The *Controllers* block is organized in two different levels. First, a high-level controller that implements an adaptive PTO control strategy defines the reference torque that the

EMU should apply to produce a certain force at the input axis and maximize the WEC power extraction. Then, a low-level controller defines the three-phase voltage to be applied at the PMSM for achieving the reference torque.

The *Electric model* block simulates the PMSM's dynamics that, according to the input three-phase voltages and EMG's rotor speed, defines the corresponding currents and thus the output torque exerted by each EMU.

To implement FW and MTPA concurrently in the *Controllers* block as described in Section 3, it is necessary to solve a nonlinear system of equations and a fourth degree polynomial for every simulation step. This task demands a reasonable amount of processing power and it is not suitable for real-time embedded systems. Therefore, a near-optimal implementation of the FW and MTPA problems was developed to achieve RT performance. This solution is inspired by [24,26] and presented in the following algorithm:

$$i_{d,MTPA} = \frac{\psi_{pm} - \sqrt{\psi_{pm}^2 + 8(L_q - L_d)^2 I_s^{*2}}}{4(L_q - L_d)}$$

$$i_{d,MTPA} = max(i_{d,MTPA}; -I_{max})$$

$$i_{q,MTPA} = sign(I_s)\sqrt{I_s^{*2} - i_{d,MTPA}^2}$$

$$\omega_{base} = \frac{V_{max}}{\sqrt{(L_d i_{d,MTPA} + \psi_{pm})^2 + (L_q i_{q,MTPA})}}$$

IF $|\omega_e| \leq \omega_{base}$

$$i_d^* = i_{d,MTPA}$$

$$i_q^* = i_{q,MTPA}$$

ELSE

$$aux = (L_d^2 - L_q^2)(\psi_{pm}^2 + (L_q I_s^*)^2 - (\frac{V_{max}}{\omega_e})^2)$$

$$i_{d,FW} = \frac{-\psi_{pm} L_d - \sqrt{(\psi_{pm} L_d)^2 - aux}}{(L_d^2 - L_q^2)}$$

$$i_d^* = max(i_{d,MTPA}; -I_{max})$$

$$i_q^* = \sqrt{I_s^{*2} - (i_d^*)^2}$$

$$T_{FW} = \frac{3}{2} p i_q^* (\psi_{pm} + (L_d - L_q) i_d^*)$$

IF $T_{FW} \geq T^*$

$$i_q^* = \frac{T^*}{T_{FW}} i_q^*$$

ENDIF

ENDIF

5. Case Study with Pretest Real-Time Model

To compare the fidelity of the pretest RT and offline models, the WEC was simulated in several operational conditions for a total of 900 s. Table 2 presents results for two representative sea states of a target site off the south-west coast of England (coordinates 49.8 N 5.8 W). These are the sea states with the highest mean annual energy production ($T_p = 7$ s, $H_s = 2.25$ m) and forces ($T_p = 8$ s, $H_s = 5.25$ m).

Table 2. NRMSE-based quantitative comparison between offline and real-time models.

Variable	$T_p = 7\,s$ $H_s = 2.25\,m$	$T_p = 8\,s$ $H_s = 5.25\,m$
PTO position	0.976	0.998
PTO velocity	0.982	0.998
PTO acceleration	0.968	0.996
PTO force	*0.873*	0.971
EMG's measured torque	*0.859*	0.967
EMG's reference torque	*0.848*	0.988
EMG's mechanical power	0.965	0.920
EMG's electrical power	0.945	0.901

The NRMSE was used as a metric to quantitatively compare the models:

$$NRMSE(x_{RT}, x_{off}) = 1 - \sqrt{\frac{\sum_{n=1}^{N}(x_{RT} - x_{off})^2}{\bar{x}_{off}}} \tag{11}$$

where N is the number of samples for the entire simulation and x_{RT}, x_{off} are the values of the pretest RT and offline models, respectively. The proposed metric compares two models without accounting for the error sign and is a measure of accuracy [27]. Its values vary between $-\infty$ (bad fit) and 1 (perfect fit).

The results from Table 2 reveal that the pretest RT model could reproduce the dynamics of the offline model with fidelity. However, discrepancies on the force, reference and output torque were present. When investigating the results closer, two main sources of deviations were found: (1) the discretization of control loops in the RT model; (2) the different field-weakening strategies in the two models.

The first source caused an overshoot of the current controllers due to the additional delay introduced by the discretization of integrators. This effect is well-documented in the control systems literature, such as in [28], and better visualized in Figure 7a. The overshoot was more prominent in sea states with a lower H_s because the EMG's current and torque controllers operated without saturation for longer periods. As consequence, the NRMSE values of the PTO force, the EMG measured, and reference torques were reduced in these sea states. To achieve the offline model performance, the current controllers in the RT model should be retuned considering the delays introduced by the discretization. For a HIL setup, measurement and communication delays must also be considered.

The second source of discrepancy is better visualized in Figure 7b. In sea states with a higher H_s, the velocity often went above the EMG's rated speed, and hence the FW algorithm was activated more frequently. By comparing the EMG's current behavior between the models, one notices that the reduction of the absolute value of i_q and the increase of i_d during FW was sharper in the RT model. This happened because the RT implementation was near-optimal and prioritized keeping voltage and current limits of the PMSM within allowed limits.

All in all, the pretest RT model was regarded as a valuable tool for the EMG's qualification tests of IMAGINE. It allowed the testing of the modular PTO system developed by UMBRAGROUP in very realistic simulation conditions but under the RT constraints of VGA's HIL test rig.

Not least, the RT model could be used for other purposes in a WEC design cycle using UMBRAGROUP's modular PTO concept. For instance, a better evaluation of a WEC performance was achieved when specifying physical constraints for each component [11]. The RT model's faster performance (up to 10 times faster than the offline model) may allow the inclusion of these constraints in the initial assessment of a target site where several sea states must be simulated to evaluate important design variables such as the energy yield, peak stroke, velocity, acceleration, forces, and power.

On the other hand, it is important to keep in mind the pretest RT model's limitations. The peak PTO force may be slightly overestimated in sea states with a lower H_s, while the average power may be slightly underestimated in sea states with a higher H_s. For a detailed evaluation of a specific sea state, including the WEC performance during faults, the use of the offline model is highly recommended.

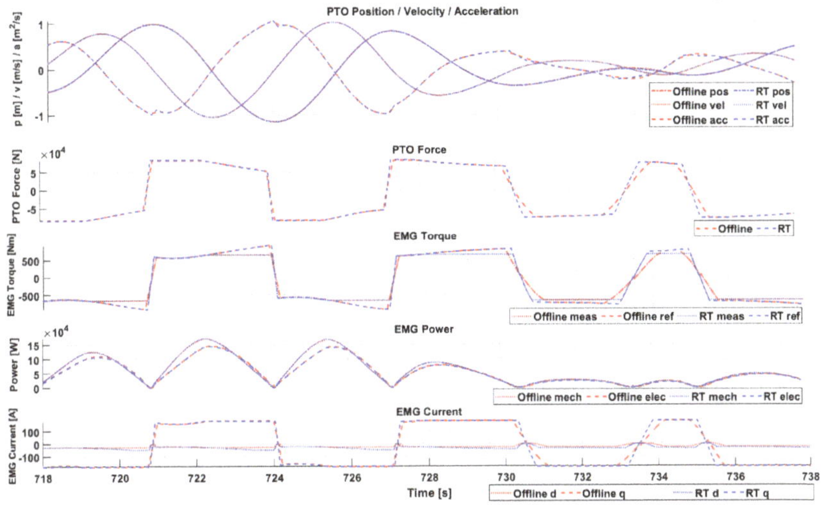

(a)

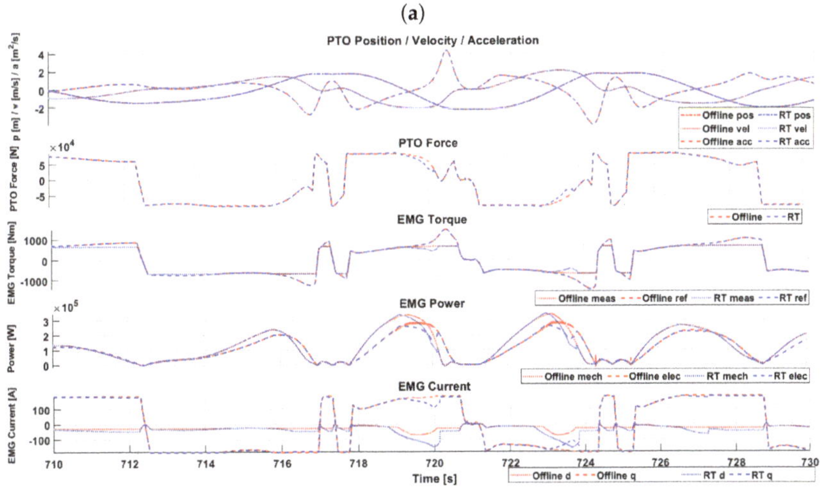

(b)

Figure 7. Visual comparison between offline and real−time models for two sea states: (**a**) $T_p = 7\,\text{s}$, $H_s = 2.25\,\text{m}$; (**b**) $T_p = 8\,\text{s}, H_s = 5.25\,\text{m}$.

6. Real-Time Model Used for Hardware-in-the-Loop Tests

In the last phase of the IMAGINE project, the HIL real-time model was updated in order to be interfaced with the real EMU and its power electronic converters. As indicated in Section 2, this configuration would allow us to test the unit up to its maximum operating limits, demonstrating the coherence with the input design data. This step was characterized by substantial changes in the model: firstly, the *Electric model* block was removed since it represents the PMSM, which is part of the EMU under test. The *CONTROLLER* block was also removed, since the respective algorithms were implemented on the programmable logic controller (PLC) of the EMU's power electronic converters. This option was preferred to the one of having an hybrid control system (where part of the controller is within the real-time model and part within the PLC) to increase the performances of the system. In fact, while the real-time model required a minimum time step of 3 ms to be run, the PLC had a cycle time of 1 ms. In particular, the high-level controller implementing the PTO damping control strategy was computed by multiplying the damping factor by the synchronous speed of the EMU's electrical generator. Finally, the FW and MTPA algorithms were already implemented within the power and control system provided by the supplier.

Figure 8 represents the updated *SM_PLANT* of the RT model, which integrates the *wecSim model* block and the interface blocks from and to the rig. The *wecSim model* connects the real hardware system (e.g., EMU with its power and control system) to the simulated OWSC concept. This block was updated by removing the simulated interface with the EMU's input axis and by connecting it to the *EMU Ball-screw*, which sends the real output force measured from the load cell on the carriage of the test bench to the *wecSim model* block.

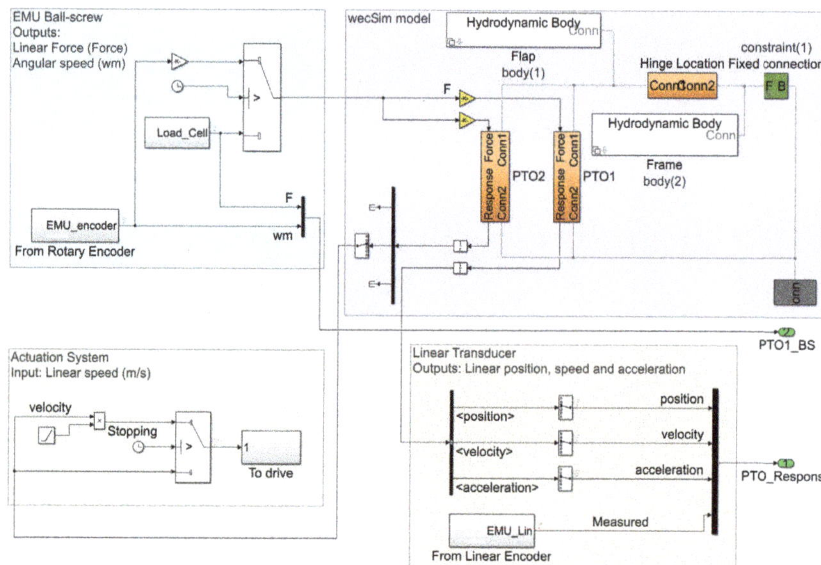

Figure 8. Matlab/Simulink Model for HIL tests including the WEC and interfaces with the test rig.

The *Actuation System* block uses the speed calculated from the *wecSim model* block and transfers it to the power and control system of the EMG to actuate the carriage. The feed-forward term accounting for the EMG's rotating inertia was implemented by increasing the motor shaft weight set on the software program of the power and control system. The other part of the feed-forward term (related to the EMG's frictions) was not implemented to reduce the computational burden of the PLC. Given the limited contribution that this term has on the overall force required at the axis, it was not expected to cause a difference in the test results.

Finally, the *Linear Transducer* block includes both simulated and measured linear kinematics, to allow for their comparison. In addition to the mentioned updates, the parameters within the controller of the actuation system had to be retuned to ensure the tests could be run smoothly. In particular, the fine-tuning of the speed derivative parameter was crucial to ensure a satisfactory performance.

7. Case Study with Real-Time Model Used for Hardware-in-the-Loop Tests

The results of the offline model and the real system were compared by running HIL tests with the EMG working as a actuator and the EMU working as a generator. The test was executed while simulating the WEC in operational conditions for a total of 60 s with the sea state having the highest mean annual energy production ($T_p = 7\,\text{s}, H_s = 2.25\,\text{m}$). Figure 9 shows a visual comparison between the model and HIL test results for the same variables considered in Section 5. The figure highlights some differences between the two sets of results.

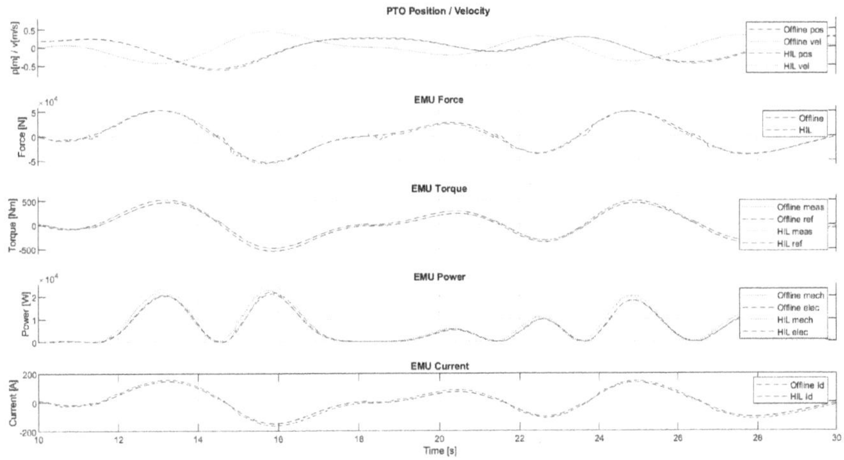

Figure 9. Comparison between offline model and HIL tests results for sea state with $T_p = 7\,\text{s}, H_s = 2.25\,\text{m}$.

In terms of kinematics, slight deviations between the respective positions (especially corresponding to the changes of direction) and speeds (corresponding to peak values) at the PTO input axis could be noticed. The input wave elevation being the same in the respective simulations (offline and real-time) and the speed output from the simulator to the actuation system being well aligned with the one measured by the encoder, the difference in terms of stroke and speed had to be related to a different behavior of the EMU. In particular, the parameters used for modeling the EMU in the offline model might not be the same as the ones of the real system.

When looking at the force profiles, a significant difference could be noticed corresponding to the stroke inversion, where the real EMU appeared to have a higher force with respect to the model. On the other hand, at maximum speed values the offline model and HIL tests reached similar values, with the EMU showing forces even beyond the ones foreseen from the offline model.

The torque did not behave like the force values: in the HIL test, the former was generally higher than the one simulated by the offline model. As a consequence of speed, torque, and (only for peaks) force having higher values in HIL tests, the input power, output power, and current profiles measured were beyond the values simulated in the offline model.

The discrepancies between simulated and real results could be corrected by updating the parameters describing the real behavior of the EMU within the offline model and rechecking the capacity of the power and control system to effectively apply the required control force.

In line with the case study described in Section 5, the NRMSE was used as metric to quantitatively compare the results; these are shown in Table 3. With respect to Table 2, the same reference parameters are shown, except for the acceleration, which was not measured during the tests. As seen in Figure 9, the correlation between the offline model and HIL tests was satisfactory. Lower values of NRMSE were, however, observed for the EMG's mechanical power and reference and measured torque, confirming an underestimation of these values by the offline model. The reason of this nonperfect correspondence was the incorrect estimation of the EMU's static and dynamic friction, whose coefficients had to be updated in the offline model to achieve more realistic results.

Table 3. NRMSE-based quantitative comparison between offline simulation and HIL real-time test.

Variable	$T_p = 7\,s$ $H_s = 2.25\,m$
PTO position	0.923
PTO velocity	0.935
PTO force	0.902
EMG's reference torque	*0.846*
EMG's measured torque	*0.856*
EMG's mechanical power	*0.845*
EMG's electrical power	0.928

8. Conclusions

This paper summarized the key findings related to the wave-to-wire models of an OWSC device connected to a PTO system (made of parallel EMUs) and to the HIL testing of the real EMU driven from a real-time simulator (implementing the same model, conveniently adapted).

First, an offline, detailed model of the overall system was created in the Matlab Simulink environment. The wave–flap interaction, its kinematics, and interface with the PTO were modeled through the use of the WEC-Sim toolbox. A block accounting for the mechanical and electrical aspects of the parallelized EMUs was included. A high-level controller and an electrical subsystem that included power converters and a dc link were modeled. The maximum torque per ampere and field-weakening techniques were also included in a low-level controller to adjust the electrical inputs to the PMSMs, according to the torque reference from the high-level controller.

Then, a real-time hardware-in-the-loop model was derived from the detailed model, to allow its execution on a RT platform, in view of the EMG's qualification tests. The simplification included the implementation of near-optimal MTPA and FW algorithms. A first comparison between the two models was presented considering the NRMSE as a quantitative metric. Values for the PTO position, velocity, and acceleration were above 0.96, which showed that the dynamics of the offline model were well represented by the RT model with execution times up to 10 times faster. However, differences in the force, reference, and output torques were present. These were mainly due to: (1) the discretization of control loops that introduced additional delays on the adjustment of the current; (2) the balance between i_d^*, i_q^* when the FW algorithm was applied. Both effects influenced the torque and force values exerted on the PMSM and ballscrew.

Finally, the real-time HIL model was adapted to allow the coupling with the real EMU under tests. The simplification included the removal of the electric model and controller block (substituted by real hardware and software). The interfaces between the simulation loop, the actuation system, and the EMU were realized by connecting the RT machine to the respective power electronic converters. The control strategies required for the actuator (feed-forward torque accounting for the mass) and generator (FW and MTPA) were implemented directly into the PLC's cyclic tasks. The overall simulation loop used the axial speed as the input of the plcs and the measured force at the carriage as the feedback to the simulator. A second comparison between the offline models and HIL test was presented; the results showed differences between the dynamics of the offline model and the ones of the tests, with the former mostly underestimating the latter.

The comparison between these three approaches is provided as a fundamental check to guarantee that the simulated environment is as similar as possible to the real space and hence to validate the product maturity albeit with a certain level of simplification.

Author Contributions: Conceptualization, E.F.A., G.A., D.M., L.C. and E.T.; methodology, G.A., E.F.A. and D.M.; software, E.F.A. and D.M.;.validation, G.A., E.F.A. and D.M.; formal analysis, E.F.A. and G.A., resources, G.A. and E.T.; data curation, G.A. and E.F.A., writing-original draft preparation, G.A. and E.F.A., writing—review and editing, B.M. and E.F.A.; supervision, E.T.; project administration, E.T., F.G. and L.C.; funding acquisition, E.T., F.G. and L.C. All authors have read and agreed to the published version of the manuscript.

Funding: The work is carried out as a part of IMAGINE project funded by the European Union's Horizon 2020 research and innovation programme under grant agreement number 764066 and "Blue for Green" project funded by the 5x1000 2021 fundraising campaign of the University of Trento (with the support of Dolomiti Energia), in cooperation with the strategic area Oceans of NTNU.

Institutional Review Board Statement: Not applicable.

Informed Consent Statement: Not applicable.

Data Availability Statement: Not applicable.

Conflicts of Interest: The authors declare no conflict of interest.

References

1. European Commission. *Committing to Climate-Neutrality by 2050: Commission Proposes European Climate Law and Consults on the European Climate Pact*; EU Commission: Brussels, Belgium, 2020.
2. European Commission. *An EU Strategy to Harness the Potential of Offshore Renewable Energy for a Climate Neutral Future*; Communication COM(2020) 741 Final; European Commission: Brussels, Belgium, 2020.
3. Ocean Energy Forum. Ocean Energy Strategic Roadmap 2016: Building Ocean Energy for Europe. In Proceedings of the Ocean Energy Europe 2016 Conference & Exhibition: Brussels, Belgium, 8–9 November 2016.
4. TP Ocean. *Strategic Research Agenda for Ocean Energy*; Ocean Energy Europe: Brussels, Belgium, 2016.
5. ETIP Ocean. *Strategic Research and Innovation Agenda for Ocean Energy*; ETIP Ocean: Brussels, Belgium, 2020.
6. Carbon Trust. *Accelerating Marine Energy*; Technical Report CTC797; Carbon Trust: London, UK, 2011.
7. Castellini, L.; D'Andrea, M.; Borgarelli, N. Analysis and Design of a Recipracating Linear Generator for a PTO. In Proceedings of the 2014 International Symposium on Power Electronics, Electrical Drives, Automation and Motion, Ischia, Italy, 18–20 June 2014; pp. 1373–1379. [CrossRef]
8. Castellini, L.; D'Andrea, M.; Martini, M.; Alessandri, G.; Coiro, D.; De Luca, F.; Calise, G.; Troise, G.; Vogler, A. Experimental Tests on a Wave-to-Wire Pivoted System for Wave Energy Exploitation. In Proceedings of the 2017 6th International Conference on Clean Electrical Power (ICCEP), Santa Margherita Ligure, Italy, 27–29 June 2017; pp. 479–487. [CrossRef]
9. Cruz, J.; Atcheson, M.; Martins, T.; Castellini, L.; Martini, M. Preliminary Load Assessment: UMBRA's 250 kW EMG Power Take-Off. In Proceedings of the European Wave and Tidal Energy Conference, Naples, Italy, 1–6 September 2019; Volume 1.
10. Scriven, J.; Cruz, J.; Atcheson, M. Upscaling Wave Energy Converters: Size vs. Modularity. In Proceedings of the International Conference on Renewable Energies Offshore, Lisbon, Portugal, 12–15 October 2020.
11. Penalba, M.; Ringwood, J. A Review of Wave-to-Wire Models for Wave Energy Converters. *Energies* **2016**, *9*, 506. [CrossRef]
12. Zhou, X.; Zou, S.; Weaver, W.W.; Abdelkhalik, O. Assessment of Electrical Power Generation of Wave Energy Converters With Wave-to-Wire Modeling. *IEEE Trans. Sustain. Energy* **2022**, *13*, 1654–1665. [CrossRef]
13. Lin, Z.; Huang, X.; Xiao, X. Fast Model Predictive Control System for Wave Energy Converters With Wave Tank Tests. *IEEE Trans. Ind. Electron.* **2022**, 1–10. [CrossRef]
14. Penalba, M.; Ringwood, J.V. A high-fidelity wave-to-wire model for wave energy converters. *Renew. Energy* **2019**, *134*, 367–378. [CrossRef]
15. Henriques, J.; Gomes, R.; Gato, L.; Falcão, A.; Robles, E.; Ceballos, S. Testing and Control of a Power Take-off System for an Oscillating-Water-Column Wave Energy Converter. *Renew. Energy* **2016**, *85*, 714–724. [CrossRef]
16. Pedersen, H.; Hansen, R.; Hansen, A.; Andersen, T.; Bech, M. Design of Full Scale Wave Simulator for Testing Power Take Off Systems for Wave Energy Converters. *Int. J. Mar. Energy* **2016**, *13*, 130–156. [CrossRef]
17. Penalba, M.; Sell, N.; Hillis, A.; Ringwood, J. Validating a Wave-to-Wire Model for a Wave Energy Converter—Part I: The Hydraulic Transmission System. *Energies* **2017**, *10*, 977. [CrossRef]
18. Penalba, M.; Cortajarena, J.A.; Ringwood, J. Validating a Wave-to-Wire Model for a Wave Energy Converter—Part II: The Electrical System. *Energies* **2017**, *10*, 1002. [CrossRef]
19. Dang, T.D.; Phan, C.B.; Ahn, K.K. Modeling and Experimental Investigation on Performance of a Wave Energy Converter with Mechanical Power Take-Off. *Int. J. Precis. Eng. Manuf.-Green Tech.* **2019**, *6*, 751–768. [CrossRef]
20. Delmonte, N.; Robles, E.; Cova, P.; Giuliani, F.; Faÿ, F.X.; Lopez, J.; Ruol, P.; Martinelli, L. An Iterative Refining Approach to Design the Control of Wave Energy Converters with Numerical Modeling and Scaled HIL Testing. *Energies* **2020**, *13*, 2508. [CrossRef]

21. Castellini, L.; Gallorini, F.; Alessandri, G.; Alves, E.; Montoya, D.; Tedeschi, E. Performance Comparison of Offline and Real-Time Models of a Power Take-Off for Qualification Activities of Wave Energy Converters. In Proceedings of the 2021 16th International Conference on Ecological Vehicles and Renewable Energies (EVER), Monte-Carlo, Monaco, 5–7 May 2021; pp. 1–8. [CrossRef]
22. Yu, Y.H.; Ruehl, K.; Rij, J.V.; Tom, N.; Forbush, D.; Ogden, D.; Keester, A.; Leon, J. *WEC-Sim v4.2*; Zenodo: Geneva, Switzerland, 2020. [CrossRef]
23. Sjolte, J.; Sandvik, C.; Tedeschi, E.; Molinas, M. Exploring the Potential for Increased Production from the Wave Energy Converter Lifesaver by Reactive Control. *Energies* **2013**, *6*, 3706–3733. [CrossRef]
24. Li, M. Flux-Weakening Control for Permanent-Magnet Synchronous Motors Based on Z-Source Inverters. Master's Thesis, Marquette, Milwaukee, WI, USA, 2014.
25. Pan, C.T.; Liaw, J.H. A Robust Field-Weakening Control Strategy for Surface-Mounted Permanent-Magnet Motor Drives. *IEEE Trans. Energy Convers.* **2005**, *20*, 701–709. [CrossRef]
26. Lim, S. *Sensorless-FOC with Flux-Weakening and MTPA for IPMSM Motor Drives*; Technical Report SPRACF3; Texas Instruments; Dallas, TX, USA, 2018.
27. Lewis-Beck, C.; Lewis-Beck, M. *Applied Regression: An Introduction*; SAGE Publications, Inc.: Thousand Oaks, CA, USA, 2016. [CrossRef]
28. Franklin, G.F.; Powell, J.D.; Emami-Naeini, A. *Feedback Control of Dynamic Systems*, 6th ed.; Pearson: Upper Saddle River, NJ, USA, 2010.

Journal of
*Marine Science
and Engineering*

MDPI

Article

Model Predictive Energy-Maximising Tracking Control for a Wavestar-Prototype Wave Energy Converter

Doudou Li and Ron Patton *

Department of Engineering, University of Hull, Cottingham Road, Hull HU6 7RX, UK; d.li-2019@hull.ac.uk
* Correspondence: r.j.patton@hull.ac.uk

Abstract: To date, one of the main challenges in the wave energy field is to achieve energy-maximizing control in order to reduce the levelized cost of energy (LCOE). This paper presents a model predictive velocity tracking control method based on a hierarchical structure for a Wavestar-like deivce in the WEC-SIM benchmark. The first part of the system structure aims to estimate the wave excitation moment (WEM) by using a Kalman filter. Then, an extended Kalman filter (EKF) is chosen to obtain the amplitude and angular frequency of the WEM in order to compute the reference velocity. Following this, a low-level model predictive control (MPC) method is designed to ensure the wave energy converter (WEC) tracks the optimal reference velocity for maximum energy extraction from irregular waves. Two Gaussian Process (GP) models are considered to predict the future wave excitation moment and future reference velocity, which are needed in MPC design. The proposed strategy can give a new vision for energy-maximizing tracking control based on MPC.

Keywords: Kalman filter; extended Kalman filter; Gaussian Process (GP) model; velocity tracking; model predictive control

Citation: Li, D.; Patton, R. Model Predictive Energy-Maximising Tracking Control for a Wavestar-Prototype Wave Energy Converter. *J. Mar. Sci. Eng.* **2023**, *11*, 1289. https://doi.org/10.3390/jmse11071289

Academic Editors: Giuseppe Giorgi and Mauro Bonfanti

Received: 15 May 2023
Revised: 15 June 2023
Accepted: 20 June 2023
Published: 25 June 2023

1. Introduction

To date, climate changes have become a significant problem and are manifested in different ways. People are starting to realize the emergence of a fossil fuel crisis and the disadvantages of using traditional fossil fuels. Hence, the concept of exploring sustainable and renewable energy has been enhanced. Wave energy shows great potential [1,2] to fulfill the growing energy demand worldwide compared with other renewable resources (wind, solar energy, etc).

Although wave energy takes the needs of commercial promotion and actual deployment into consideration, the technologies involved are still immature and lead to a high levelized cost of energy (LCOE) [3] compared with the other marine renewable energy sources, e.g., compared with wind energy. Hence, lowering the LCOE has been a primary task in the wave energy field. Excluding the reduction of operation and maintenance costs [4] to reduce LCOE, the other elegant way is to achieve energy maximization through a WEC device, attempting to harvest energy from the ocean waves and convert the absorbed mechanical energy into electrical energy [5] using a control system.

For the research target, the point absorber wave energy converter (PAWEC) type of device is popular as it has has a relatively simple structure and can be arranged into an array [6] of suitable size to absorb as much power as possible. For example, there have been numbers of related works in the literature based on a special form of point absorber device known as the Wavestar-prototype WEC [7–10]. This was selected in an open competition [11] to compare different WEC control systems on a standard benchmark, the WEC-SIM software and experimental tests. This paper focuses on the use of a wavestar-like system to perform energy-maximizing control design [12] based on a very good WEC-SIM numerical model.

For the purpose of energy maximization, a considerable number of control strategies can be found in the literature, such as reactive control [13], model predictive control

(MPC) [14], moment-matching based control [15], Spectral/Pseudospectral-based (MPC-like type) control [16], LTI (LiTe-Con) controller [17], etc. The most well known of these is reactive control [18], based on linear resonance theory, and it has been discussed and tested in many references. It is usually necessary to amplify the WEC motion to capture as much wave energy as possible. At first glance, this seems to be because the use of a "complex conjugate" method may violate the WEC linear model assumption of the hydrodynamics, involving many power fluctuations (both positive and negative), and can be limited to application for regular wave conditions, due to its frequency-dependent characteristic. The other choice is MPC [14,16,19], which selects the average absorbed mechanical power as an objective function and maximizes the power by solving an optimization problem. The MPC can reach the optimal solution for energy extraction and deal with system constraints elegantly. This is why the MPC approach has a significant research following. Additionally, an alternative choice of energy-maximizing design is based on a hierarchical structure [20] with a low-level controller, e.g., a robust controller, for optimal reference velocity tracking. This tracking system idea can achieve a near-resonance operation to bring acceptable results for energy maximization production.

On the basis of MPC and the hierarchical structure strategies, this paper combines them and produces a model predictive velocity tracking control approach for maximum energy extraction. As for the second or high level part of the hierarchical structure, the WEM is estimated by a Kalman filter, and reference velocity profile calculation is set in it. On the other hand, the reference velocity computation requires the instantaneous amplitude and frequency of the WEM, which are estimated by the EKF in this paper. Theoretically, the WEM is assumed to be a narrow-band harmonic process and modeled as a single cyclical component based on Harvey's structural model [21]. Following this, on-line estimation can be performed to obtain the angular amplitude and frequency of WEM by the designed EKF in order to compute the reference velocity. Next, model predictive tracking control is placed in the low-level part of the whole designed structure. The objective function comprises the velocity error signal and control input and is transferred into a quadratic index to reach its minimum with input constraint. However, the future excitation moment and future reference velocity information are required to be predicted and used in the MPC objective function. Usually, an autoregressive model is used for the prediction of the WEM [22]. However, for the real-time forecasting requirement, the AR model needs to be updated, which will increase the prediction time. Here, two GP models are chosen instead to provide the short-term forecasting [23]. The hyperparameters can be preselected before the training process to reduce the prediction time with no need for them to be updated during real-time forecasting. Besides this, since the GP is a kernel-based and nonparametric learning method, it has the advantages of modeling flexibility, as well as prediction with learning smoothness and the use of noise parameters based on a training set.

The remainder of this paper is organized as follows. Section 2 describes the Wavestar-prototype WEC modeling and the overall hierarchical tracking system structure. Section 3 presents the design of a Kalman filter for the WEM estimation and EKF for reference velocity computation. Section 4 shows the design of the model predictive velocity tracking control with GP model for short-term forecasting. Section 5 gives the simulation results and discussions. Finally, Section 6 presents the concluding discussion.

2. Modeling and the Whole System Scheme

This section presents the description of the dynamics of a Wavestar-like device (the scaled 1:20 benchmark) based on the well-known Cummins equation and the expression of the hierarchical tracking control structure. The parameters of the WEC dynamics [19] are provided by Wavestar experiments and a hydrodynamic database (WAMIT computation). The numerical simulation model has been developed on the WEC-SIM simulator and verified in [24].

As for the hierarchical tracking control structure, this was first proposed by [20] and adopted for energy-maximizing controller design. This paper focuses on this strategy and gives a new version of velocity tracking design based on MPC to reach a near resonance condition. The control input constraint can be handled in MPC, which shows an advantage compared with some other control methods.

2.1. WEC Dynamics

The scaled Wavestar-prototype device is a kind of wave-activated body WEC, as shown in Figure 1. A hemispherical float is mechanically connected to an arm that can rotate around a fixed hinge point A [11], which has three independent motions (surge, heave and pitch). At the other side of the arm, a linear motor (power take-off system) is attached on the rotating arm to provide the power take-off force, and it only has one degree of freedom.

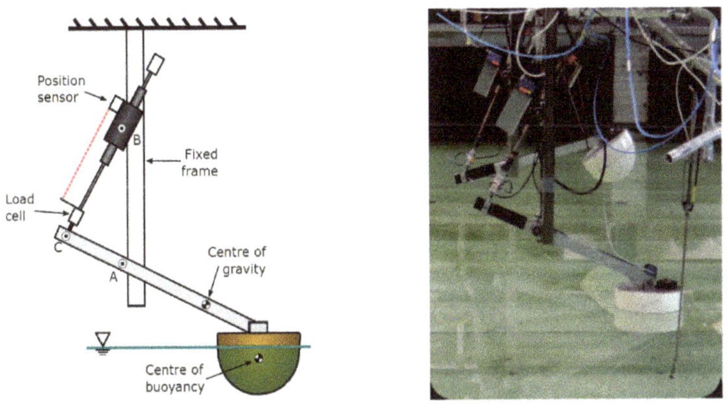

(a) The sketch model. (b) The physical model in laboratory.

Figure 1. The scaled Wavestar-prototype WEC device [25].

In order to reduce the design complexity for estimation and control, the hydrodynamic response of the float-arm can be equivalent to pitch moment only around the fixed hinge point. This means that the linear position and force measurements can be converted to the rotational displacement and moment. Then, the float rotational dynamics at the hinge point A [19] can be treated as the equivalent pitch moment:

$$
\begin{aligned}
(J + J_\infty)\ddot{\theta}(t) &= -K_{hs}\theta(t) - K_v\dot{\theta}(t) + M_{ex}(t) - M_{ra}(t) - M_{PTO}(t) \\
\dot{r}_a(t) &= A_{ra}r_a(t) + B_{ra}\dot{\theta}(t) \\
M_{ra}(t) &= C_{ra}r_a(t) + D_{ra}\dot{\theta}(t)
\end{aligned}
\tag{1}
$$

where J is the inertia of the float and arm, J_∞ is the added inertia, $\ddot{\theta}$ is the rotational angular acceleration of the float, K_{hs} and K_v are the hydrostatic stiffness coefficient and linear damping coefficient respectively, and M_{ra}, M_{ex} *and* M_{PTO} are the equivalent radiation moment, the equivalent wave excitation moment and power take-off moment around the hinge point. The radiation moment $M_{ra} = \int_0^t h_r(t-l)\dot{\theta}(l)dl$ is a convolution integral term, which can dramatically increase computational burden and bring difficulties in estimation and control work design. To overcome these problems, the convolution term can be converted into an order-two linear state space model by using system identification according to Prony's method according to the realization theory. The internal variable $r_a(t)$ in the identified order-two state space model does not have physical meaning. $(A_{ra}; B_{ra}; C_{ra}; D_{ra})$ are the state space identified matrices of the convolution term of M_{ra}.

Theoretically, the equivalent wave excitation moment around the hinge point based on [19] can be computed as

$$
M_{ex} = -F_{ex,x}sin(\theta_0 + \theta)l_{arm} - F_{ex,z}cos(\theta_0 + \theta)l_{arm} + M_{ex,\theta}
\tag{2}
$$

where θ_0 is the initial angular displacement of the float when it is located at the equilibrium point, l_{arm} is the length of the arm, and $F_{ex,x}$, $F_{ex,z}$, $M_{ex,\theta}$ are the surge, heave and pitch direction components of the wave excitation force acting on the float.

The state space model of the WEC system can be expressed as

$$\begin{aligned} \dot{x} &= Ax + Bu + BM_{ex} \\ y &= Cx \end{aligned} \tag{3}$$

where $A = \begin{bmatrix} 0 & 1 & \mathbf{0}_{1\times 2} \\ -\frac{K_{hs}}{J_t} & -\frac{K_v + D_{ra}}{J_t} & -\frac{C_{ra}}{J_t} \\ \mathbf{0}_{2\times 1} & B_{ra} & A_{ra} \end{bmatrix}$, $x = \begin{bmatrix} \theta \\ \dot{\theta} \\ r_a \end{bmatrix}$, $B = \begin{bmatrix} 0 \\ \frac{1}{J_t} \\ \mathbf{0}_{2\times 1} \end{bmatrix}$, $u = -M_{PTO}$,
$C = \begin{bmatrix} 1 & 0 & \mathbf{0}_{1\times 2} \\ 0 & 1 & \mathbf{0}_{1\times 2} \end{bmatrix}$.

The state variables θ and $\dot{\theta}$ are the angular displacement and velocity of the float. r_a is the internal variable of the identified state space model in Equation (1). Concerning the model parameters, $J_t = J + J_\infty$ is the total inertia, where $\mathbf{0}_{p\times q}$ represents a zero matrix with r rows and q columns.

The electrical energy E_e absorbed by the grid [14] can be defined as

$$E_e(t) = -\int_t^{t+T} P_e(r)dr = -\int_t^{t+T} \Gamma(r)P_m(r)dr \tag{4}$$

where P_m is the absorbed mechanical power by the PTO system, Γ is the conversion efficiency, P_e is the extracted electrical power, and r is the integration variable.

The relationship between P_e and P_m is given below:

$$P_e(t) = \Gamma(t)P_m(t) = \Gamma(t)M_{PTO}(t)\dot{\theta}(t), \begin{cases} \Gamma(t) = \mu_{gen} & if\ P_m(t) \geq 0 \\ \Gamma(t) = \mu_{mot} & if\ P_m(t) < 0 \end{cases} \tag{5}$$

where μ_{gen} is the efficiency when the PTO system is assumed to be working in generator mode and μ_{mot} is the motor mode efficiency.

2.2. The Overall Hierarchical Control Structure

For model predictive velocity tracking control design, the proposed hierarchical structure is based on previous work [20] as shown in Figure 2.

The high-level part of the system structure consists of WEM estimation and reference angular velocity generation. The Kalman filter is selected to estimate the WEM, and an EKF is chosen to estimate the instantaneous amplitude and frequency of the WEM. From this, the velocity reference $\dot{\theta}_{ref}$ can be calculated.

The low-level part of the control design is based on the MPC, which aims to force the scaled Wavestar-prototype device to track the optimal angular velocity trajectory for the power maximization purpose. Two Gaussian process models are adopted to predict both the future WEM and the future reference velocity, respectively, which are needed in the MPC tracking design.

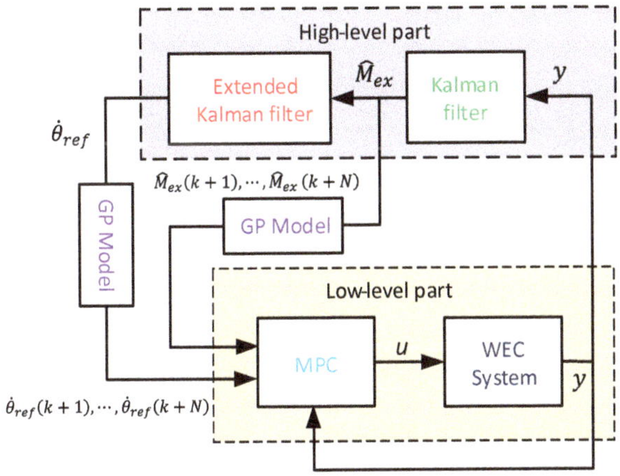

Figure 2. Overall estimation and tracking control structure.

3. Wave Excitation Moment Estimation and Reference Velocity Computation

For the Wavestar-prototype device, the wave excitation information is a physically unmeasurable quantity, although it is required for optimal control or energy-maximizing control design. The equivalent M_{ex} around the hinge point is an external term (unknown input) of the WEC system and can be estimated by using a Kalman filter. According to the reference velocity calculation, the WEM instantaneous amplitude $\hat{A}_{ex}$ and frequency $\hat{\omega}$ are required, and they are computed using an EKF.

3.1. Kalman Filter with Random Walk

The discretized system of Equation (3) has the following form:

$$\begin{aligned} x(k+1) &= A_d x(k) + B_d u(k) + B_d M_{ex}(k) + \varepsilon_x(k) \\ y(k) &= C_d x(k) + \mu(k) \end{aligned} \tag{6}$$

where $\varepsilon_x(k)$ denotes the unmodeled dynamics and $\mu(k)$ is the measurement noise. $M_{ex}(k)$ is considered as an external moment term acting on the WEC float, and it can be treated as an unknown input term of System (6). Then, the discrete time dynamics of the WEM [26] can be considered as

$$M_{ex}(k+1) = M_{ex}(k) + \varepsilon_m(k) \tag{7}$$

where $\varepsilon_m(k)$ is a Gaussian distributed random variable. It means that the next value of WEM will conduct a random step away from the present value after a sampling time, and all moving steps are considered independent. $M_{ex}(k)$ can be estimated by a Kalman filter when it is being considered as one of the system states. Hence, the augmented system for estimating the WEM is

$$\begin{aligned} \bar{x}(k+1) &= \bar{A}\bar{x}(k) + \bar{B}u(k) + \varepsilon(k) \\ y(k) &= \bar{C}\bar{x}(k) + \mu(k) \end{aligned} \tag{8}$$

where $\bar{A} = \begin{bmatrix} A_d & B_d \\ \mathbf{0}_{1\times 4} & 1 \end{bmatrix}, \bar{x} = \begin{bmatrix} x \\ M_{ex} \end{bmatrix}, \bar{B} = \begin{bmatrix} B_d \\ 0 \end{bmatrix}, \bar{C} = \begin{bmatrix} C_d & \mathbf{0}_{2\times 1} \end{bmatrix}, \varepsilon(k) = \begin{bmatrix} \varepsilon_x(k) \\ \varepsilon_m(k) \end{bmatrix}$.

The dimensions of system matrices are $\bar{A} \in \mathfrak{R}^{5\times 5}$, $\bar{B} \in \mathfrak{R}^{5\times 1}$, $\bar{C} \in \mathfrak{R}^{2\times 5}$. $\varepsilon(k)$ and $\mu(k)$ are uncorrelated zero-mean white noise sequences with covariance matrices Q_f and R_f.

Then, the Kalman filter prediction update equation [26] is

$$\begin{aligned}\hat{x}(k|k-1) &= \bar{A}\hat{x}(k-1|k-1) + \bar{B}u(k-1) \\ P_f(k|k-1) &= \bar{A}P_f(k-1|k-1)\bar{A}^T + Q_f\end{aligned}$$

The Kalman filter correction update equation [26] becomes

$$\begin{aligned}K_f(k) &= P_f(k|k-1)\bar{C}^T(\bar{C}P_f(k|k-1)\bar{C}^T + R_f)^{-1} \\ \hat{x}(k|k) &= \hat{x}(k|k-1) + K_f(k)(y(k) - \bar{C}\hat{x}(k|k-1)) \\ P_f(k|k) &= (I - K_f(k)\bar{C})P_f(k|k-1)\end{aligned}$$

Hence, the estimated $\hat{M}_{ex}$ can be obtained from the optimal estimation of state vector $\hat{x}$. Note the following:

$\hat{x}(k|k-1)$ is a predicted prior state estimate given the observations at time k−1.
$P_f(k|k-1)$ is a predicted priori covariance matrix given the observations at time k−1.
$\hat{x}(k|k)$ is an updated posteriori state estimate given the observations at time k.
$P_f(k|k)$ is an updated posteriori covariance matrix given the observations at time k.

3.2. Extended Kalman Filter

On the basis of the high-level part of the hierarchical strategy in Figure 2, the instantaneous amplitude and frequency of WEM must be obtained in order to compute the reference velocity $\dot{\theta}_{ref}$. Hence, an efficient EKF method for the recursive estimation is required. Additionally, it is assumed that the WEM signal is a narrow-band process, and its harmonic model can be expressed as

$$M_{ex}(t) = A_{ex}(t)cos(\omega(t) \cdot t + \beta(t)) \tag{9}$$

where $A_{ex}(t), \omega(t)$ and $\beta(t)$ are the time-varying amplitude, angular frequency and phase of the WEM signal, respectively. Based on Harvey's structural model presented in [21], the WEM can be modeled as a single cyclical component:

$$\begin{aligned}\begin{bmatrix}\psi(k+1) \\ \psi^*(k+1) \\ \omega(k+1)\end{bmatrix} &= \begin{bmatrix}cos(\omega(k)T_s) & sin(\omega(k)T_s) & 0 \\ -sin(\omega(k)T_s) & cos(\omega(k)T_s) & 0 \\ 0 & 0 & 1\end{bmatrix}\begin{bmatrix}\psi(k) \\ \psi^*(k) \\ \omega(k)\end{bmatrix} + \begin{bmatrix}\varsigma(k) \\ \varsigma^*(k) \\ \kappa(k)\end{bmatrix} \\ M_{ex}(k) &= \psi(k) + \zeta(k)\end{aligned} \tag{10}$$

where $d_e(k) = [\varsigma(k) \quad \varsigma^*(k) \quad \kappa(k)]^T$ and $\zeta(k)$ are random process and measurement noise, and $\psi(k)$ and $\psi^*(k)$ are related states of the WEM amplitude and phase. The state vector $x_e(k) = \begin{bmatrix} x_{e,1}(k) & x_{e,2}(k) & x_{e,3}(k) \end{bmatrix}^T = \begin{bmatrix} \psi(k) & \psi^*(k) & \omega(k) \end{bmatrix}^T$ and $x_e(k) \in \Re^{3\times1}$, corresponding to the sampling time T_s.

The non-linear time-varying Model (10) can be formulated as follows:

$$\begin{aligned}x_e(k) &= f(x_e(k-1), d_e(k-1)) \\ z_e(k) &= h(x_e(k), \zeta(k))\end{aligned} \tag{11}$$

where the estimated excitation moment $\hat{M}_{ex}$ is treated as the actual measurement $z_e(k)$. Based on the above definition and without knowing the noise information, the a *priori* state and measurement from time $k-1$ are

$$\begin{aligned}\hat{x}_e(k|k-1) &= f(\hat{x}_e(k-1|k-1), 0) \\ \hat{z}_e(k|k-1) &= h(\hat{x}_e(k|k-1), 0)\end{aligned} \tag{12}$$

where $\hat{x}_e(k|k-1)$ is the estimate of $x_e(k)$ and $\hat{z}_e(k|k-1)$ is the estimate of $z_e(k)$ based on measurements from time $k-1$.

On application of a first-order Taylor series expansion of Equation (12), the linearized system time-varying Jacobian matrix $F(k)$ [27] is obtained as

$$F(k) = \frac{\partial f}{\partial x_e}\bigg| \hat{x}_e(k-1|k-1) = \begin{pmatrix} cos(\omega(k)T_s) & sin(\omega(k)T_s) & T_s(-sin(\omega(k)T_s)\psi(k) + cos(\omega(k)T_s)\psi'(k)) \\ -sin(\omega(k)T_s) & cos(\omega(k)T_s) & T_s(-cos(\omega(k)T_s)\psi(k) - sin(\omega(k)T_s)\psi'(k)) \\ 0 & 0 & 1 \end{pmatrix}$$

The observed Jacobian matrix is

$$H(k) = \frac{\partial h}{\partial x_e}\bigg| \hat{x}_e(k|k-1) \triangleq [1 \quad 0 \quad 0]$$

Thus, the EKF time-update equations are

$$\begin{aligned} \hat{x}_e(k|k-1) &= f(\hat{x}_e(k-1|k-1), 0) \\ P_e(k|k-1) &= F(k)P_e(k-1|k-1)F^T(k) + Q_e \end{aligned}$$

The use of Jacobian matrices $F(k)$ and $H(k)$ to perform the model and measurement updates leads to

$$\begin{aligned} K_e(k) &= P_e(k|k-1)H^T(k)(H(k)P_e(k|k-1)H^T(k) + R_e)^{-1} \\ \hat{x}_e(k|k) &= \hat{x}_e(k|k-1) + K_e(k)(z_e(k) - h(\hat{x}_e(k|k-1), 0)) \\ P_e(k|k) &= (I - K_e(k)H(k))P_e(k|k-1) \end{aligned}$$

where Q_e and R_e are suitably chosen variance process and measurement noise matrices. After the on-line estimation, the estimated angular amplitude and frequency [21] are

$$\begin{aligned} \hat{A}_{ex}(k|k) &= \sqrt{\hat{x}_{e,1}(k|k)^2 + \hat{x}_{e,2}(k|k)^2} \\ \hat{\omega}(k|k) &= \hat{x}_{e,3}(k|k) \end{aligned} \tag{13}$$

Next, consider the forcing of the float to reach a near resonance condition. The reference velocity [20] can then be described as

$$\dot{\theta}_{ref}(t) = \frac{1}{T(t)}\hat{M}_{ex}(t) \tag{14}$$

$$where \quad \frac{1}{T(t)} = \begin{cases} \dfrac{1}{2B(\omega) + 2K_v^0}, & if \quad \dfrac{\hat{\omega}\theta_{lim}}{\hat{A}_{ex}} > \dfrac{1}{2B(\omega) + 2K_v^{(0)}} \\ \dfrac{\hat{\omega}\theta_{lim}}{\hat{A}_{ex}}, & otherwise \end{cases} \tag{15}$$

$B(\omega)$ is the radiation damping coefficient, and K_v^0 is the extra viscous damping coefficient. θ_{lim} is the maximum angular displacement within the allowable range. Hence, the reference velocity can be calculated from Equations (14) and (15) since the instantaneous amplitude $\hat{A}_{ex}$ and frequency $\hat{\omega}$ of the WEM signal have been obtained using the EKF.

4. Model Predictive Tracking Control with Short-Term Forecasting

In general, MPC is often used as an optimal strategy for the purpose of energy-maximizing by setting the average mechanical power as its objective function with control input and physical position constraints [16]. This paper tests the value of MPC in another way by a hierarchical velocity tracking structure based on the scheme in Figure 2. The velocity error signal is considered in the quadratic objective function for the purpose of tracking. In addition, the future WEM and future reference velocity are predicted in the short term by two separate Gaussian Process models. Furthermore, it is desirable to adopt the short-term forecasting of M_{ex} and $\dot{\theta}_{ref}$ using uncorrelated Gaussian Process models. The forecasting time needs to be considered carefully, based on a trade-off between prediction accuracy and computation complexity.

4.1. Model Predictive Control

By defining $v = \dot{\theta}$ and with $C_v = [0 \quad 1 \quad 0 \quad 0]$, and by considering the angular velocity as the WEC system output, this leads to a discrete-time state space model:

$$\begin{aligned} x(k+1) &= A_d x(k) + B_d u(k) + B_d \hat{M}_{ex}(k) \\ v(k) &= C_v x(k) \end{aligned} \tag{16}$$

After iterating Model (16) with the prediction horizon N, a prediction model will have the following form:

$$V_N = S_{av} x(k) + S_{bv} u_N + S_{bv} M_{ex,N} \tag{17}$$

with $V_N = \begin{bmatrix} v(k+1) \\ v(k+2) \\ \cdots \\ v(k+N) \end{bmatrix}, S_{av} = \begin{bmatrix} C_v A_d \\ C_v A_d^2 \\ \cdots \\ C_v A_d^N \end{bmatrix}, u_N = \begin{bmatrix} u(k) \\ u(k+1) \\ \cdots \\ u(k+N-1) \end{bmatrix},$

$M_{ex,N} = \begin{bmatrix} \hat{M}_{ex}(k) \\ \hat{M}_{ex}(k+1) \\ \cdots \\ \hat{M}_{ex}(k+N) \end{bmatrix}, S_{bv} = \begin{bmatrix} C_v B_d & 0 & 0 & \cdots & 0 \\ C_v A_d B_d & C_v B_d & 0 & \cdots & 0 \\ \vdots & \vdots & \vdots & \ddots & \vdots \\ C_v A_d^{N-1} B_d & C_v A_d^{N-2} B_d & \cdots & C_v A_d B_d & C_v B_d \end{bmatrix}.$

Note that these stacked predictions are used in the objective function of the next optimization problem.

By defining $v_{ref} = \dot{\theta}_{ref}$, the constrained discrete time optimization problem can be considered as follows:

$$\min_{\substack{V_N, u_N \\ M_{eN}}} \sum_{i=1}^{N} [v(k+i) - v_{ref}(k+i)]^T Q_m [v(k+i) - v_{ref}(k+i)] + \sum_{i=0}^{N-1} [u(k+i)]^T R_m u(k+i) \tag{18}$$

$$subject\ to \quad |u(k+i)| \le u_{max} \tag{19}$$

According to Equations (17) and (18), the MPC objective function is selected as

$$J_m = \frac{1}{2} V_N^T \bar{Q} V_N - v_{ref,N}^T \bar{Q} V_N + \frac{1}{2} v_{ref,N}^T \bar{Q} v_{ref,N} + \frac{1}{2} u_N^T \bar{R} u_N \tag{20}$$

where $\bar{Q} = \begin{pmatrix} Q_m & 0 & 0 \\ 0 & \ddots & 0 \\ 0 & 0 & Q_m \end{pmatrix}, \bar{R} = \begin{pmatrix} R_m & 0 & 0 \\ 0 & \ddots & 0 \\ 0 & 0 & R_m \end{pmatrix}, v_{ref,N} = \begin{pmatrix} v_{ref}(k+1) \\ \vdots \\ v_{ref}(k+N) \end{pmatrix}.$

By substituting Equations (17) into (20), the objective function is transformed into

$$\begin{aligned} J_m =& \frac{1}{2}(S_{av} x(k) + S_{bv} u_N + S_{bv} M_{ex,N})^T \bar{Q} (S_{av} x(k) + S_{bv} u_N + S_{bv} M_{ex,N}) + \frac{1}{2} u_N^T \bar{R} u_N \\ & - v_{ref,N}^T \bar{Q} S_{av} x(k) - v_{ref,N}^T \bar{Q} S_{bv} u_N - v_{ref,N}^T \bar{Q} S_{bv} M_{ex,N} + \frac{1}{2} v_{ref,N}^T \bar{Q} v_{ref,N} \end{aligned} \tag{21}$$

Considering the velocity tracking and dropping some bias terms, the objective function J_m can be converted into its quadratic form:

$$J_m = \frac{1}{2} u_N^T H_m u_N + f_m^T u_N \tag{22}$$

where $H_m = S_{bv}^T \bar{Q} S_{bv} + \bar{R}$, $f_m^T = x^T(k) S_{av}^T \bar{Q} S_{bv} + M_{ex,N}^T S_{bv}^T \bar{Q} S_{bv} - v_{ref,N}^T \bar{Q} v_{ref,N}$.

Thereby, the velocity tracking problem is transformed into a Quadratic Programming (QP) optimization problem, and the energy-maximizing generation can be achieved when the scaled Wavestar-like device can track the reference angular velocity $\dot{\theta}_{ref}$. The qpOASES QP solver [28] is utilized to solve the QP optimization. It is particularly well suited for MPC applications that need fast QP solving rates with high reliability. As for the sequences $M_{ex,N}$ and $v_{ref,N}$, these can be obtained from the GP model to obtain short-term forecasting.

4.2. Gaussian Process Model

To specify the a *priori* expression of a GP model [29], the mean function $m(a)$ and covariance function $k(a, a^*)$ are defined as follows:

$$\begin{aligned} f(a) &\sim \mathcal{GP}(m(a), k(a, a^*)) \\ m(a) &= \mathbb{E}[f(a)] \\ k(a, a^*) &= cov(f(a), f(a^*)) \end{aligned} \tag{23}$$

where $a \in \Re^{\mathcal{D}}$ is a vector of the dynamics input with dimension $\mathcal{D}$, and $f(a)$ and $f(a^*)$ are arbitrary Gaussian scalar variables indexed by a and a^*.

Next, define a training set $\mathcal{D} = (\boldsymbol{a}, \boldsymbol{z})$, where the matrix $\boldsymbol{a} = [a_1, a_2, \cdots, a_n]$ contains all input vectors, and $\boldsymbol{z} = [z_1, z_2, \cdots, z_n]$ is a corresponding vector with all scalar outputs. Then, a GP *posterior* model can be determined as

$$z_i = f(a_i) + \epsilon_i \quad \epsilon_i \sim \mathcal{N}(0, \sigma^2) \tag{24}$$

where z is the observed output values, $f(a)$ is the GP model values, and ϵ is a zero mean white Gaussian noise.

Here, a spectral mixture (SM) kernel is chosen as the covariance function for excitation moment and reference velocity forecasting. The form is given by [30] as

$$k_{SM}(\tau) = \sum_{s=1}^{S} w_s cos(2\pi\tau^T\mu_s) \prod_{o=1}^{O} \exp(-2\pi^2\tau_o^2 v_s^{(o)}) \tag{25}$$

Regarding the one-dimensional input $O = 1$, S is a selected number of the wave frequency components, and τ is the distance between two arbitrary points a_i and a_j. According to the view of wave reconstruction, the parameters of the SM kernel hyperparameter vector $\Theta = (\mu_s, w_s, v_s)^T$ can denote the the amplitude, period and evolutionary-scale of each wave component. This means that this SM kernel [23] can be applied as the covariance function of a GP model for wave forecasting directly owing to its automatic discovery capability.

The hyperparameters of Θ in the SM kernel can be calculated by optimizing the marginal likelihood function [29] as

$$logp(\boldsymbol{z}|\boldsymbol{a}, \Theta) = -\frac{1}{2}log|K + \sigma^2 I| - \frac{1}{2}\boldsymbol{z}^T(K + \sigma^2 I)^{-1}\boldsymbol{z} - \frac{n}{2}log(2\pi) \tag{26}$$

Note that the choice of highly suitable initial hyperparameters before the training process can promote the convergence rate of optimization and avoid reaching an unsatisfactory local optimum. After the training process, the posterior joint distribution of the prediction f^* with given input vector $\boldsymbol{a}$ has the following form:

$$\begin{bmatrix} f^* \\ \boldsymbol{z} \end{bmatrix} \sim \left(\begin{bmatrix} m(\boldsymbol{a}^*) \\ m(\boldsymbol{a}) \end{bmatrix}, \begin{bmatrix} k(\boldsymbol{a}^*, \boldsymbol{a}^*) & k(\boldsymbol{a}^*, \boldsymbol{a}) \\ k(\boldsymbol{a}, \boldsymbol{a}^*) & K + \sigma^2 I \end{bmatrix} \right) \tag{27}$$

with $k(a^*, a) = k(a, a^*) = [k(a_1, a^*), \cdots, k(a_N, a^*)]$. Then, based on the Joint Gaussian Distribution Theorem, the predicted result about f^* can be described as

$$\begin{aligned} \mu(f^*) = & \quad m(a^*) + k(a^*, a)[K + \sigma^2 I]^{-1}(z - m(a)) \\ \sigma(f^*) = & \quad k(a^*, a^*) - k(a^*, a)[K + \sigma^2 I]^{-1}k(a, a^*) \end{aligned} \tag{28}$$

The Gaussian Process for Machine Learning (GPML) package [29] is applied to design the GP model and perform the training and test prediction work.

5. Results and Discussion

The simulation works are conducted using the open-source software WEC-Sim, developed in Matlab/Simulink using the multi-body dynamics solver Simscape Multibody—refer to WEC-Sim documentation [31] and applications [32–36]. The main contribution focuses on a Kalman filter approach to WEM estimation; in addition to optimal reference computation based on EKF, GP modeling is used for short-term forecasting and velocity tracking through MPC and tested on a WEC-Sim model of the 1:20 scaled Wavestar device in WEC-Sim. The simulation parameters are listed in Table 1. A fixed-step size ode8 (Dormand-Prince) solver is selected to conduct the simulation works for the WEC-Sim numerical model in Matlab/Simulink software.

Table 1. The simulation parameters.

Parameters	Values
Simulation sampling time T_s	0.05 s
Inertia of float and arm J	1.0 kg m^2
Added inertia J_∞	0.4805 kg m^2
Hydrostatic coefficient K_{hs}	92.33 Nm rad^{-1}
Linear damping coefficient K_v	1.8 Nm rad^{-1}s^{-1}
Length of the arm l_{arm}	0.54875 m
Efficiency of generator mode μ_{gen}	0.7
Efficiency of motor mode μ_{gen}	0.7^{-1}
KF coefficient Q_f	diag(0.01 0.1 0.01 0.01 200)
KF coefficient R_f	diag(0.01; 0.01)
EKF initial state x_e	$[1, 1, 5]^T$
EKF coefficient Q_e	$diag(0.2, 0.2, 0.001)$
EKF coefficient R_e	0.1
Angular displacement limit θ_{lim}	0.4 rad
Control limit u_{max}	±12 Nm
GP coefficient S	12

The matrices for the approximation of radiation moment are $A_{ra} = \begin{bmatrix} -13.59 & -13.35 \\ 8.0 & 0 \end{bmatrix}$, $B_{ra} = \begin{bmatrix} 8.0 \\ 0 \end{bmatrix}$, $C_{ra} = [4.739 \quad 0.5]$, $D_{ra} = -0.1586$.

The JONSWAP (JS) wave Spectrum is adopted to generate three irregular waves: Seastate1, Seastate2 and Seastate3. The significant wave height H_{m0} and peak wave period T_p of irregular waves, MPC prediction horizon N_p and MPC coefficients Q_m, R_m are shown in Table 2. The wave peak enhancement factor $\varkappa = 1$.

Table 2. The irregular wave information and MPC coefficients.

Case	H_{m0}	T_p	N_p	Q_m	R_m
Seastate1	0.0208	0.988	20	0.5	2×10^{-3}
Seastate2	0.0625	1.412	30	0.5	5×10^{-4}
Seastate3	0.1042	1.836	40	0.5	2×10^{-4}

Figure 3 presents the estimated $\hat{M}_{ex}$ from the KF over the time interval 50–75 s. It can be seen that some biases arise in the crests of the estimated WEM due to the WEC-Sim numerical model having some additional static moments and the offset between the center of bouyancy (CoB) and center of gravity (CoG). Additionally, a small lag of about 0.05 s occurs between the estimated $\hat{M}_{ex}$ and the calculated M_{ex}. The reason for this lag is that a relatively large sampling time has to be utilized for the requirement of the QP solving time in MPC due to its computational complexity. The sampling time is 0.05 s, which is larger than the most frequently used sampling rate (0.001 s). This means that the KF will not produce the most accurate estimation performance with zero lag, but the estimation performance is acceptable.

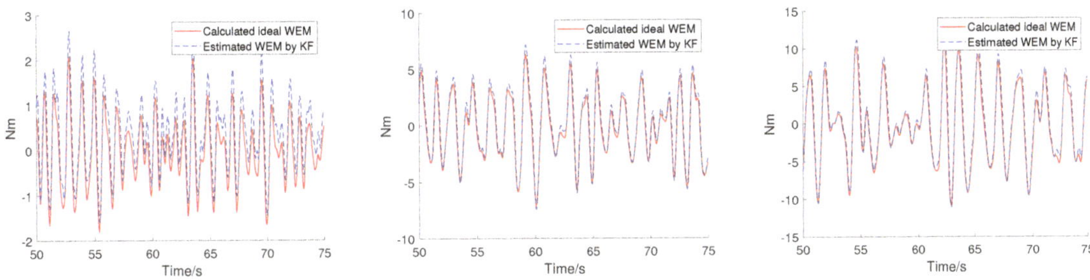

(**a**) Seastate1. (**b**) Seastate2. (**c**) Seastate3.

Figure 3. The estimated $\hat{M}_{ex}$ under three irregular waves by KF.

The estimated instantaneous amplitude $\hat{A}_{ex}$ is shown in Figure 4 along with the estimated excitation moment $\hat{M}_{ex}$. The estimated instantaneous frequency $\hat{\omega}$ is given in Figure 5. The EKF convergence time is usually long, and the initial values need to be chosen reasonably close to the expected values to avoid a long convergence time.

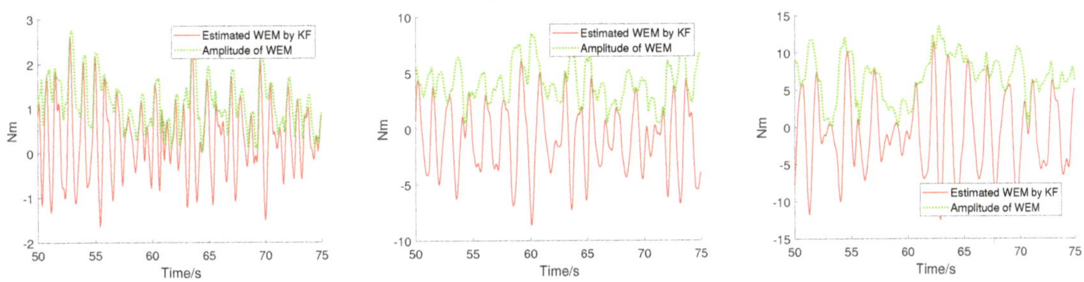

(**a**) Seastate1. (**b**) Seastate2. (**c**) Seastate3.

Figure 4. The amplitude $\hat{A}_{ex}$ of $\hat{M}_{ex}$ under three irregular waves by EKF.

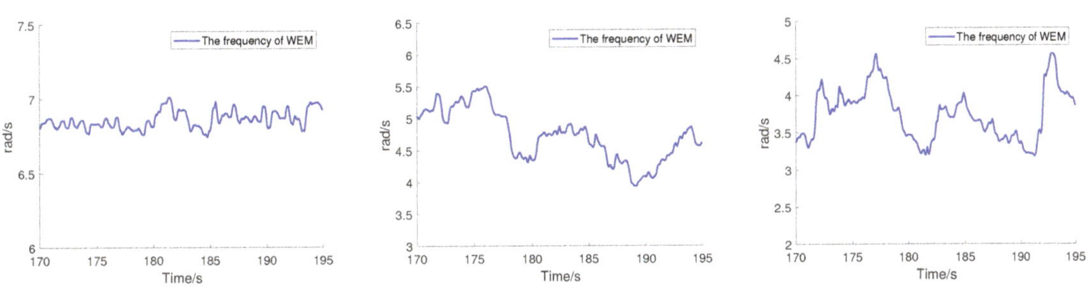

(**a**) Seastate1. (**b**) Seastate2. (**c**) Seastate3.

Figure 5. The frequency $\hat{\omega}$ of $\hat{M}_{ex}$ under three irregular waves by EKF.

Two GP models are employed to provide the future WEM and velocity information for MPC tracking design. In other words, the predicted future WEM sequence and future reference velocity sequence are used in the MPC objective function. It is clear that, most of the time, the predicted WEM based on the GP model is matched well with the calculated ideal WEM in Figure 6. The GP modeling method shows good learning smoothness and accuracy for WEM prediction. The suitable hyperparameters of the GP model can be preselected before the training process to reduce the prediction time and promote the convergence rate of optimization. There is no need to update the hyperparameters during the real-time forecasting. From Equations (14) and (15), the wave excitation moment $\hat{M}_{ex}$ and reference angular velocity $\dot{\theta}_{ref}$ share the same phase but a different amplitude when the tracking control is achieved for the WEC system. Hence, the predicted future reference velocity shows similar trends to the predicted future WEM in Figure 7.

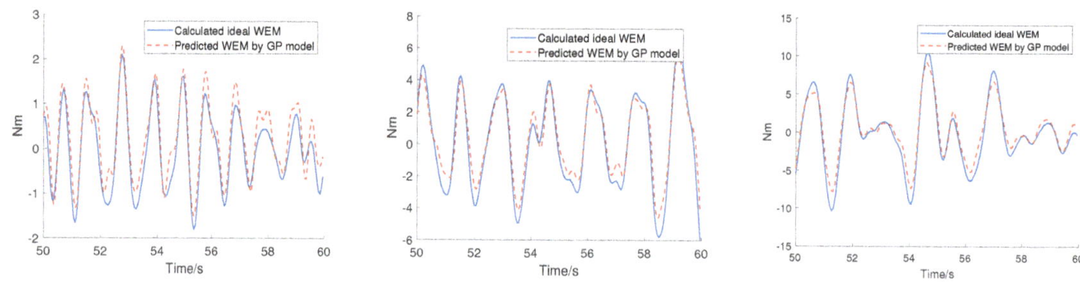

(**a**) Seastate1. (**b**) Seastate2. (**c**) Seastate3.

Figure 6. The predicted WEM by GP model under three irregular waves.

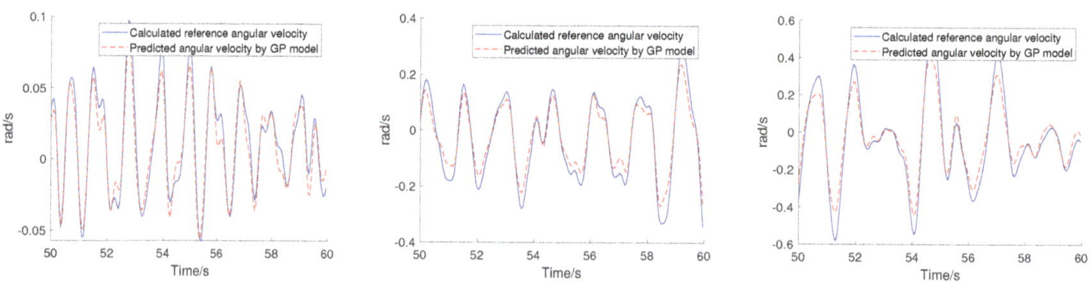

(**a**) Seastate1. (**b**) Seastate2. (**c**) Seastate3.

Figure 7. The predicted reference angular velocity by GP model under three irregular waves.

The MPC angular velocity tracking result is shown in Figure 8. To start from the MPC standpoint, the overall tracking performance is good, even if the float velocity shows occasional amplitude differences with the reference velocity. However, the tracking errors between the reference velocity and float velocity are small and acceptable. Most of the time, they share the same phase. The primary role of the reference velocity is to capture the same phase with the estimated $\hat{M}_{ex}$ so as to reach a near resonance condition for energy-maximizing purposes. This means that the WEC system is forced to follow the incident waves by MPC when the tracking mission is obtained. In total, the MPC method lacks some robustness to a certain extent, which is unlike the other robust methods, which can result in very small tracking errors for the WEC system doing the tracking work. However, the MPC has a good advantage in terms of input constraint handling.

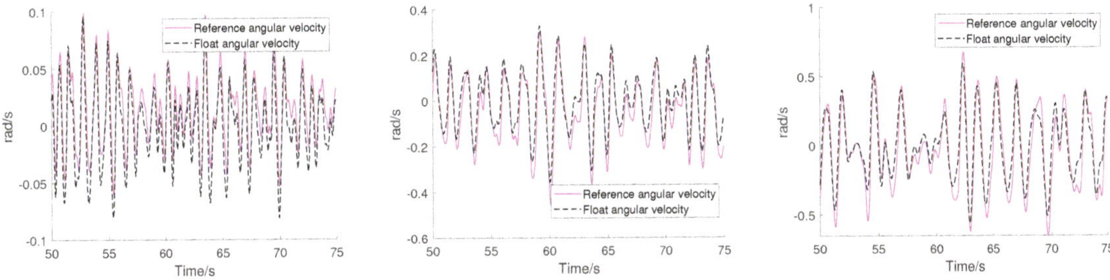

(**a**) Seastate1. (**b**) Seastate2. (**c**) Seastate3.

Figure 8. The float angular velocity $\dot{\theta}$ based on tracking control under three irregular waves.

The generated instantaneous power and extracted energy through the power take-off system are shown in Figures 9 and 10. It is clear that the proposed MPC tracking approach only causes very few negative power excursions, which is a benefit for the PTO system solving the large bidirectional flow problem and reducing the energy loss during the motor mode.

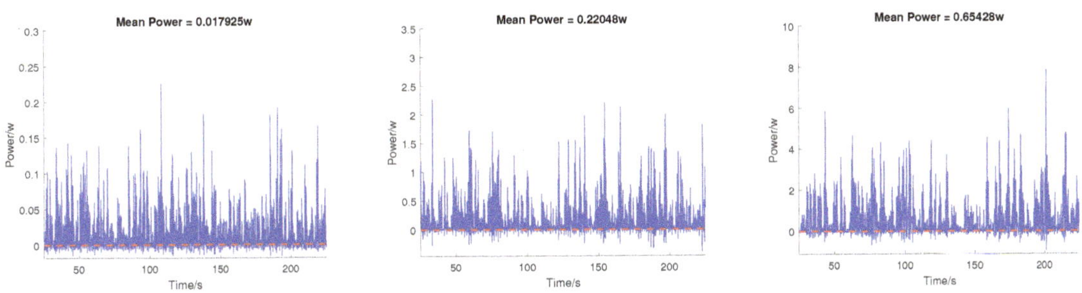

(**a**) Seastate1. (**b**) Seastate2. (**c**) Seastate3.

Figure 9. The generated instantaneous power from three irregular waves.

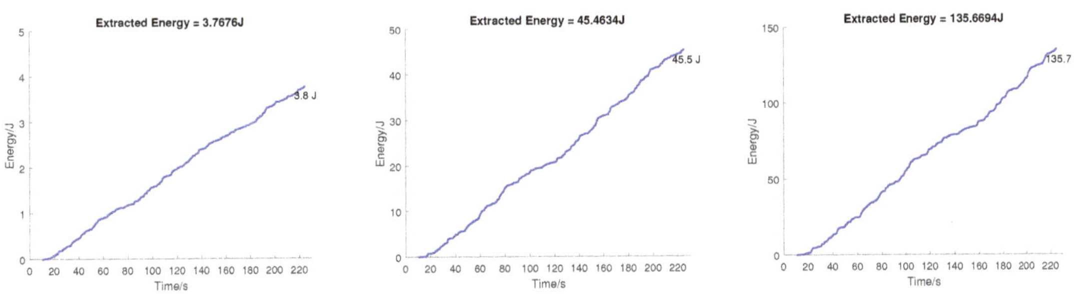

(**a**) Seastate1. (**b**) Seastate2. (**c**) Seastate3.

Figure 10. The extracted energy from three irregular waves.

The PTO moment control input is given in Figure 11. The red lines represent the input maximum limits of ± 12 Nm, showing that the MPC control input satisfies the input constraint conditions. The MPC tracking performance can be improved if the MPC coefficient R_m is decreased, but this can bring a large PTO moment in the control input. Conversely, if a large R_m is adopted and the control input can be small, the MPC tracking performance will certainly decline.

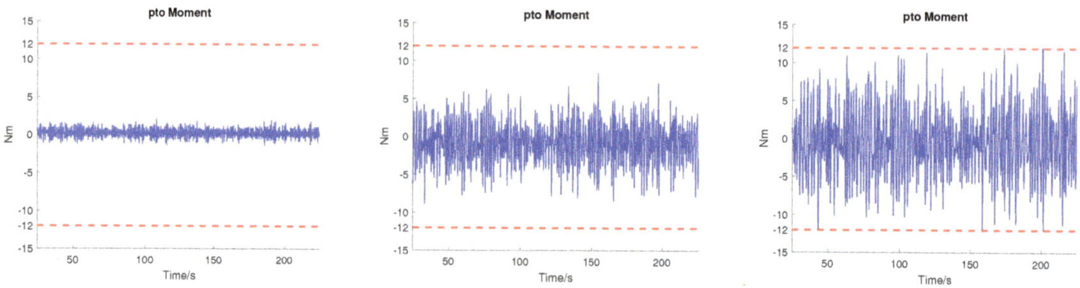

(**a**) Seastate1. (**b**) Seastate2. (**c**) Seastate3.

Figure 11. The PTO moment under three irregular waves.

6. Conclusions

This paper concentrates on model predictive velocity tracking control design based on a hierarchical structure to reach energy-maximizing generation for a Wavestar-like device in the WEC-SIM benchmark. The high-level part of the overall structure includes a KF for WEM estimation and an EKF for providing the instantaneous amplitude and frequency of WEM to calculate the optimal reference velocity. Two Gaussian Process models are selected to provide the future information, multiple steps ahead of the reference velocity and wave excitation moment for MPC design with accurate prediction accuracy and good smoothness. The low-level model predictive controller shows good tracking performance with small errors and pushes the WEC system into a near-resonance condition for power maximization extraction. Furthermore, it is shown that the MPC tracking system approach can take the input constraint into consideration as well and generate a low level of negative power, which is beneficial for the PTO system to avoid large bidirectional energy flow. Considering that the KF and MPC still lack robustness to a certain extent, future work may focus on how to design a robust controller under a large sampling time and how to improve the robustness of the designed MPC.

Author Contributions: Conceptualization, D.L. and R.P.; methodology, D.L. and R.P.; software, D.L.; validation, D.L.; formal analysis, D.L. and R.P.; writing—original draft preparation, D.L.; writing—review and editing, D.L. and R.P.; supervision, R.P. All authors have read and agreed to the published version of the manuscript.

Funding: This research received no external funding.

Institutional Review Board Statement: Not applicable.

Informed Consent Statement: Not applicable.

Data Availability Statement: Not applicable.

Conflicts of Interest: The authors declare no conflict of interest.

References

1. Lopez, I.; Andreu, J.; Ceballos, S.; Alegria, I.M.D.; Kortabarria, I. Review of wave energy technologies and the necessary power-equipment. *Renew. Sustain. Energy Rev.* **2013**, *27*, 413–434. [Crossref] [CrossRef]
2. Widen, J.; Carpman, N.; Castellucci, V.; Lingfors, D.; Olauson, J.; Remouit, F.; Bergkvist, M.; Grabbe, M.; Waters, R. Variability assessment and forecasting of renewables: A review for solar, wind, wave and tidal resources. *Renew. Sustain. Energy Rev.* **2015**, *44*, 356–375. [Crossref] [CrossRef]
3. Guo, B.; Ringwood, J.V. A review of wave energy technology from a research and commercial perspective. *IET Renew. Power Gener.* **2021**, *15*, 3065–3090. [Crossref] [CrossRef]
4. Mérigaud, A.; Ringwood, J.V. Condition-based maintenance methods for marine renewable energy. *Renew. Sustain. Energy Rev.* **2015**, *66*, 53–78. [Crossref] [CrossRef]
5. Sheng, W. Wave energy conversion and hydrodynamics modelling technologies: A review. *Renew. Sustain. Energy Rev.* **2015**, *109*, 482–498. [Crossref] [CrossRef]

6. Jakobsen, M.M. Wave-Structure Interactions on Point Absorbers—An Experimental Study. Ph.D. Dissertation, Aalborg University, Aalborg, Denmark, 2015.
7. Francesco, F. Wave-to-wire Modelling of Wave Energy Converters: Critical Assessment, Developments and Applicability for Economical Optimisation. Ph.D. Dissertation, Aalborg University, Aalborg, Denmark, 2014.
8. Windt, C.; Davidson, J.; Ringwood, J.V. Numerical analysis of the hydrodynamic scaling effects for the wavestar wave energy converter. *J. Fluids Struct.* **2021**, *105*, 103328. [Crossref] [CrossRef]
9. Garcia, V.D.; Pena-Sanchez, Y.; Faedo, N.; Windt, C.; Ringwood, J.V. Experimental implementation and validation of a broadband lti energy-maximizing control strategy for the wavestar device. *IEEE Trans. Control Syst. Technol.* **2021**, *99*, 1–13. [Crossref]
10. Heo, S.; Koo, W. Dynamic Response Analysis of a Wavestar-Type Wave Energy Converter Using Augmented Formulation in Korean Nearshore Areas. *Processes* **2021**, *9*, 1721. [Crossref] [CrossRef]
11. Ringwood, J.; Ferri, F.; Tom, N.; Ruehl, K.; Coe, R.G. The Wave Energy Converter Control Competition: Overview. In Proceedings of the ASME 2019 38th International Conference on Ocean, Offshore and Arctic Engineering, Glasgow, Scotland, UK, 9–14 June 2019.
12. Ringwood, J.V.; Bacelli, G.; Fusco, F. Energy-maximizing control of wave-energy converters. *IEEE Control Syst.* **2014**, *34*, 30–55. [Crossref]
13. Hansen, R.H.; Kramer, M. Modelling and control of the wavestar prototype. In Proceedings of the 9th European Wave and Tidal Conference, Southampton, UK, 5–9 September 2011.
14. Juan, L.G.; Oscar, J.; González, V.; John,A.R. Efficiency-aware nonlinear model-predictive control with real-time iteration scheme for wave energy converters. *Int. J. Control* **2022**, *preprint*. [Crossref]
15. Faedo, N.; Scarciotti, G.; Astolfi, A.; Ringwood, J.V. Energy-maximising control of wave energy converters using a moment-domain representation. *Control Eng. Pract.* **2018**, *81*, 85–96. [Crossref] [CrossRef]
16. Faedo, N.; Olaya, S.; Ringwood, J.V. Optimal control, mpc and mpc-like algorithms for wave energy systems: An overview. *IFAC J. Syst. Control* **2017**, *1*, 37–56. [Crossref] [CrossRef]
17. Violini, D.G.; Pena-Sanchez, Y.; Faedo, N.; Ringwood, J.V. An energy-maximising linear time invariant controller (lite-con) for wave energy devices. *IEEE Trans. Sustain. Energy* **2020**, *11*, 2713–2721. [Crossref] [CrossRef]
18. Falnes, J. *Ocean Waves and Oscillating Systems: Linear Interactions Including Wave-Energy Extraction*, 1st ed.; Cambridge University Press: Cambridge, UK, 2002; pp. 196–222.
19. Tona, P.; Sabiron, G.; Nguyen, H.N. An Energy-Maximising MPC Solution to the WEC Control Competition. In Proceedings of the ASME 2019 38th International Conference on Ocean, Offshore and Arctic Engineering, Glasgow, Scotland, UK, 9–14 June 2019.
20. Fusco, F.; Ringwood, J.V. Hierarchical robust control of oscillating wave energy converters with uncertain dynamics. *IEEE Trans. Sustain. Energy* **2014**, *5*, 958–966. [Crossref] [CrossRef]
21. Francesco, F. Real-Time Forecasting and Control for Oscillating Wave Energy Devices. Ph.D. Dissertation, National University of Ireland, Maynooth, Ireland, 2012.
22. Fusco, F.; Ringwood, J.V. Short-term wave forecasting for real-time control of wave energy converters. *IEEE Trans. Sustain. Energy* **2010**, *1*, 99–106. [Crossref] [CrossRef]
23. Shi, S.; Patton, R.J.; Liu, Y. Short-term wave forecasting for real-time control of wave energy converters. *IFAC Pap.* **2018**, *51*, 44–49. [Crossref] [CrossRef]
24. Tom, N.; Ruehl, K.; Ferri, F. Numerical Model Development and Validation for the WECCCOMP Control Competition. In Proceedings of the 37th International Conference on Ocean, Offshore, and Arctic Engineering, Madrid, Spain, 17–22 June 2018.
25. Yerai, P.S. Hydrodynamic Excitation Force Estimation and Forecasting for Wave Energy Applications. Ph.D. Dissertation, National University of Ireland, Maynooth, Ireland, 2020.
26. Nguyen, H.N.; Tona, P. Wave excitation force estimation for wave energy converters of the point-absorber type. *IEEE Trans. Control Syst. Technol.* **2018**, *26*, 2173–2181. [Crossref] [CrossRef]
27. Becker, V.S.P. Theoretical and Practical Aspects of Control Strategies for Wave Energy Converters of the Point Absorber Type. Master's Thesis, University of Kassel, Kassel, Germany, 2014.
28. Potschka, A.; Ferreau, H.J.; Diehl, M.; Kirches, C.; Bock, H.G. qpOASES: A parametric active-set algorithm for quadratic programming. *Math. Program. Comput. A Publ. Math. Program. Soc.* **2014**, *26*, 2173–2181. [Crossref]
29. Rasmussen, C.E.; Williams, C.K.I. *Gaussian Processes for Machine Learning*, 1st ed.; MIT Press: Cambridge, MA, USA, 2006; pp. 7–30.
30. Wilson, A.G.; Adams, R.P. Gaussian Process Kernels for Pattern Discovery and Extrapolation. In Proceedings of the 30th International Conference on Machine Learning, Atlanta, GA, USA, 16–21 June 2013.
31. WEC-Sim (Wave Energy Converter SIMulator). Available online: https://wec-sim.github.io/WEC-Sim/master/index.html (accessed on 12 June 2023).
32. WECCCOMP. Available online: https://github.com/WEC-Sim/WEC-Sim_Applications/tree/master/WECCCOMP (accessed on 12 June 2023).
33. Guerrero-Fernandez, J.L.; Tom, N.M.; Rossiter, J. Nonlinear Model Predictive Control Using Real-Time Iteration Scheme for Wave Energy Converters Using Wecsim Platform. In Proceedings of the ASME 41st International Conference on Ocean, Offshore and Arctic Engineering (OMAE2022), Hamburg, Germany, 5–10 June 2022.
34. García-Violini, D.; Faedo, N.; Jaramillo-Lopez, F.; Ringwood, J.V. Simple Controllers for Wave Energy Devices Compared. *J. Mar. Sci. Eng.* **2020**, *8*, 793. [Crossref] [CrossRef]

35. Guo, B.; Wang, T.; Jin, S.; Duan, S.; Yang, K.; Zhao, Y. A Review of Point Absorber Wave Energy Converters. *J. Mar. Sci. Eng.* **2022**, *10*, 1534. [Crossref] [CrossRef]
36. Nicolás, F.; Yerai, P.S.; Demián, G.V.; Francesco, F.; Giuliana, M.; John, V.R. Experimental assessment and validation of energy-maximising moment-based optimal control for a prototype wave energy converter. *Control Eng. Pract.* **2023**, *133*, 105454. [Crossref]

Journal of
Marine Science and Engineering

MDPI

Article

On the Effect of Wave Direction on Control and Performance of a Moored Pitching Wave Energy Conversion System

Bruno Paduano *, Nicolás Faedo and Giuliana Mattiazzo

Marine Offshore Renewable Energy Lab, Politecnico di Torino, 10128 Turin, Italy; nicolas.faedo@polito.it (N.F.); giuliana.mattiazzo@polito.it (G.M.)
* Correspondence: bruno.paduano@polito.it

Abstract: In the pathways towards the commercialisation of wave energy systems, the need for reliable mathematical models is of paramount importance for the design and synthesis of model-based control techniques to maximise the performance of wave energy converters (WECs). Furthermore, these offshore marine systems are held in position by the use of mooring systems, which have recently been analysed beyond survivability conditions to investigate their influence on control synthesis and device performance. In this study, we delve into the complex challenge of incorporating relevant mooring dynamics in defining a representative control action while also examining the influence of wave directionality on the overall procedure. For the specific case of a spread mooring system, where the hull cannot weathervane and operates based on directionality, control synthesis must be performed taking into account this characteristic of the resource. In this context, because it is able to harvest energy from only the bow-directed waves, the PeWEC is considered as a representative case study. The control synthesis is realised using a tailored data-based model, and device performance is evaluated across different site conditions while accounting for wave direction. Among our overall conclusions, we show that neglecting the directionality of the wave resource for the PeWEC case study can lead to an overestimation of device performance of up to 50%, even though a prevalent wave direction exists at the site.

Keywords: wave energy; pitching device; PeWEC; moorings; productivity; performances; control; data-based model

Citation: Paduano, B.; Faedo, N.; Mattiazzo, G. On the Effect of Wave Direction on Control and Performance of a Moored Pitching Wave Energy Conversion System. *J. Mar. Sci. Eng.* **2023**, *11*, 2001. https://doi.org/10.3390/jmse11102001

Academic Editor: Eugen Rusu

Received: 13 September 2023
Revised: 8 October 2023
Accepted: 12 October 2023
Published: 17 October 2023

1. Introduction

The evolution of renewable energies has paved the way for the development of various technologies, with ocean energy still remaining largely untapped. In particular, ocean energy encompasses different harvesting methods with wave energy holding significant promise, as it possesses the second-largest energy potential [1,2]. While the ocean energy sector has made remarkable strides in recent years, it remains in its early stages, with several advanced prototypes undergoing current testing. However, the global installed capacity remains limited to just a few megawatts. The industry faces pressing challenges, including the need for further technological development to establish reliability and robustness and cost reduction, as well as overcoming deployment hurdles and mitigating risks [3–6]. Evaluating the importance of an energy source involves considering both its potential and the capacity of the available technology to extract power from it. Recent studies indicate that wave energy extraction could reach around 10–20% of its total potential [5,7–9]. This level of power extraction has the capabilities of contributing with a substantial portion of the overall energy consumption [10].

As the energy stored in a wave diminishes with increasing water depth and distance from the shore [11], most wave energy converters (WECs) are classified as offshore floating devices [12]. Consequently, these WECs require specific locations for deployment. A crucial component in ensuring the proper functioning of these devices is the mooring system, which addresses the station-keeping problem.

Compared to traditional offshore structures, like floating production storage and offloading units, wave energy technologies present more complex systems for station keeping. This complexity arises from the need to maintain the device energy extraction characteristics, i.e., minimising the effect of the mooring dynamics in operating conditions, while ensuring its structural integrity [12].

The optimisation of energy extraction typically involves employing control techniques derived from optimal control theory [13]. In wave energy systems, control actions are virtually always computed using a model-based approach [14]. Therefore, accounting for the effect of the mooring system on the overall system response becomes crucial in the process of synthesising effective and representative control actions. Designing a reliable controller necessitates a thorough understanding of the system dynamics, including the relevant mooring dynamics. However, modelling moorings accurately within a tractable WEC model is inherently challenging due to their potentially nonlinear behaviour [15], which can (at least partially) explain why the vast majority of control studies available for WEC systems do not incorporate any information on the effect of the station-keeping component. As a matter of fact, while the influence of moorings on device dynamics is acknowledged in the existing literature, the vast majority of the state-of-the-art studies in the wave energy domain primarily focus on mooring system design under survivability conditions (see, e.g., [16–18], to cite a few).

If the station-keeping system significantly influences the dynamics of WECs, synthesising control actions without considering the mooring system can lead to non-representative and suboptimal results [19–22].

Furthermore, though, in general, performance assessments omit the influence of the wave direction on the device performances [23,24], and the associated device dynamics can be significantly influenced if a directional device is investigated, (i.e., a device capable of harvesting energy from a single direction), because the mooring system reduces the device's weathervaning capabilities. In addition, it is not clear in the state of the art how the wave directionality directly influences the response of a moored system and, consequently, the definition of a representative control action.

With the intention of providing insight into wave energy systems modelling, this study expands upon the control synthesis approach proposed in [20]. In particular, we incorporate wave directionality into the analysis and the assessment of a representative wave energy system performance based on representative site conditions. Acknowledging the impact of nonlinear factors on the overall dynamics, such as mooring restoring forces and wave second-order forces, among others (see Section 3), the results are evaluated using a nonlinear numerical model. By overcoming the limitations associated with wave directionality, this study comprehensively assesses the productivity of the pendulum wave energy converter (PeWEC) across a wide spectrum of scenarios. Among the available WECs, the PeWEC has been chosen as a representative case study due to its multi-degrees-of-freedom (DoFs) nature, comprehensive PTO conversion mechanism, and the relevance of the mooring system for device station keeping and alignment.

The remainder of this study is organised as follows: In Section 2, the device under investigation is presented alongside its main characteristics, working principle, and corresponding mooring system. In Section 3, the numerical models adopted to evaluate both the data-based models and device productivity are presented. In Section 4, the control synthesis of the device is analysed and the application of such a synthesis by means of a tailored data-based model is described. Section 5 outlines the simulation process and describes the main results of the productivity analysis and, finally, in Section 6 the main findings of this study are reported.

Notation

With $\mathbb{R}^+$, the set of positive real numbers is represented. $F(\omega) = \mathcal{F}(f(t))$ represents the Fourier transform of the function f. Moreover, whenever a function is represented by a Greek letter, its Fourier transform is indicated with a *tilde*, e.g., the Fourier transform of the

η is expressed with $\tilde{\eta}$. Given a matrix $A \in \mathbb{C}^{n\times m}$, $A^* \in \mathbb{C}^{n\times m}$ represents the Hermitian of A. Additionally, given a square non-singular matrix $B \in \mathbb{C}^{n\times n}$, $B^{-*} \in \mathbb{C}^{n\times n}$ denotes the inverse Hermitian of B. Considering $A \in \mathbb{C}^{n\times m}$, $\{\Re(A), \Im(A)\} \subset \mathbb{R}^{n\times m}$ represents the real and the imaginary parts of the matrix A, respectively.

2. The PeWEC Case

With the purpose of keeping this paper reasonably self-contained, a brief introduction to the PeWEC underlying working principle, mooring, and associated mechanical system is presented within the following paragraphs.

The PeWEC is a floating offshore pendulum-based WEC, which harvests energy by means of the wave-induced pitch motion. Its working principle is outlined in Figure 1. The PeWEC pitch motion, excited by the incoming wave, induces a pendulum rotation around its axis (ε), which is connected, by means of a gearbox, to a PTO system.

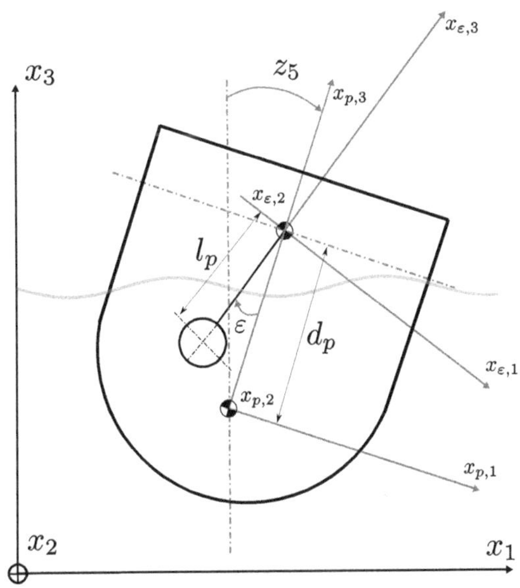

Figure 1. The PeWEC working principle in a schematic two-dimensional fashion.

Hull geometries and inertial properties are the result of an optimisation algorithm, which evaluates the device performance and minimises a corresponding cost function (i.e., capex over productivity) on a specific site. In particular, within this study, the PeWEC device has been optimised, taking into account the environmental conditions of Pantelleria (Sicily, Italy), by leveraging a genetic-based algorithm [25].

The PeWEC station-keeping problem is solved by adopting symmetrical spread mooring, formed by four catenary lines (see Figure 2), which are effectively "common" within the wave energy field [26,27]. Each line has a corresponding jumper attached to reduce the vertical load on the device and hence minimise any undesired effects on the device response.

A spread mooring system can significantly affect the harvested energy of a pitching device like the PeWEC, because it can restrain the weathervaning capabilities of the wave energy system. Moreover, Pantelleria represents a peculiar, environmentally speaking, site, because waves have a prevalent direction, as exposed in Figure 3. The PeWEC properties and the associated installation site data are both available in Table 1.

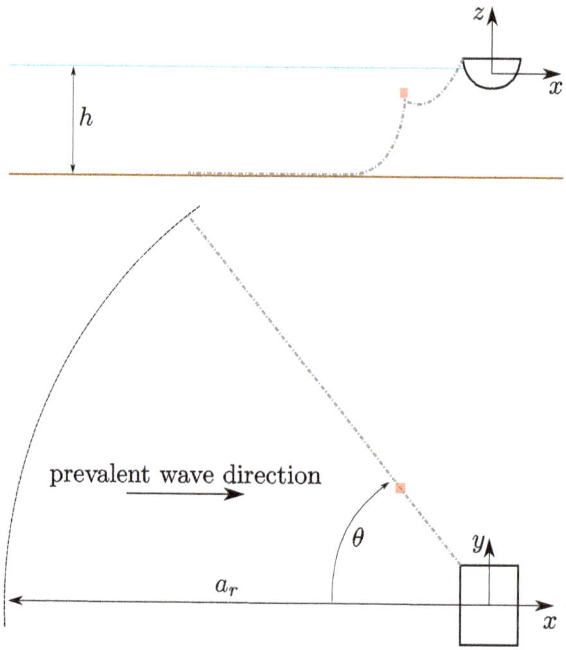

Figure 2. The PeWEC mooring system layout: schematic representation of one of the four catenary lines. Please note that, the sub-surface jumper is figured in red.

Table 1. PeWEC properties.

Property	Symbol	Unit	Value
	Device properties		
device—mass	$m_{1,1}$	(kg)	$1.11 \cdot 10^7$
device—roll inertial moment	$m_{4,4}$	(kg·m^2)	$5.5 \cdot 10^7$
device—pitch inertial moment	$m_{5,5}$	(kg·m^2)	$3 \cdot 10^7$
device—yaw inertial moment	$m_{6,6}$	(kg·m^2)	$7.2 \cdot 10^7$
hydrostatic stiffness—heave	$h_{k,3,3}$	(N/m)	$3.1 \cdot 10^6$
hydrostatic stiffness—roll	$h_{k,4,4}$	(Nm/rad)	$1.2 \cdot 10^8$
hydrostatic stiffness—pitch	$h_{k,5,5}$	(Nm/rad)	$4 \cdot 10^7$
	Pendulum properties		
pendulum—mass	m_p	(kg)	$7.17 \cdot 10^4$
pendulum—extent	d_p	(m)	2.44
pendulum—length	l_p	(m)	2.40
	Site and mooring properties		
site water depth	w_d	(m)	38
mooring anchor radius	a_r	(m)	175
mooring line length	–	(m)	190
mooring line angle	θ	(deg)	60
chain nominal diameter	–	(mm)	80
chain axial stiffness	–	(N)	$546 \cdot 10^6$
chain linear mass	–	(kg/m)	127
jumper net buoyancy	–	(kg)	4000

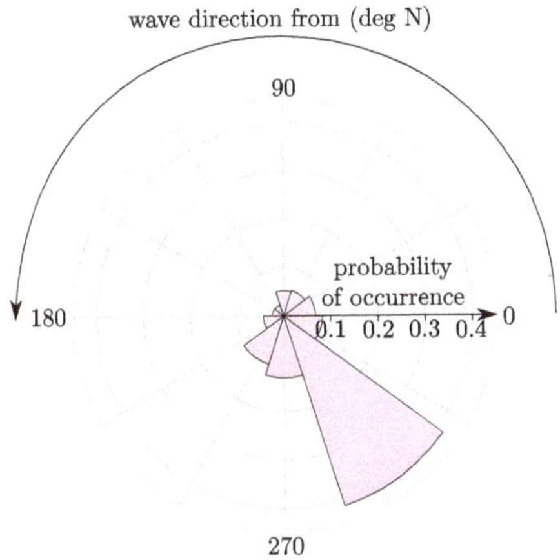

Figure 3. Rose diagram of the Pantelleria wave conditions: wave occurrences related to the coming-from direction.

3. Numerical Modelling

In this study, various numerical models are employed to capture the underlying system dynamics. While a frequency-domain approach can be utilised within linear assumptions (also by linearising nonlinear terms, see [28]), it is recognised that the use of a dynamic model is necessary when modelling a mooring system [29]. The following numerical models are employed in this study:

- Device hydrodynamics: The hydrodynamic effects acting on the device are incorporated using the assumptions corresponding with potential-flow theory [30]. Time-domain simulations are performed by computing the corresponding impulse response functions based on the well-known Cummins' equation [31].
- Mooring system: The mooring system is modelled using a dynamic lumped-mass model described in [32], and the numerical solver adopted is OrcaFlex (OF), being one of the most used in the offshore field [33]. External forces acting on the mooring system are included within the OF environment using the procedure proposed in [20].

Accordingly, the equation of motions of the device is

$$(m + m_\infty)\ddot{z}(t) + \int_{-\infty}^{t} h_r(\tau)\dot{z}(t-\tau)\mathrm{d}\tau + h_k z(t) = f_w(t) - f_{\text{pend}}(t) - f_m(t), \quad (1)$$

in which $\{m, m_\infty, h_k\} \subset \mathbb{R}^{6\times 6}$ represent the system inertia matrix, the so-called infinite-frequency added mass [31], and the hydrostatic stiffness, respectively. $\{f_w, f_{\text{pend}}, f_m\} : \mathbb{R}^+ \to \mathbb{R}^6, t \mapsto \{f_w(t), f_{\text{pend}}(t), f_m(t)\}$ are the wave exciting force, the pendulum reaction on the device, and the mooring restoring force, respectively. Finally, $h_r : \mathbb{R}^+ \to \mathbb{R}^{6\times 6}, t \mapsto h_r(t)$ represents the radiation impulse response function. Furthermore, it is possible to appreciate the integration of the following nonlinear terms:

- Wave force: The excitation force on the hull is evaluated by including the second-order terms (slow varying forces) defined in [34], precisely $f_w(t) = f_e(t) + f_{drift}(t)$, in which $f_{drift}(t)$ represents the drift forces acting on the device, given by the superposition of the mean drift forces and different-frequencies forces [34].
- Mooring force: The mooring problem is solved by means of OF as exposed in [32].
- Mechanism force: The integration of the mechanism and the associated control (the pendulum and the PTO) is achieved by means of an external library, following the approach proposed in [20]. The corresponding nonlinear equations are discussed within the following paragraph.

A graphical (schematic) representation of the equation of motion is presented in Figure 4. The pendulum equations are derived in a 3-DoFs model, by considering the surge, heave, and pitch motions. Note that the presented mathematical model, and its associated parameters, have been validated previously, see, e.g., [35,36]. The pendulum reaction on the hull, $f_{\text{pend}} : \mathbb{R} \to \mathbb{R}^6, t \mapsto f_{\text{pend}}(t)$, can be written as

$$\begin{aligned} f_{\text{pend}} &= [f_{\text{pend},1}, 0, f_{\text{pend},3}, 0, f_{\text{pend},5}, 0,]^{\intercal}, \\ f_{\text{pend},1} &= -m_p d_p \cos(z_5)\ddot{z}_5 - m_p l_p \cos(z_5 + \varepsilon)(\ddot{z}_5 + \ddot{\varepsilon}) \\ &+ m_p d_p \sin(z_5)\ddot{z}_5 - m_p l_p \sin(z_5 + \varepsilon)(\dot{z}_5 + \dot{\varepsilon})^2, \\ f_{\text{pend},3} &= m_p d_p \sin(z_5)\ddot{z}_5 - m_p l_p \sin(z_5 + \varepsilon)(\ddot{z}_5 + \ddot{\varepsilon}) \\ &+ m_p d_p \cos(z_5)\ddot{z}_5 - m_p l_p \cos(z_5 + \varepsilon)(\dot{z}_5 + \dot{\varepsilon})^2, \\ f_{\text{pend},5} &= f_{\text{ctrl}} + f_{\text{pend},1} d_p \cos(z_5) - f_{\text{pend},3} d_p \sin(z_5), \end{aligned} \tag{2}$$

where $\{m_p, l_p, d_p\} \subset \mathbb{R}^+$ represent the pendulum mass, pendulum length, and vertical extent between the pendulum fulcrum and the device centre of gravity (CoG), respectively. Moreover, $z_5 : \mathbb{R} \to \mathbb{R}, t \mapsto z_5(t)$ is the pitch motion, defined as the fifth entry of the device motion vector z, and $\varepsilon : \mathbb{R} \to \mathbb{R}, t \mapsto \varepsilon(t)$ is the rotation of the PTO axis. Finally, $f_{\text{ctrl}} : \mathbb{R} \to \mathbb{R}, t \mapsto f_{\text{ctrl}}(t)$ is the control torque applied to the PTO axis. The relation between the PeWEC pendulum rotation and its associated pitch motion can be described by the following equation:

$$\begin{aligned} &(m_{5,5} + m_p l_p^2)\ddot{\varepsilon} - m_p l_p \cos(z_5 + \varepsilon)\ddot{z}_1 + m_p l_p \sin(z_5 + \varepsilon)\ddot{z}_3 + \\ &(m_{5,5} + m_p l_p^2 - m_p d_p l_p \cos(\varepsilon))\ddot{z}_5 - m_p d_p l_p \sin(\varepsilon)\dot{z}_5^2 + \\ &m_p g l_p \sin(z_5 + \varepsilon) + f_{\text{ctrl}} = 0, \end{aligned} \tag{3}$$

with $m_{5,5} \in \mathbb{R}^+$ the hull pitch inertial moment.

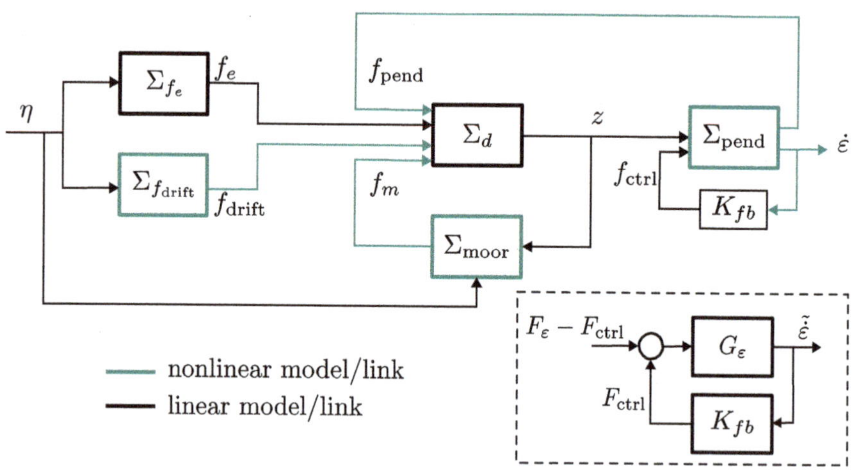

Figure 4. Schematic representation of the PeWEC system. The controlled system (under linear assumptions) is represented as the linear map G_ε.

As discussed previously in this section, the overall model experiences nonlinear actions, which are highlighted in Figure 4 using green lines. According to Cummins' Equation (1), it is possible to define the following subsystems: $\{\Sigma_{f_e}, \Sigma_{f_{\text{drift}}}\} : \mathbb{R} \to \mathbb{R}^6$, the linear and nonlinear maps related to the wave forces (f_e and f_{drift}, respectively); $\Sigma_{\text{moor}} : \mathbb{R}^6 \to \mathbb{R}^6$, the mooring nonlinear map; $\Sigma_{\text{pend}} : \mathbb{R}^6 \to \mathbb{R}^7$, the pendulum numerical representation; and, finally, $\Sigma_d : \mathbb{R}^6 \to \mathbb{R}^6$, the linear map associated to the device hydrodynamics. Regarding the control action itself, $K_{fb} : \mathbb{R} \to \mathbb{R}$ represents the controller acting on the mechanical system.

Please note that, under linear assumptions, it is possible to represent the controlled system separately, in terms of an equivalent input/output map $G_\varepsilon : \mathbb{R} \to \mathbb{C}, \omega \mapsto G_\varepsilon(\omega)$ which defines the link between the total force acting on the PTO axis (F_ε) and the associated velocity ($\dot{\varepsilon}$).

We summarise, in the following, the main assumptions adopted within the mathematical modelling procedure:

- Hydrodynamics: The model is based on linear potential theory, which is expanded using nonlinear second-order forces. Linear potential models rely on the assumption of small motions and are generally unsuitable when excitation forces and device hydrodynamic properties (e.g., stiffness, among others) vary significantly. In most cases, this issue is limited to extreme wave scenarios [37], and the use of a linear potential-based model has already been validated for the PeWEC case [38].
- Mechanism: The model assumes that the pendulum effect primarily affects the pitch, surge, and heave motions. Note that the overall model is validated against experimental data in [36].
- Mooring: This model is based on a dynamic lumped-mass model, which effectively represents the state of the art in mooring modelling [33].

4. Control Synthesis for the PeWEC

As suggested in Paduano et al. [20], the control problem of the moored PeWEC can be faced by leveraging impedance-matching theory [39]. To keep this study reasonably self-contained, the description of the control synthesis and the associated theoretical background are briefly recalled. For a complete description, the interested reader is referred to [20,39–41].

4.1. Impedance-Matching Application

Figure 5 shows the PeWEC system under linear assumptions feedback-controlled in terms of the map $K_{fb}^{opt} : \mathbb{R} \to \mathbb{C}, s \mapsto K_{fb}^{opt}(s)$. The control action F_{ctrl}^{opt} can be interpreted in terms of an electric load, which needs to be designed with the objective of maximising the power extraction from the source, i.e., the force acting on the ε axis. Note that, as per the discussion provided previously within this section, $\dot{\varepsilon}$ represents the velocity of the PTO axis, and the map $G_\varepsilon : \mathbb{R} \to \mathbb{C}, s \mapsto G_\varepsilon(s)$ is the system to be controlled.

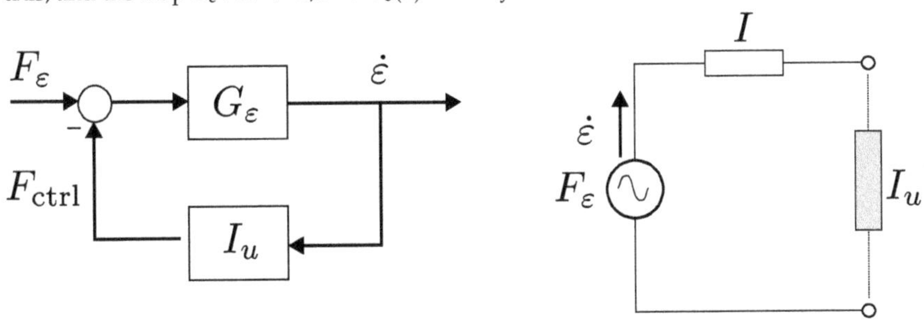

Figure 5. Impedance-matching principle: equivalent electrical circuit representation.

The resolution of this problem, under its electrical representation, can be addressed by means of the so-called impedance-matching theorem [42]. In particular, in order to maximise the power transfer, the load impedance K_{fb}^{opt} must be designed as the complex conjugate of the system impedance, which means, in the PeWEC case,

$$K_{fb}^{opt}(\omega) = I^*(\omega) = G_\varepsilon^{-*}(\omega). \tag{4}$$

The direct application of the condition in Equation (4), for every possible ω, cannot be achieved in practice due to the non-causality induced by the Hermitian operator (see [39]). To address this problem, it is possible to apply the impedance-matching principle by interpolating the optimal condition in (4) with an implementable (i.e., causal) controller structure on a matching frequency ω_m, i.e.,

$$K_{fb}(\omega_m) = K_{fb}^{opt}(\omega_m). \tag{5}$$

Though the choice of the interpolating frequency can be straightforward in monochromatic exciting conditions, when a panchromatic force excites the controlled system, the definition of ω_m is not trivial, compromising the performance of the adopted controller [39].

Therefore, in order to synthesise a suitable controller, the interpolating frequency can be selected by analysing, according to a given input wave spectrum, the spectrum of the exciting force $S_{f_\varepsilon} : \mathbb{R} \to \mathbb{R}^+, \omega \mapsto S_{f_\varepsilon}(\omega)$ [41]. To evaluate the force spectrum, the map $\tilde{\eta} : \mathbb{R} \to \mathbb{C}, \omega \mapsto \tilde{\eta}(\omega)$, which links the wave elevation with the total force acting on the PTO axis, can be leveraged.

As such, the uncontrolled system can be represented, under linear assumptions, as exposed in Figure 6. Finally, the spectrum of the exciting input acting on the PeWEC system can be defined as

$$S_{f_\varepsilon} = S_\eta G_{\eta 2 f_\varepsilon} G^*_{\eta 2 f_\varepsilon}, \tag{6}$$

in which $S_\eta : \mathbb{R} \to \mathbb{R}^+, \omega \mapsto S_\eta(\omega)$ represents the wave spectrum.

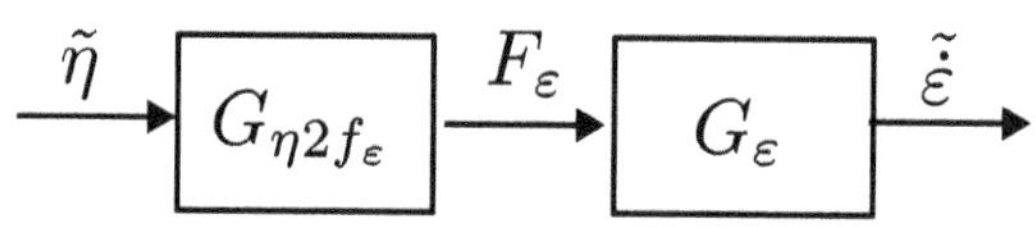

Figure 6. Identification of the pendulum axis excitation force.

4.2. Data-Based Modelling of the Controlled System

Bearing in mind the numerical tools available (see Section 3), it is clear that a significant limitation arises in the use of impedance-matching-based control synthesis, because most of the used models are nonlinear. This problem can be circumnavigated by leveraging a data-based model of the controlled system. The identification of the map G_ε, for the PeWEC case, is achieved by imposing a set of $N_j \in \mathbb{N}$ multisine input signals F^j_{ctrl} [43]. By applying this set of signals, it is straightforward to define the I/O empirical transfer function estimate for G_ε as

$$G_\varepsilon(\omega) = \sum_{j=1}^{N_j} \frac{1}{N_j} \frac{F^j_{\text{ctrl}}(\omega)}{\tilde{\ddot{\varepsilon}}^j(\omega)}, \tag{7}$$

where $\tilde{\ddot{\varepsilon}}^j$ denotes each output signal corresponding with the input F^j_{ctrl}.

The process described above is applied to the PeWEC-moored configuration so as to evaluate the corresponding frequency response map. In particular, this is exposed explicitly in Figure 7.

With the identified response (7), we proceed to synthesise an energy-maximising controller, following the impedance-matching theory presented in Section 4.1. In particular, the structure of the feedback controller adopted in this paper, which is used to interpolate the optimal impedance-matching response, is

$$K_{fb}(\omega) = \frac{k_\alpha \jmath\omega}{\jmath\omega + k_\beta}, \tag{8}$$

with $\{k_\alpha, k_\beta\} \subset \mathbb{R}$. Note that the proposed controller structure is an alternative of the classic proportional–integral (PI) controller, well known in the wave energy field. This particular structure is adopted for the intrinsic stability condition that can be guaranteed in the controller closed-loop form, taking advantage of the WEC dissipativity property [44].

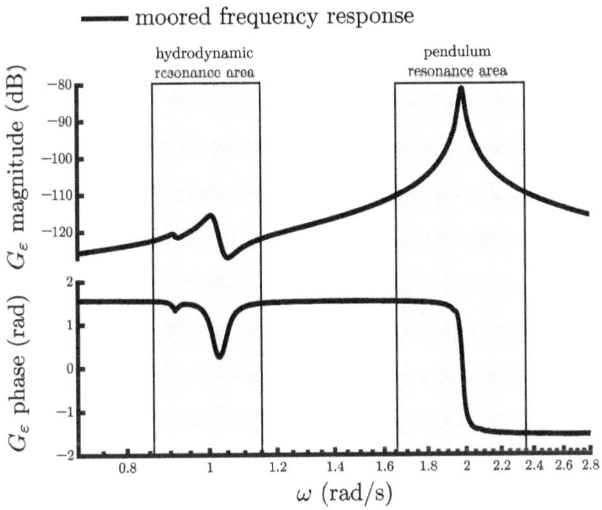

Figure 7. Averaged frequency responses (G_ε) for the moored and unmoored models.

Finally, it is possible to define, for a given interpolating frequency $\omega_m \in \mathbb{R}$, the control parameters for (8) as follows:

$$\begin{aligned} k_\alpha(\omega_m) &= \Re\left(\frac{1}{G_\varepsilon^*(\omega_m)}\right)\frac{\omega_m^2 + k_{\beta,m}^2}{\omega_m^2}, \\ k_\beta(\omega_m) &= \Im\left(\frac{1}{G_\varepsilon^*(\omega_m)}\right)\frac{\omega_m}{\Re\left(\frac{1}{G_\varepsilon^*(\omega_m)}\right)}. \end{aligned} \tag{9}$$

Within this study, irregular sea states are experienced by the pitching device, which means that the interpolating frequency can be achieved, as suggested in Section 4.1, by taking advantage of the map between the wave elevation and the PTO force, i.e., by identifying map $G_{\eta 2 f_\varepsilon}$ (see Figure 6), which can be achieved for a nonlinear system by computing the best linear approximation of the system [43]:

- Known wave spectra (i.e., Schroder phase multisine [43]) are simulated on the high fidelity model. For any simulation, the overall response is computed and averaged, i.e.,

$$G_\star = \frac{\tilde{\varepsilon}}{\tilde{\eta}}. \tag{10}$$

- The map $G_{\eta 2 f_\varepsilon}$ can be computed as

$$G_{\eta 2 f_\varepsilon} = \frac{G_\star}{G_\varepsilon}, \tag{11}$$

  with G_ε the frequency response evaluated as exposed in Section 4.
- The spectrum of the force on the ε axis can now be computed, i.e.,

$$S_{f_\varepsilon} = S_\eta\left(G_{\eta 2 f_\varepsilon} G_{\eta 2 f_\varepsilon}^*\right). \tag{12}$$

The map $G_{\eta 2 f_\varepsilon}$ is evaluated, and the associated amplitude is exposed in Figure 8. The system can be divided into two frequency regions, one defined by the device resonance (close to 1 rad/s) and one defined by the pendulum resonance (close to 2 rad/s). It can be noticed that the output of the system $G_{\eta 2 f_\varepsilon}$ is more susceptible to exciting input with a frequency closer to the device resonance one. Once the force spectrum is computed, the matching frequency ω_m can be selected. Among the characteristic frequencies of the force exciting spectrum, within this study, the median is considered.

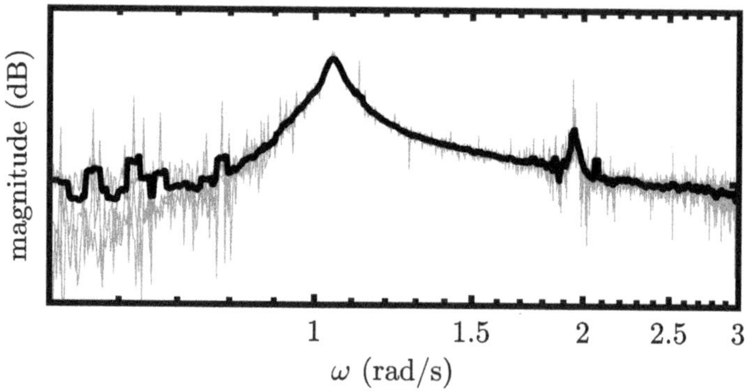

Figure 8. Identification of the $G_{\eta 2 f_\varepsilon}$ map. The averaged value of the several multisine signals is exposed in black.

5. Performance Assessment

The assessment of device productivity requires considering the performance across the entire scatter diagram of the site, which necessitates the inclusion of directionality. This aspect is particularly relevant for the PeWEC, as the device's ability to harvest energy is dependent on the incident wave direction. Including the directionality in the performance assessment of the PeWEC requires considerations on the deployment of the associated mooring system, because the orientation of the spread mooring (see Figure 2) will definitely impact the device performances. In the next paragraphs, the problem is faced by analysing a large set of environmental data in order to evaluate by means of a representative analysis the productivity of the PeWEC by analysing the directional scatter diagrams. The obtained productivity result is compared to the productivity achieved in an omnidirectional scenario. This comparison allows for the power loss resulting from the waves' directionality to be quantified and emphasises their significance in the overall performance assessment of the device.

To gather a comprehensive dataset, a total of 20 years' worth of hourly data are downloaded from the ERA5 online database [45]. This dataset serves as the foundation for the subsequent analysis. The downloaded data are utilised to investigate the distribution of the wave directions over the designated time period. This analysis provides valuable insights into the prevailing wave directions at the chosen site.

In Figure 9, the wave energy and occurrences (O_{en} and O_w, respectively) are represented as a function of the wave directionality. It can be noted that almost 70% of the site total energy is carried by waves coming from 300° N. Nevertheless, the waves from the same directions represent only 40% of the site conditions.

By leveraging the directionality characteristic of the site, and taking advantage of the device symmetry[1], the following steps can be followed in order to define and organise the site environmental conditions:

- Waves are divided into four main directions, which means four different scatters are built according to the direction between the waves and the device bow: $0 \pm 15°$, $30 \pm 15°$, $60 \pm 15°$, and $90 \pm 15°$.
- The 0° scatter is defined according to the most energetic directions (which actually is 300° N, see Figure 9).
- The bin size is set to 0.75 m and 0.75 s in order to define a representative set of waves.

- The waves simulated represent 95% of the site occurrence and 98% of the site energy: it means that most energetic waves on directional scatters are taken until 98% of the site energy and 95% of the site occurrence is reached[2].

According to the proposed discretisation, it is possible to minimise the number of simulated waves for non-prevalent directions. The scatters discretised according to the four main directions are reported in Appendix A.

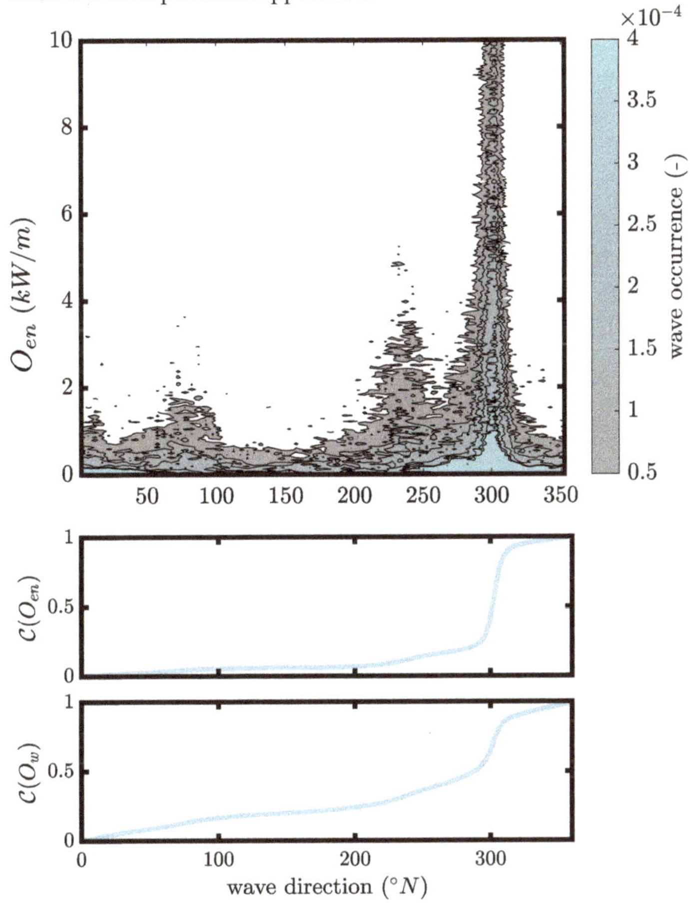

Figure 9. On top, wave energy and occurrences as a function of the wave directionality; on bottom, the cumulate of the wave energy and occurrence.

Once scatters are defined, a representative linear model of the moored system needs to be computed (such as the one exposed in Section 3). Please note that the directionality significantly influences the device response in the corresponding map $G_{\eta 2 f_\varepsilon}$. Therefore, by varying the directionality of the input signal $\tilde{\eta}$, it is possible to define the map $G_{\star,dd}$ (see Equation (11)), in which $dd \in [0, 30, 60, 90]$ represents the wave–bow orientation.

The maps of $G_{\eta 2 f_\varepsilon, dd}$ for different wave directions are shown in Figure 10. The influence of the wave direction on the equivalent force acting on the ε axis can be observed. The map associated with a 90°-oriented wave is not exactly zero due to the small yaw angles allowed by the mooring system. However, the force acting on the pendulum axis is lower than 30 dB when a wave comes from the side of the device, compared to a bow-oriented wave.

Based on this, the following simulations are conducted by omitting the 90° scatter.

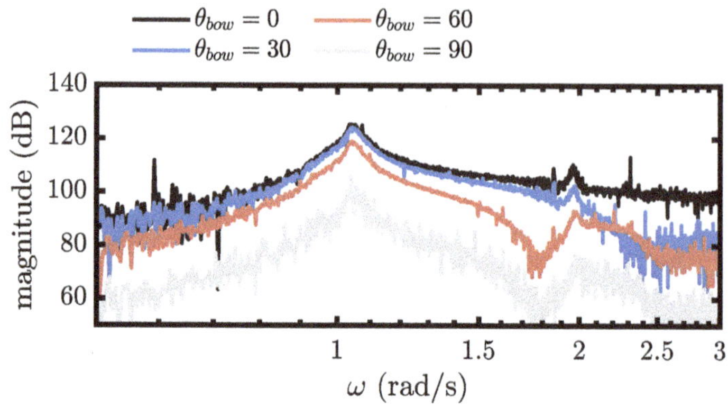

Figure 10. Identification of the $G_{\eta 2 f_e}$ map for each direction.

Results

A total of 122 waves are simulated by means of the OF-based proposed tool.

Harvested mechanical energy is evaluated within the proposed case of study. The average mechanical power is calculated for a single wave condition as the mean of the instantaneous power, i.e.,

$$\mu_p = \frac{1}{N_k} \sum_{k=1}^{N_k} p_k(t), \tag{13}$$

in which $p : \mathbb{R} \to \mathbb{R}, t \mapsto p(t)$ is the instantaneous mechanical power evaluated as $p(t) = \dot{\varepsilon}(t) f_{\mathrm{ctrl}}(t)$. Consequently, the device power matrix, for each direction, is constructed by averaging the mechanical performance across all scenarios. The resultant power matrices are presented in Figure 11. Furthermore, Figure 12 illustrates the device response under one of the most energetic waves in the 0-degree scatter.

Once the device power matrices are available, the assessment of the overall productivity can be conducted by incorporating wave occurrences. This is achieved by multiplying the power matrix by the occurrences matrices, specifically for each direction:

$$\Gamma = N_{\mathrm{hours}} \sum_{ii} \sum_{jj} \mu_{p,ii,jj} O_{w,ii,jj}, \tag{14}$$

where $N_{\mathrm{hours}} \in \mathbb{N}$ represents the total number of hours in a year. The productivity is reported in Table 2.

Therefore, the productivity of the device on the Pantelleria site results in being equal to 46.8 $\frac{\mathrm{MWh}}{\mathrm{year}}$.

Omnidirectional Analysis

The omnidirectional productivity is evaluated to quantify the energy loss due to the monodirectional nature of the PeWEC. In this evaluation, it is assumed that all the waves at the site are directed towards the bow of the device. This allows for the definition of the omnidirectional scatter, which represents the overall performance of the device under these conditions.

Please note that the omnidirectional scatter is equivalent to the $\theta_{\mathrm{bow}} = 0$ scatter in terms of the wave conditions. This means that the power matrix of the device is exactly the same in both cases, and the resulting productivity needs to be multiplied by the occurrences from the omnidirectional scatter.

The resulting productivity for the omnidirectional analysis results in being

$$\Gamma_{omni} = 70.3 \frac{\mathrm{MWh}}{\mathrm{year}}. \tag{15}$$

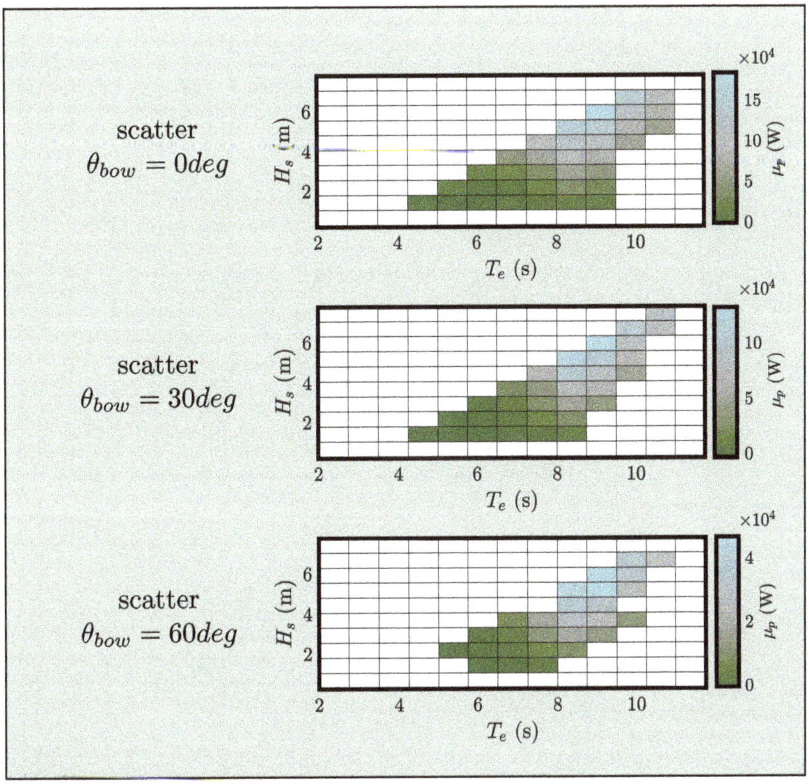

Figure 11. Directional power matrices for the moored PeWEC device.

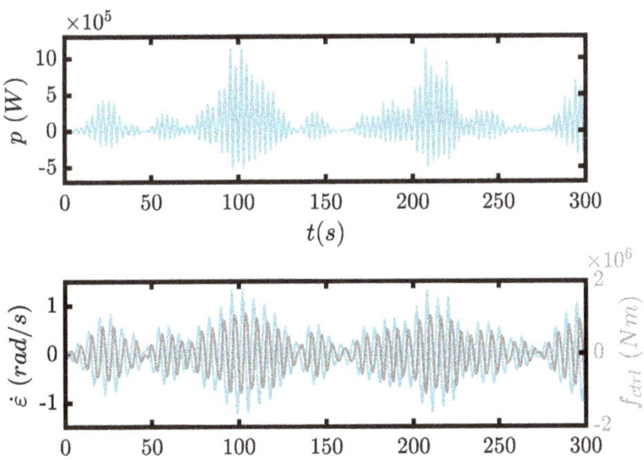

Figure 12. Device response when subject to a panchromatic wave condition, with the instantaneous power displayed at the top, and the PTO velocity $\dot{\varepsilon}$ (in blue) and control force f_{ctrl} (in grey) at the bottom. The wave spectrum is characterised by the following properties: $T_p = 6.5$ s, $H_s = 2.375$ m, and dir = 0 deg.

Table 2. Device productivity.

θ_{bow} (°)	$\Gamma\ (\frac{MWh}{year})$
0	31.7
30	11.4
60	3.7

The analysis of the omnidirectional productivity reveals that the PeWEC overall energy extraction capability is reduced by 34% due to the waves' directionality at the Pantelleria site. This finding highlights the significant impact that wave direction can have on the performance of the device. Despite the PeWEC's ability to harvest energy from stern-coming waves as well as bow-coming waves, the loss of energy due to the restricted wave approach direction is significant. For the sake of completeness, the device capture width ratio (CWR) is evaluated on the overall 0 deg scatter, exposed in Figure 13. The CWR, also called hydrodynamic efficiency [46], expresses the fraction of the wave power absorbed by the device, i.e.,

$$\mathrm{CWR} = 100\frac{\mu_p}{O_{en}B}, \tag{16}$$

in which $B = 22.5$ m represents the device width.

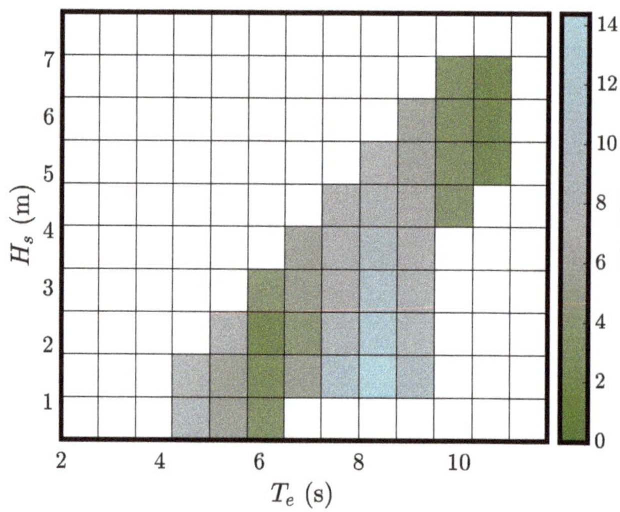

Figure 13. Device CWR 0 deg scatter.

Please note that the CWR of the PeWEC is consistent with that of other point-absorber devices [46].

6. Conclusions

Moorings influence WEC performances by affecting the overall motion and restraining the device's weathervaning capability. Although the first problem is analysed in the state of the art [20], the inclusion of the wave directionality within the control synthesis and performance assessment of a moored device still results in being a novelty. This study aims to address this problem by analysing the PeWEC representative case study, because its performances can be influenced by the waves' direction.

Site environmental conditions are pre-processed in order to define a comprehensive set of waves, defined in four directional scatter diagrams. The problem of the control synthesis is then addressed including the wave directionality and, consequently, performances of the moored PeWEC on the directional scatters are evaluated. Finally, the omnidirectional conditions are compared in terms of productivity to remark on the importance of directionality in the device performance assessment.

Although the site conditions result in having a prevalent wave direction, by neglecting the directionality problem for the PeWEC case study, device performances can be overestimated up to 50%. To synthesise, because a significant part of wave energy systems can be influenced by the wave directionality [12], such information needs to be included in the system design stage in order to provide reliable results.

While the control action synthesised using impedance-matching theory has limitations in practical applications, such as the absence of inherent constraint handling, this study emphasises the importance of assessing WEC performance comprehensively, even when analysing sites with prevalent wave directions. It is worth noting that the significant dynamics of the mooring system can be incorporated by linearising mooring actions to synthesise controllers using numerical techniques [22]. Furthermore, although the optimal control action remains unaffected by wave directions in unconstrained conditions, the inclusion of a mooring system can also influence the synthesised controller, also when considering heaving devices [21]. In general, to pave the way for the commercialisation of wave energy systems, careful consideration of the mooring system and wave directionality should be performed in the initial optimisation stages of device development. This study focuses on addressing the current challenges in wave energy system modelling to advance representative holistic optimisation strategies.

Author Contributions: Conceptualization, B.P. and G.M.; Methodology, B.P. and N.F.; Investigation, B.P.; Writing—original draft, B.P.; Writing—review & editing, B.P. and N.F.; Supervision, N.F. and G.M. All authors have read and agreed to the published version of the manuscript.

Funding: This research received no external funding.

Data Availability Statement: Not applicable.

Conflicts of Interest: The authors declare no conflict of interest.

Appendix A

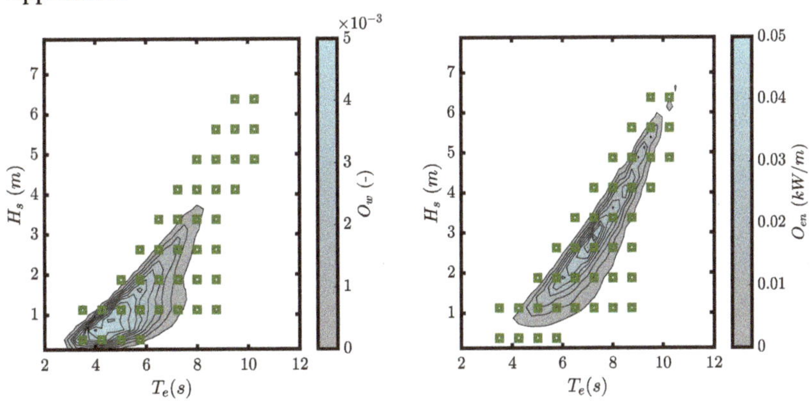

Figure A1. Wave directional scatter for the Pantelleria site, Italy. Occurrences scatter on the left, and energetic scatter on the right. The scatter is defined according to the wave directionality of 0° to the device's bow, and the chosen waves are represented with green square markers.

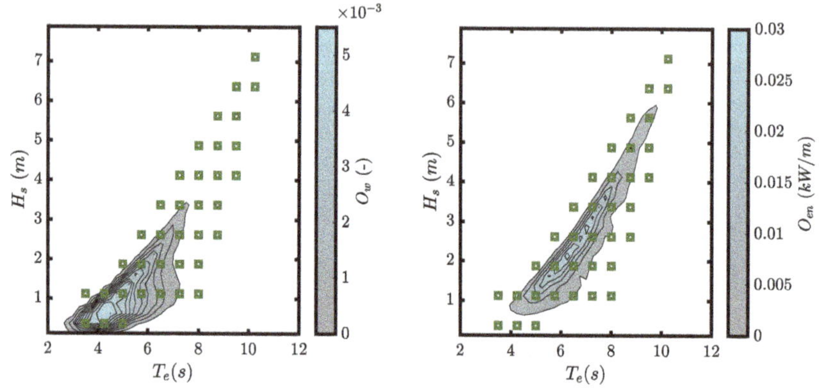

Figure A2. Wave directional scatter for the Pantelleria site, Italy. Occurrences scatter on the left, and energetic scatter on the right. The scatter is defined according to the wave directionality of 30° to the device's bow, and the chosen waves are represented with green square markers.

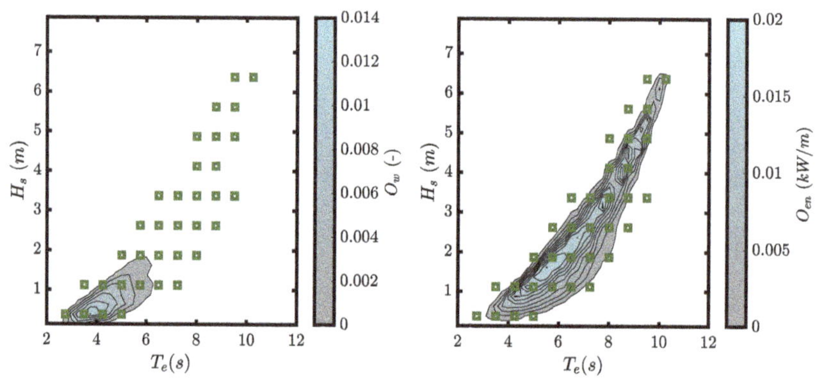

Figure A3. Wave directional scatter for the Pantelleria site, Italy. Occurrences scatter on the left, and energetic scatter on the right. The scatter is defined according to the wave directionality of 60° ± 15° according to the device's bow, and the chosen waves are represented with green square markers.

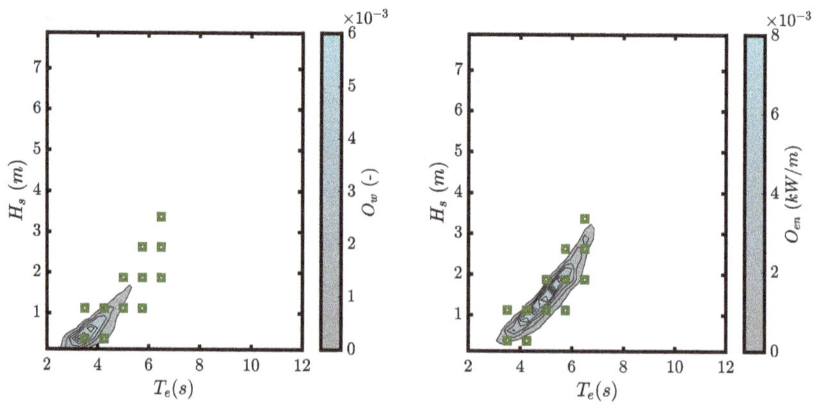

Figure A4. Wave directional scatter for the Pantelleria site, Italy. Occurrences scatter on the left, and energetic scatter on the right. The scatter is defined according to the wave directionality of 90° to the device's bow, and the chosen waves are represented with green square markers.

Notes

1 The PeWEC takes advantage of a double symmetry. It is able to harvest energy from stern-coming waves, such as the bow-coming ones.

2 Please note that cutting off 2% of the total energy and 5% of the total occurrences halves the number of waves.

References

1. Thorpe, T.W. A Brief Review of Wave Energy Harwell Laboratory, Energy Technology Support Unit, 1999. Available online: https://books.google.com/books/about/A_Brief_Review_of_Wave_Energy.html?id=9NeZswEACAAJ (accessed on 11 October 2023).
2. Ilyas, A.; Kashif, S.A.; Saqib, M.A.; Asad, M.M. Wave electrical energy systems: Implementation, challenges and environmental issues. *Renew. Sustain. Energy Rev.* **2014**, *40*, 260–268. [CrossRef]
3. Aderinto, T.; Li, H. Ocean Wave energy converters: Status and challenges. *Energies* **2018**, *11*, 1250. [CrossRef]
4. Sasaki, K. Ocean wave energy: current status and future perspectives. *J. Soc. Mech. Eng.* **2008**, *45*, 45–6810. [CrossRef]
5. Cruz, J. *Ocean Wave Energy: Current Status and Future Prespectives*; Springer: Berlin/Heidelberg, Germany, 2008. [CrossRef]
6. Lehmann, M.; Karimpour, F.; Goudey, C.A.; Jacobson, P.T.; Alam, M.R. Ocean wave energy in the United States: Current status and future perspectives. *Renew. Sustain. Energy Rev.* **2017**, *74*, 1300–1313. [CrossRef]
7. Mustapa, M.A.; Yaakob, O.B.; Ahmed, Y.M.; Rheem, C.K.; Koh, K.K.; Adnan, F.A. Wave energy device and breakwater integration: A review. *Renew. Sustain. Energy Rev.* **2017**, *77*, 43–58. [CrossRef]
8. Wahyudie, A.; Jama, M.A.; Susilo, T.B.; Saeed, O.; Nandar, C.S.; Harib, K. Simple bottom-up hierarchical control strategy for heaving wave energy converters. *Int. J. Electr. Power Energy Syst.* **2017**, *87*, 211–221. [CrossRef]
9. Mørk, G.; Barstow, S.; Kabuth, A.; Pontes, M.T. Assessing the global wave energy potential. *Proc. Int. Conf. Offshore Mech. Arct. Eng.-OMAE* **2010**, *3*, 447–454. [CrossRef]
10. European Commission. Energy Rodmap 2050, Impact Assessment and Scenario Analysis. Brussels, 2011. Available online: https://energy.ec.europa.eu/system/files/2014-10/roadmap2050_ia_20120430_en_0.pdf (accessed on 11 October 2023).
11. World Meteorological Organization. *Guide to Wave Analysis and Forecasting*; Secretariat of the World Meteorological Organization: Geneva, Switzerland, 1988; ISBN 92-63-12702-6.
12. Czech, B.; Bauer, P. Wave Energy Converter Concepts: Design Challenges and Classification. *IEEE Ind. Electron. Mag.* **2012**, *6*, 4–16. [CrossRef]
13. Faedo, N. Optimal Control and Model Reduction for Wave Energy Systems: A Moment-Based Approach. Ph.D. Thesis, Maynooth University, Maynooth, Ireland, 2020. Available online: https://mural.maynoothuniversity.ie/13541/ (accessed on 11 October 2023).
14. Pasta, E.; Faedo, N.; Mattiazzo, G.; Ringwood, J. Towards data-driven and data-based control of wave energy systems: Classification, overview, and critical assessment. *Renew. Sustain. Energy Rev.* **2023**, 188, 113877. [CrossRef]
15. Bhinder, M.; Karimirad, M.; Weller, S.; Debruyne, Y.; Guerinel, M.; Sheng, W. Modelling mooring line non-linearities (material and geometric effects) for a wave energy converter using AQWA, SIMA and Orcaflex. In Proceedings of the 11th European Wave and Tidal Energy Conference (EWTEC 2015), Nantes, France, 6–11 September 2015. Available online: https://tethys-engineering.pnnl.gov/publications/modelling-mooring-line-non-linearities-material-geometric-effects-wave-energy (accessed on 11 October 2023).
16. Depalo, F.; Wang, S.; Xu, S.; Soares, C.G. Design and analysis of a mooring system for a wave energy converter. *J. Mar. Sci. Eng.* **2021**, *9*, 782. [CrossRef]
17. Mao, Y.; Wang, T.; Duan, M. A DNN-based approach to predict dynamic mooring tensions for semi-submersible platform under a mooring line failure condition. *Ocean. Eng.* **2022**, *266*, 112767. [CrossRef]
18. Wei, H.; Xiao, L.; Liu, M.; Kou, Y. Data-driven model and key features based on supervised learning for truncation design of mooring and riser system. *Ocean. Eng.* **2021**, *224*, 108743. [CrossRef]
19. Paduano, B.; Pasta, E.; Papini, G.; Carapellese, F.; Bracco, G. Mooring Influence on the Productivity of a Pitching Wave Energy Converter. In Proceedings of the OCEANS 2021, San Diego, CA, USA, 20–23 September 2021 ; pp. 1–6. [CrossRef]
20. Paduano, B.; Carapellese, F.; Pasta, E.; Sergej, S.; Faedo, N.; Mattiazzo, G. Data-based control synthesis and performance assessment for moored wave energy conversion systems: The PeWEC case. *IEEE Trans. Sustain. Energy* **2023**, 1–12. [CrossRef]
21. Paduano, B.; Pasta, E.; Carapellese, F.; Papini, G.; Baltazar, J.; Faedo, N.; Mattiazzo, G. Control co-design mooring optimisation for wave energy systems: a three-tethered point absorber case. In Proceedings of the IFAC World Congress, Yokohama, Japan, 9–14 July 2023.
22. Papini, G.; Paduano, B.; Pasta, E.; Carapellese, F.; Mattiazzo, G.; Faedo, N. On the influence of mooring systems in optimal predictive control for wave energy converters. *Renew. Energy* **2023**, *218*, 119242. [CrossRef]
23. Carapellese, F.; Pasta, E.; Faedo, N.; Giorgi, G. Dynamic analysis and performance assessment of the Inertial Sea Wave Energy Converter (ISWEC) device via harmonic balance. *IFAC-PapersOnLine* **2022**, *55*, 439–444. [CrossRef]
24. Rusu, L.; Onea, F. Assessment of the performances of various wave energy converters along the European continental coasts. *Energy* **2015**, *82*, 889–904. [CrossRef]
25. Sirigu, S.A.; Foglietta, L.; Giorgi, G.; Bonfanti, M.; Cervelli, G.; Bracco, G.; Mattiazzo, G. Techno-Economic Optimisation for a Wave Energy Converter via Genetic Algorithm. *J. Mar. Sci. Eng.* **2020**, *8*, 482. [CrossRef]

26. Harris, R.; Johanning, L.; Wolfram, J. Mooring systems for wave energy converters: A review of design issues and choices. *PRoceedings Inst. Mech. Eng. Part B J. Eng. Manuf.* **2006**, *220*, 159–168.
27. Qiao, D.; Haider, R.; Yan, J.; Ning, D.; Li, B. Review of wave energy converter and design of mooring system. *Sustainability* **2020**, *12*, 8251. [CrossRef]
28. Bonfanti, M.; Giorgi, G. Improving Computational Efficiency in WEC Design: Spectral-Domain Modelling in Techno-Economic Optimization. *J. Mar. Sci. Eng.* **2022**, *10*, 1468. [CrossRef]
29. Paduano, B.; Carapellese, F.; Pasta, E.; Faedo, N.; Mattiazzo, G. Optimal controller tuning for a nonlinear moored wave energy converter via non-parametric frequency-domain techniques. In *Trends in Renewable Energies Offshore*, 1st ed.; CRC Press: Boca Raton, FL, USA, 2022; 8p. [CrossRef]
30. Journee, J.M.J.; Massie, W.W. *Offshore Hydromechanics*, 1st ed.; Delft University of Technology: Delft, The Netherland, 2001.
31. Cummins, W.E. The Impulse Response Function and Ship Motions. *Schiffstechnik* **1962**, *9*, 101–109.
32. Orcina. *OrcaFlex Manual*; Orcina Ltd.: Ulverston, UK, 2020. Available online: https://www.orcina.com/webhelp/OrcaFlex/ (accessed on 11 October 2023).
33. Davidson, J.; Ringwood, J.V. Mathematical modelling of mooring systems for wave energy converters—A review. *Energies* **2017**, *10*, 666. [CrossRef]
34. Faltinsen, O. *Sea Loads on Ships and Offshore Structures*; Cambridge University Press: Cambridge, UK, 1993.
35. Pozzi, N.; Bonfanti, M.; Mattiazzo, G. Mathematical Modeling and Scaling of the Friction Losses of a Mechanical Gyroscope. *Int. J. Appl. Mech.* **2018**, *10*. [CrossRef]
36. Pozzi, N.; Bracco, G.; Passione, B.; Sirigu, S.A.; Mattiazzo, G. PeWEC: Experimental validation of wave to PTO numerical model. *Ocean. Eng.* **2018**, *167*, 114–129. [CrossRef]
37. Penalba, M.; Giorgi, G.; Ringwood, J.V. Mathematical modelling of wave energy converters: a review of nonlinear approaches. *Renew. Sustain. Energy Rev.* **2017**, *78*, 1188–1207. [CrossRef]
38. Niosi, F.; Begovic, E.; Bertorello, C.; Rinauro, B.; Sannino, G.; Bonfanti, M.; Sirigu, S.A. Experimental validation of Orcaflex-based numerical models for the PEWEC device. *Ocean. Eng.* **2023**, *281*, 114963. [CrossRef]
39. Faedo, N.; Carapellese, F.; Pasta, E.; Mattiazzo, G. On the principle of impedance-matching for underactuated wave energy harvesting systems. *Appl. Ocean. Res.* **2022**, *118*, 102958. [CrossRef]
40. Carapellese, F.; Pasta, E.; Paduano, B.; Faedo, N.; Mattiazzo, G. Intuitive LTI energy-maximising control for multi-degree of freedom wave energy converters: the PeWEC case. *Ocean. Eng.* **2022**, 256, 111444. [CrossRef]
41. Paduano, B.; Edoardo, P.; Faedo, N.; Mattiazzo, G. Control synthesis via Impedance-Matching in panchromatic conditions: A generalised framework for moored systems. In Proceedings of the European Wave and Tidal Energy Conference, Bilbao, Spain, 3–7 September 2023; Volume 15. [CrossRef]
42. Floyd, T.L.; Pownell, E. *Principles of Electric Circuits: Conventional Current*; Pearson Education, 2021. Available online: https://books.google.com/books?id=KskQzgEACAAJ (accessed on 11 October 2023).
43. Pintelon, R.; Schoukens, J. *System Identification: A Frequency Domain Approach*, 2nd ed.; Wiley: Hoboken, NJ, USA, 2012; pp. 1–787.
44. Faedo, N.; Peña-Sanchez, Y.; Carapellese, F.; Mattiazzo, G.; Ringwood, J.V. LMI-based passivation of LTI systems with application to marine structures. *IET Renew. Power Gener.* **2021**, *15*, 3424–3433. [CrossRef]
45. Hersbach, H.; Bell, B.; Berrisford, P.; Hirahara, S.; Horányi, A.; Muñoz-Sabater, J.; Nicolas, J.; Peubey, C.; Radu, R.; Schepers, D.; et al. The ERA5 global reanalysis. *Q. J. R. Meteorol. Soc.* **2020**, *146*, 1999–2049. [CrossRef]
46. Babarit, A. A database of capture width ratio of wave energy converters. *Renew. Energy* **2015**, *80*, 610–628. [CrossRef]

Journal of
Marine Science and Engineering

MDPI

Article

Comparison of Advanced Control Strategies Applied to a Multiple-Degrees-of-Freedom Wave Energy Converter: Nonlinear Model Predictive Controller versus Reinforcement Learning

Ali S. Haider [1,*], Kush Bubbar [1] and Alan McCall [2]

[1] System-Model Development Engineering Lab, University of New Brunswick, Fredericton, NB E3B 5A3, Canada; kush.bubbar@unb.ca
[2] Dehlsen Associates, LLC, Santa Barbra, CA 93101, USA; amccall@ecomerittech.com
* Correspondence: ali.haider@unb.ca

Abstract: Achieving energy maximizing control of a Wave Energy Converter (WEC) not only needs a comprehensive dynamic model of the system—including nonlinear hydrodynamic effects and nonlinear characteristics of Power Take-Off (PTO)—but to treat the entire system using an integrated approach, i.e., as a cyber–physical system considering the WEC dynamics, control strategy, and communication interface. The resulting energy-maximizing optimization formulation leads to a non-quadratic and nonstandard cost function. This article compares the (1) Nonlinear Model Predictive Controller (NMPC) and (2) Reinforcement Learning (RL) techniques as applied to a class of multiple-degrees-of-freedom nonlinear WEC–PTO systems subjected to linear as well as nonlinear hydrodynamic conditions in simulation, using the WEC-Sim™ toolbox. The results show that with an optimal choice of RL agent and hyperparameters, as well as suitable training conditions, the RL algorithm is more robust under more stringent operating requirements, for which the NMPC algorithm fails to converge. Further, RL agents are computationally efficient on real-time target machines with a significantly reduced Task Execution Time (TET).

Keywords: energy maximizing control; nonlinear model predictive control; cyber–physical modeling; wave energy converter; reinforcement learning; nonlinear viscous drag; non-ideal power take-off

Citation: Haider, A.S.; Bubbar, K.; McCall, A. Comparison of Advanced Control Strategies Applied to a Multiple-Degrees-of-Freedom Wave Energy Converter: Nonlinear Model Predictive Controller versus Reinforcement Learning. *J. Mar. Sci. Eng.* **2023**, *11*, 2120. https://doi.org/10.3390/jmse11112120

Academic Editors: Giuseppe Giorgi and Mauro Bonfanti

Received: 2 October 2023
Revised: 30 October 2023
Accepted: 1 November 2023
Published: 6 November 2023

1. Introduction

Renewable energy technologies offer a feasible, sustainable, and green solution to increasing global energy needs, and the ocean offers an immense, untapped resource of energy with the potential to become an integral part of the world's energy mix [1,2]. The prospect of ocean wave energy has triggered researchers to explore techniques to maximize energy capture [3] for wave energy converters under operating conditions deviating from ideality, to include practical PTO system constraints [4] and the nonlinear hydrodynamics effects of ocean waves. Energy maximization for a WEC system is in practice a multi-objective optimization problem, requiring considerations of the physical geometry of the WEC, the PTO system design, the mooring system design, the ocean conditions of the deployment site, the communication interface, and the control methodology.

On the control front, Model Predictive Control (MPC) yields superior overall system performance for wave energy converters because it optimizes energy capture while enforcing the electro-mechanical operating limits of the system [5]. MPC is a constrained online optimal control strategy that forecasts future trajectories of the system dynamics to solve an optimization program over a receding horizon window and determine the best instantaneous control action to maximize the output power of the WEC. The MPC algorithm uses an internal model of the plant to predict the system's future states. However, WEC systems are increasingly growing in complexity [6], and there is a need for the control algorithm to handle the resulting non-ideal operating conditions. MPC algorithms suffer convergence issues under stringent non-ideal operating conditions, due to the limitations of the complex online optimization algorithm. These convergence issues become more prominent as the complexity of the optimization problem increases, due to the inclusion of multiple-DoF PTO mechanisms. The performance of the MPC algorithm is also vulnerable when incorporating nonlinearities such as viscous drag effects and nonlinear hydrodynamic forces.

Moreover, the MPC controller typically does not consider cyber-related issues, such as communication latency and packet loss between the real-time target machine that implements the controller and the WEC hardware. These factors contribute not only to the degree of optimality of the MPC solution but also to the degree of convergence of the optimal control problem. There is always an intrinsic limitation of the mathematical model of a WEC when simulating a real-world system, and if the internal plant prediction is too simple, the MPC optimization algorithm generates a poor solution under non-ideal conditions and may even become unstable.

Reinforcement Learning (RL) is a data-driven, goal-oriented, computational technique. In the RL approach, a computer interacts with a given unknown dynamic system through the RL inputs (i.e., observations and reward) and RL outputs (i.e., actions). During these interactions, the RL approach trains an agent to perform a task based on a reward from the environment [7]. Given a suitable training environment and RL agent structure, the agent can be trained for any practical environment. For a given energy-maximizing problem for a WEC, if the training environment includes effects such as non-ideal PTO behavior, communications interface latencies, and nonlinear hydrodynamic responses, then the trained RL agent learns to maximize the reward (i.e., optimization objective) in the presence of these effects. The adaptability of the RL approach to a given environment has led to an uptake in the usage of this technique for energy-maximizing problems for WEC systems. For example, an RL approach based on the Q-learning approach is presented to maximize the energy extraction in regular and irregular sea states for a point-absorber-type wave energy converter in [8], where the controller damping and stiffness are adjusted based on a reward function. Resistive control of a realistic WEC model using an RL approach based on a least-squares policy iteration is presented in [9]. A nonlinear reactive control strategy for a two-body point-absorber wave energy converter using the Q actor–critic learning method is presented for a two-body 1-DoF point absorber in [10]. A deep-RL-agent-based real-time control is presented in [11] for a 1-DoF heaving point absorber under a linear environment and is compared with a linear MPC.

This work presents the energy-maximizing control of a 2-DoF WEC array device for the digital twin of Dehlsen Associates' three-pod CENTIPOD™ device [12]. The optimizing objective is to maximize the energy harnessed by PTO machines in heave and pitch axes, subject to the electro-mechanical constraints of the system. The objective function is a nonlinear and non-quadratic function of PTO current, heave velocity, and pitch velocity, considering the practical electric machine loss characteristics of the PTOs. Moreover, the wave energy converter model includes nonlinear hydrodynamic effects due to the quadratic drag of fluid, yielding a WEC model with nonlinear dynamics. To enact the energy-maximizing control of the WEC plant, we designed two controllers: (1) a Nonlinear MPC (NMPC) and (2) an RL-agent-based controller. For NMPC design, we extended the approach in [13] to two degrees of freedom, exploiting the technique of pseudo-quadratization using the ACADO Toolkit [14]. The WEC plant is modeled in surge–heave–pitch degrees of freedom using Cummin's equation, where the radiation force convolution terms are approximated by state-space models [15]. For array devices, more thorough energy-based modeling approaches are possible, such as the port-Hamiltonian approach [16]; however, for this study, array effects and body-to-body interactions are neglected. On the RL side, we trained a Deep Deterministic Policy Gradient (DDPG) RL agent for the heave and pitch degrees of freedom. The simulation results of (1) NMPC and (2) RL are compared under the operation of the device in linear sea conditions as well as with the nonlinear hydrodynamics effects enabled in WECSim™ [17]. The WEC digital twin is simulated on an emulator machine and interfaced with the controller/training machine over EtherCAT and Universal Datagram Port (UDP) buses.

2. Developing Time-Domain Equations of the WEC

This work is related to the investigation of the power capture performance of advanced controllers for three-pod Centipod devices made by Dehlsen Associates, LLC (the multi-pod CENTIPOD) [12]. Figure 1 shows a 35th-Froude-scaled model of the WEC. However, for this work, a full-scale WECSIM [17] based digital twin of the Centipod device is considered, as shown in Figure 2; this is an array of three floating bodies (pods) that are free to heave and pitch against reaction bodies (spars) attached to a single submerged backbone structure, which is moored with three taut lines. The backbone structure is the main contributor of reaction damping to the PTO, as well as providing a stable common junction point for multiple pod–spars mechanisms. The backbone is taut-moored to the seabed, as shown in Figure 2. The pods have linear direct-drive permanent magnetic AC generator PTO machines in the heave axis and rotatory direct-drive permanent magnetic AC generator PTO machines in the pitch axes. For this study, body-to-body radiation coupling between pods is ignored; it will be evaluated in future research.

Figure 1. 35th-scale Centipod wave energy converter by Dehlsen Associates, LLC.

As per the multi-body dynamics convention for the floating pods, subscripts "1", "3", and "5" denote surge, heave, and pitch axes, respectively. Table 1 lists the variables and their descriptions, which are used in WEC dynamics.

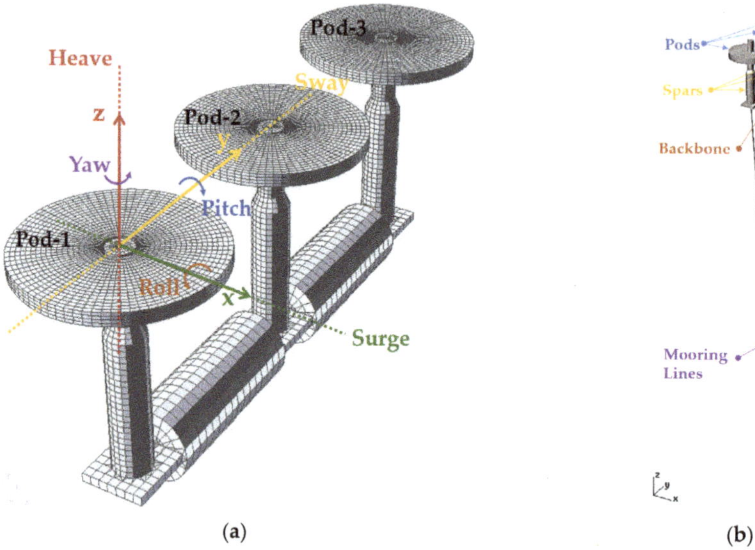

Figure 2. Degrees of freedom for dynamic modeling of Centipod WEC: (**a**) baseline configuration and (**b**) model with mooring lines.

Table 1. Nomenclature for WEC dynamics.

Symbol	Unit	Description
$v_i(t)$	m/s, rad/s	Generalized velocity
$x_i(t)$	m, rad	Generalized displacement
$\xi_i(t)$	–	Auxiliary state variable for radiation force dynamics
$F_{r,pq}(t)$	N, Nm	Force from wave radiation in p axis due to velocity in q axis
$F_{hs,i}(t)$	N, Nm	Buoyancy restoring force
$F_{v,i}(t)$	N, Nm	Fluid damping force
$F_{e,i}(t)$	N, Nm	Force due to wave excitation
$F_{p,i}(t)$	N, Nm	Power take-off actuation force
m	Kg	Pod mass
$A_{pq}(\infty)$	Kg, Kg m, Kg m^2	Infinite frequency generalized radiation added mass in p axis due to acceleration in q axis.
C_i	N/m, Nm/rad	Buoyancy restoring constant
$C_{vd,i}$	N/(m/s)^2, Nm/(rad/s)^2	Fluid quadratic damping constant
$A_{qp}(\omega)$	Kg, Kg m, Kg m^2	Added mass due to wave radiation in p axis due to acceleration in q axis
$B_{qp}(\omega)$	N/m/s, Nm/rad/s	Radiation damping due to wave radiation in p axis due to velocity in q axis
$K_{pq}(t)$	N/m, Nm/rad	Impulse response function for wave radiation
$Z_{qp}(\omega)$	N/m/s, Nm/rad/s	Mechanical impedance
g	m/s^2	Gravity constant
ρ	Kg/m^3	Water density

2.1. Dynamic Model of WEC in Surge, Heave, and Pitch Axes

The orientation of the Centipod device in Figure 2 with respect to incoming waves in the surge direction results in negligible roll, sway, and yaw displacements of the pods; hence, it is adequate to consider the surge–pitch–heave model of each pod for energy capture considerations. The floating pods in Figure 2 are modeled as point-absorber bodies. Heave motion is very weakly coupled to surge and pitch; hence, this coupling effect can be ignored. In the local frame of reference, the Cummins equations for the three axes of freedom (surge–pitch–heave) are

$$M_{11}\dot{v}_1 + A_{15}(\infty)\dot{v}_5 = -F_{r,11}(t) - F_{r,15}(t) - F_{v,1}(t) + F_{e,1}(t), \tag{1}$$

$$M_{33}\dot{v}_3(t) = -F_{r,33}(t) - F_{hs,3}(t) - F_{v,3}(t) - F_{p,3}(t) + F_{e,3}(t), \tag{2}$$

$$M_{55}\dot{v}_5 + A_{51}(\infty)\dot{v}_1 = -F_{r,55}(t) - F_{r,51}(t) - F_{v,5}(t) - F_{hs,5}(t) - F_{p,5}(t) + F_{e,5}(t) \tag{3}$$

Here, $M_{ii} = (m + A_{ii}(\infty))$. The radiation force, buoyancy restoring force, and fluid quadratic damping terms in (1) through (3), respectively, are given by

$$F_{r,ij}(t) = \int_{-\infty}^{t} K_{ij}(t-\tau)v_j d\tau, \tag{4}$$

$$F_{hs,i}(t) = C_i x_i, \tag{5}$$

$$F_{v,i}(t) = C_{d,i} v_i |v_i|. \tag{6}$$

The time-domain convolution integral term in (4) can be transformed into the frequency-domain expression $Z_{pq}(j\omega)V_q(j\omega)$ through the application of the Fourier transform. The frequency-domain hydrodynamic parameters for the Centipod hull geometry without the mooring system are determined using the WAMIT™ (Version 7.201-x64) software package [18]. A single-body WEC intrinsic impedance $Z_{pq}(j\omega)$ [19] is calculated using these hydrodynamic parameters plotted in Figures 3–5. A minimal-order transfer function in Laplace space is approximated for $Z_{pq}(j\omega)$ using system identification techniques, and an equivalent state-space representation can be formulated as in [15,20]. After performing algebraic manipulations, the final state-space model of the plant developed in [20] is given by

$$\dot{\mathrm{X}} = \mathrm{AX} + \mathrm{B_p F_p} + \mathrm{B_v F_v} + \mathrm{B_e F_e}, \tag{7}$$

where X is the state vector and

$$\mathrm{F_p} = \begin{bmatrix} F_{p,5} & F_{p,3} \end{bmatrix}^T \tag{8}$$

$$\mathrm{F_v} = \begin{bmatrix} F_{v,1} & F_{v,5} & F_{v,3} \end{bmatrix}^T \tag{9}$$

$$\mathrm{F_e} = \begin{bmatrix} F_{e,1} & F_{e,5} & F_{e,3} \end{bmatrix}^T \tag{10}$$

The state matrix and input matrices in (7) are given by (11) and (12) with appropriate systems parameter constants m_{ij}, a_i, and b_i.

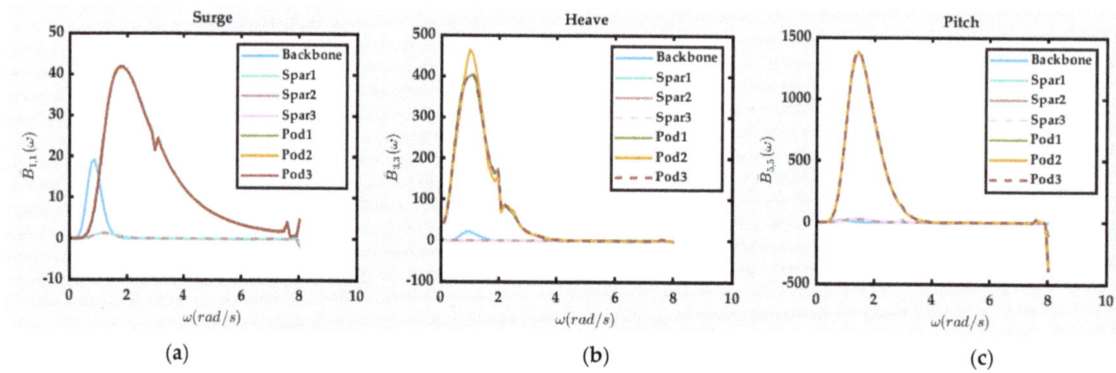

Figure 3. Normalized radiation damping $B/\omega\rho$ of the Centipod WEC: (**a**) surge axis, (**b**) heave axis, and (**c**) pitch axis.

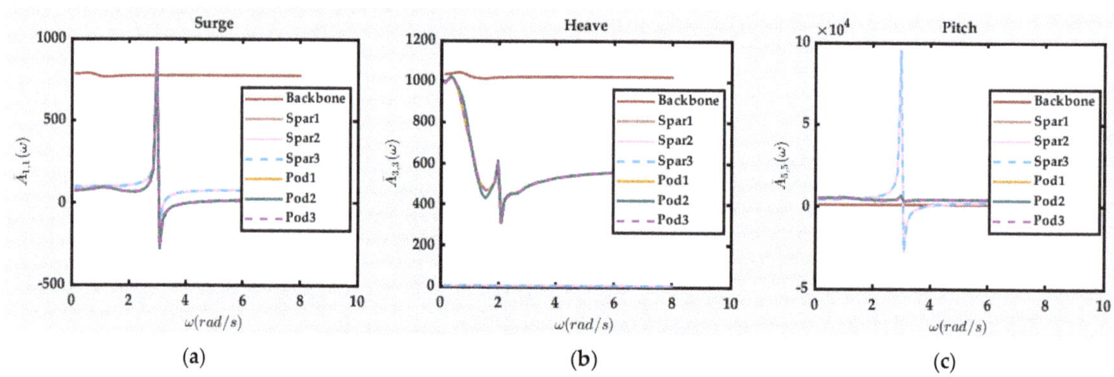

Figure 4. Normalized added mass A/ρ of the Centipod WEC: (**a**) surge axis, (**b**) heave axis, and (**c**) pitch axis.

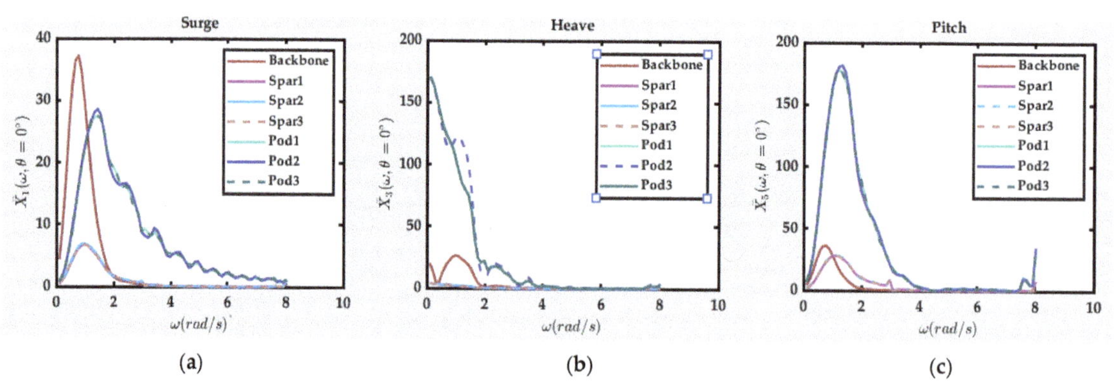

Figure 5. Normalized excitation amplitude $X/\rho g$ of the Centipod WEC: (**a**) surge axis, (**b**) heave axis, and (**c**) pitch axis.

2.2. Non-Ideal Power Take-Off Model

The power take-off machine for the heave axis is a Linear Universal Modular Actuator/Absorber (LUMA) machine [21]. For the pitch-axis, PTO comprises a direct-drive permanent magnet AC generator. The non-ideal power take-off model is taken from the case study in [20], where the PTO power capture is a function of PTO force and velocity with system parameter constants c_i given by (13).

$$
A=\begin{bmatrix}
\begin{matrix}
0 & 0 & -m_{15}C_5 & -m_{11} & 0 & -m_{11} & 0 & -m_{15} & 0 & -m_{15} & 0\\
0 & 0 & -m_{55}C_5 & -m_{51} & 0 & -m_{51} & 0 & -m_{55} & 0 & -m_{55} & 0\\
0 & 1 & 0 & 0 & 0 & 0 & 0 & 0 & 0 & 0 & 0\\
b_3 & 0 & 0 & 0 & 1 & 0 & 0 & 0 & 0 & 0 & 0\\
b_4 & 0 & 0 & a_3 & a_4 & 0 & 0 & 0 & 0 & 0 & 0\\
0 & b_5 & 0 & 0 & 0 & 0 & 1 & 0 & 0 & 0 & 0\\
0 & b_6 & 0 & 0 & 0 & a_5 & a_6 & 0 & 0 & 0 & 0\\
0 & b_7 & 0 & 0 & 0 & 0 & 0 & 0 & 1 & 0 & 0\\
0 & b_8 & 0 & 0 & 0 & 0 & 0 & a_7 & a_8 & 0 & 0\\
b_9 & 0 & 0 & 0 & 0 & 0 & 0 & 0 & 0 & 0 & 1\\
b_{10} & 0 & 0 & 0 & 0 & 0 & 0 & 0 & 0 & a_9 & a_{10}
\end{matrix} & 0_{11\times 4}\\
0_{4\times 11} & \begin{matrix}
0 & \frac{-C_3}{M_{33}} & \frac{-1}{M_{33}} & 0\\
1 & 0 & 0 & 0\\
b_1 & 0 & 0 & 1\\
b_2 & 0 & a_1 & a_2
\end{matrix}
\end{bmatrix}
$$

$$
=\begin{bmatrix}
0 & 0 & 1.4 & -3.8e^{-6} & 0 & -3.8e^{-6} & 0 & 6.7e^{-8} & 0 & 6.7e^{-8} & 0 & 0 & 0 & 0 & 0\\
0 & 0 & -2.8 & 6.7e^{-8} & 0 & 6.7e^{-8} & 0 & -1.3e^{-7} & 0 & -1.3e^{-7} & 0 & 0 & 0 & 0 & 0\\
0 & 1 & 0 & 0 & 0 & 0 & 0 & 0 & 0 & 0 & 0 & 0 & 0 & 0 & 0\\
2.2e^{5} & 0 & 0 & 0 & 1 & 0 & 0 & 0 & 0 & 0 & 0 & 0 & 0 & 0 & 0\\
-4.7e^{5} & 0 & 0 & -5.3 & -2.2 & 0 & 0 & 0 & 0 & 0 & 0 & 0 & 0 & 0 & 0\\
0 & 5.4e^{5} & 0 & 0 & 0 & 0 & 1 & 0 & 0 & 0 & 0 & 0 & 0 & 0 & 0\\
0 & -7e^{5} & 0 & 0 & 0 & -3.3 & -1.3 & 0 & 0 & 0 & 0 & 0 & 0 & 0 & 0\\
0 & 2.4e^{6} & 0 & 0 & 0 & 0 & 0 & 0 & 1 & 0 & 0 & 0 & 0 & 0 & 0\\
0 & -2.4e^{6} & 0 & 0 & 0 & 0 & 0 & -2.6 & -1.1 & 0 & 0 & 0 & 0 & 0 & 0\\
5.4e^{5} & 0 & 0 & 0 & 0 & 0 & 0 & 0 & 0 & 0 & 1 & 0 & 0 & 0 & 0\\
-6.9e^{5} & 0 & 0 & 0 & 0 & 0 & 0 & 0 & 0 & -3.3 & -1.3 & 0 & 0 & 0 & 0\\
0 & 0 & 0 & 0 & 0 & 0 & 0 & 0 & 0 & 0 & 0 & 0 & -2.1 & -1.2e^{-6} & 0\\
0 & 0 & 0 & 0 & 0 & 0 & 0 & 0 & 0 & 0 & 0 & 1 & 0 & 0 & 0\\
0 & 0 & 0 & 0 & 0 & 0 & 0 & 0 & 0 & 0 & 0 & 6.3e^{5} & 0 & 0 & 1\\
0 & 0 & 0 & 0 & 0 & 0 & 0 & 0 & 0 & 0 & 0 & -8.9e^{5} & 0 & -1.5 & -1.5
\end{bmatrix} \tag{11}
$$

$$
B_p=\begin{bmatrix}
-m_{15} & 0\\
-m_{55} & 0\\
0 & 0\\
0 & 0\\
0 & 0\\
0 & 0\\
0 & 0\\
0 & 0\\
0 & 0\\
0 & 0\\
0 & 0\\
0 & \frac{-1}{M_{33}}\\
0 & 0\\
0 & 0\\
0 & 0
\end{bmatrix}=\begin{bmatrix}
6.7e^{-8} & 0\\
-1.3e^{-7} & 0\\
0 & 0\\
0 & 0\\
0 & 0\\
0 & 0\\
0 & 0\\
0 & 0\\
0 & 0\\
0 & 0\\
0 & 0\\
0 & 1.2e^{-6}\\
0 & 0\\
0 & 0\\
0 & 0
\end{bmatrix}, B_e=-B_v=\begin{bmatrix}
m_{11} & m_{15} & 0\\
m_{51} & m_{55} & 0\\
0 & 0 & 0\\
0 & 0 & 0\\
0 & 0 & 0\\
0 & 0 & 0\\
0 & 0 & 0\\
0 & 0 & 0\\
0 & 0 & 0\\
0 & 0 & 0\\
0 & 0 & 0\\
0 & 0 & \frac{1}{M_{33}}\\
0 & 0 & 0\\
0 & 0 & 0\\
0 & 0 & 0
\end{bmatrix}=\begin{bmatrix}
3.8e^{-6} & -6.7e^{-8} & 0\\
-6.8e^{-8} & 1.3e^{-7} & 0\\
0 & 0 & 0\\
0 & 0 & 0\\
0 & 0 & 0\\
0 & 0 & 0\\
0 & 0 & 0\\
0 & 0 & 0\\
0 & 0 & 0\\
0 & 0 & 0\\
0 & 0 & 0\\
0 & 0 & 1.2e^{-6}\\
0 & 0 & 0\\
0 & 0 & 0\\
0 & 0 & 0
\end{bmatrix} \tag{12}
$$

$$
P_{E,i}=c_{0,i}F_{p,i}v_i-\left(c_{1,i}F_{p,i}^6+c_{2,i}F_{p,i}^5+c_{3,i}F_{p,i}^4+c_{4,i}F_{p,i}^3+c_{5,i}F_{p,i}^2+c_{6,i}F_{p,i}+c_{7,i}\right), \tag{13}
$$

3. Nonlinear MPC Design for WEC

A given NMPC problem optimizes a manipulated variable $\mathrm{u} \subseteq \mathrm{w}$ to maximize some cost function P of a set of system variables "w" while respecting the given system constraints. A general class of NMPC problems has been formulated in [13], in which the cost function takes on a nonlinear piecewise polynomial form. Considering the case of finite-horizon optimization, we can mathematically describe the NMPC problem of such a class as (Tables 2 and 3)

$$\underset{\mathrm{u}}{max}\mathrm{P}(\mathrm{w}) = \begin{cases} \mathrm{P}_1(\mathrm{w}) + \rho_{N,1}(\mathrm{w}), & w_k < R_1 \\ \mathrm{P}_2(\mathrm{w}) + \rho_{N,2}(\mathrm{w}), & R_1 \leq w_k \leq R_2 \\ \vdots & \vdots \\ \mathrm{P}_\mathrm{j}(\mathrm{w}) + \rho_{N,j}(\mathrm{w}), & R_{j-1} \leq w_k \leq R_j \end{cases}, \quad (14)$$

Table 2. Nomenclature for nonlinear MPC formulation.

Symbol	Description
w	Set of system variables
N	Prediction horizon
$\mathrm{X} \subseteq \mathrm{w}$	State vector of WEC dynamics
$\mathrm{u} \subseteq \mathrm{w}$	Manipulated variable vector, PTO force/torque $\mathrm{F_p}(N)$
$\rho_{N,i}$	Finite-horizon terminal cost penalty
P_i	Polynomial of system variables
Ψ_i	Constant weighting matrices
$\mathrm{B_i}$	Constant column vectors
Υ_i	Column vectors of nonlinear functions of state variables
q	Column vectors of nonlinear functions of state variables
d	Excitation force disturbance vector, $\mathrm{F_e}(N)$
R_i	Some real number

Table 3. WEC system parameters.

Parameter	Value
m	2.36×10^5 Kg
$A_{11}(\infty)$	2.88×10^4 Kg
$A_{33}(\infty)$	5.71×10^5 Kg
$A_{55}(\infty)$	4.4×10^6 Kg m^2
$A_{15}(\infty)$	1.33×10^5 Kg m
$A_{51}(\infty)$	1.33×10^5 Kg m
C_3	1.69×10^6 N/m
C_5	2.12×10^7 Nm/rad
$C_{d,1}$	1.48×10^5 N/(m/s)^2
$C_{d,3}$	1.73×10^5 N/(m/s)^2
$C_{d,5}$	1.29×10^7 Nm/(rad/s)^2
M_{11}	2.65×10^5 Kg
M_{33}	8.07×10^5 Kg
M_{55}	7.53×10^6 Kg m^2
Water depth	212 m
Pod volume	400 m^3
Pod immersed volume	359 m^3
Mooring line length	192.32 m
Mooring line type	Chain
Mooring no of lines	3
Mooring line diameter	0.175 m
Mooring mass density in air	18.375 kg/m
Mooring damping ratio	0.8
Mooring stiffness	1.11×10^4 MN
Mooring transverse drag coefficient	1.6
Mooring transverse added mass coefficient	1
Mooring tangential drag coefficient	0.05
Mooring tangential added mass coefficient	0

4. RL Agent Design for 2-DoF Heave–Pitch PTOs

Similar to Section 3, we propose a method to optimize the overall electrical power captured by the 2-DoF PTO, in this case through designing an appropriate RL agent for our problem. For continuous action and observation spaces, the typical options for the candidate agents are

1. Deep Deterministic Policy Gradient (DDPG).
2. Twin-Delayed Deep Deterministic policy gradient (TD3).
3. Proximal Policy Optimization (PPO).
4. Soft Actor-Critic (SAC).

Regarding increasing complexity, the DDPG is the simplest compatible agent, followed by TD3, PPO, and SAC. TD3 is an improved, more complex version of DDPG, while PPO has more stable updates but requires more training [22]. On the other hand, SAC is an improved and more complex version of DDPG that generates stochastic policies. We utilize the DDPG for our problem, typically the first choice for problems with continuous action and observation spaces [7].

4.1. RL DDPG Agent Reward Function and Properties

DDPG-based control aims to maximize the PTO power capture while respecting the PTO velocity limits. The observation consists of the pod velocity, and the action is the PTO force. Training is performed offline for this study. To specify a reward function to train our RL DDPG, we propose using a modified version of (14) below, which includes a penalty term for the agent for exceeding the velocity limits of the PTO mechanisms:

$$Reward_i = \mathrm{k}_p \frac{1}{2} \mathbf{h}_\mathbf{i}^\mathrm{T} (2\mathbf{W}_\mathbf{i}) \mathbf{h}_\mathbf{i} - \mathrm{k}_{v_i} (|v_i| > v_{i,max}) \tag{15}$$

Here, k_p and k_{v_i} are some appropriate scaling factors. We have designed two separate RL agents for the pitch and heave control because these DoFs are decoupled. The DDPG agent options for both DoFs are given in Table 4.

Table 4. RL DDPG agent properties for heave and pitch control.

RL Agent Options	Value
Sample time	0.1
Target smooth factor	1×10^{-6}
Discount factor	0.95
Mini batch size	512
Length of experience buffer	1×10^{6}
Noise variance	0.3
Variance decay rate	1×10^{-5}
Target update frequency	1

4.2. Design of Actor and Critic Deep Networks for RL Training

The DDPG-based RL algorithm requires the critic and actor neural networks to implement the optimal policy by generating actions in response to the given observations. A critic neural network predicts the discounted value of the cumulative long-term reward by looking at the observations and actions, and an agent neural network implements the RL policy to produce actions to maximize the predicted discounted cumulative long-term reward [22]. An experience-based design choice of the deep network structures for RL actor and critic is shown in Figure 6. Based on extensive training trials, an RL-Q-value representation for the critic network is finally selected for both heave and pitch, and the other hyperparameter choice for the critic network is given in Table 5.

An RL-deterministic representation is chosen for the actor network for both heave and pitch, based on extensive training trials; the other hyperparameters chosen for the actor network are provided in Table 6.

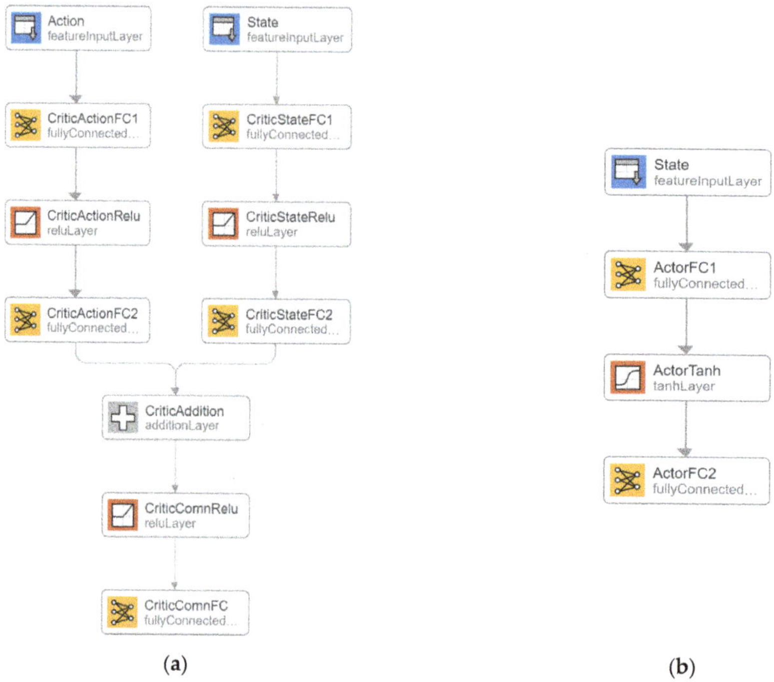

Figure 6. Deep network design for RL actor and critic in MATLAB for the heave and pitch PTOs: (**a**) deep network for the critic and (**b**) deep network for the actor.

Table 5. RL critic properties for heave and pitch control.

RL Critic Options	Value
Representation	RL Q-value
Learn rate	0.1
Gradient threshold	inf
Action feature input layer size	1
State feature input layer size	1
Action and critic fully connected layer-1 size (for unconstrained F_{pto})	64
Action and critic fully connected layer-2 size (for constrained F_{pto})	50
Critic common fully connected layer-2 (FC1) size	1

Table 6. RL actor properties for heave and pitch control.

RL Critic Options	Value
Representation	RL deterministic
Learn rate	0.1
Gradient threshold	inf
Optimizer momentum	0.95
State feature input layer size	1
Actor fully connected layer-1 size (unconstrained F_p)	32 (heave), 16 (pitch)
Actor fully connected layer-1 size (constrained F_p)	25 (heave), 16 (pitch)
Actor fully connected layer-2 size	1
Optimizer for pitch (constrained or unconstrained F_p)	Root mean square propagation (RMS-Prop)
Optimizer for heave (for constrained F_p)	Root mean square propagation (RMS-Prop)
Optimizer for heave (for unconstrained F_p)	Stochastic gradient descent with momentum (SGDM)

4.3. RL Agent Training

To train the RL agent, we generate training data by simulating a full-scale version of the 2-DoF version of Dehlsen's three-pod CENTIPOD WEC of Figure 1 in an emulator machine, by using its WEC-Sim model, as shown in Figure 7. The mean of a cluster of sea states used to execute the WEC-Sim simulation are described in Table 7, which were selected based on the geographic location and wave resources at the deployment site at PacWave [23,24]; the corresponding wave spectrum is shown in Figure 8.

Table 7. Mean of the cluster of sea states for Centipod digital twin simulation in WECSim.

Sea-State Parameter in WECSim	Value
Significant wave height [m]	2.5
Peak wave period [s]	7.35
Spectrum type of ocean waves	Pierson Moskowitz (PM)
Class of waves	Irregular

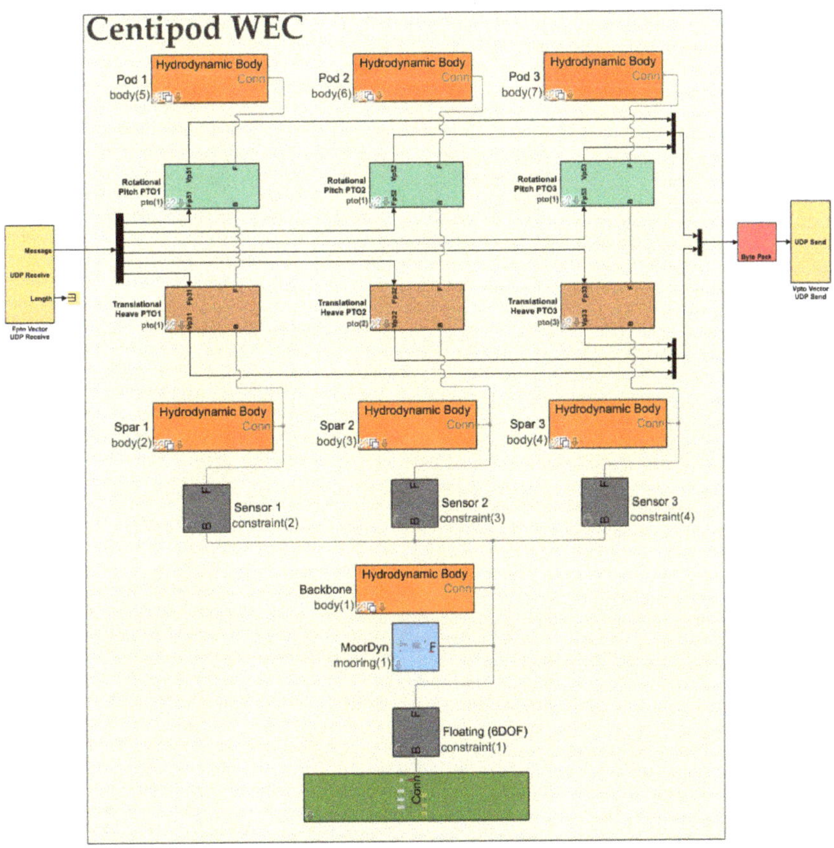

Figure 7. WECSim digital twin of 3-pod Centipod WEC with PTOs in heave and pitch axes.

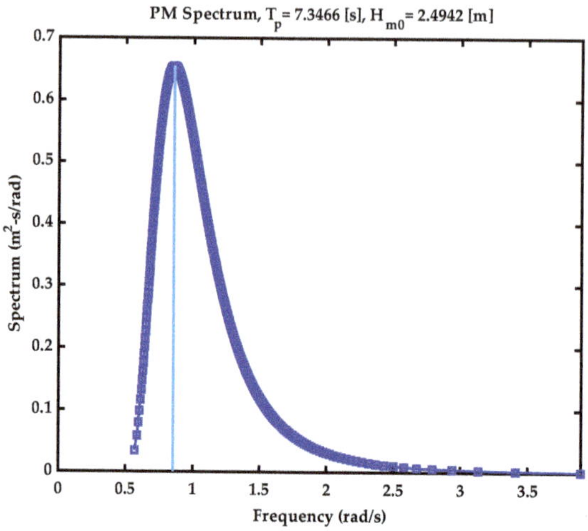

Figure 8. Wave spectrum of the mean of a cluster of sea states for WEC-Sim simulation.

The strategy to train the RL agent is shown schematically in Figure 9. An emulator machine simulates the real-time digital twin of the plant in Simulink/WEC-Sim., which is connected to another real-time controller target machine via an Ethernet/Universal Datagram Packet (UDP) link which runs the RL training algorithm in MATLAB/Simulink. A MATLAB script to train the RL algorithm in the controller machine establishes the connection between the two machines for each training episode and trains the agent for the environment marked in Figure 9.

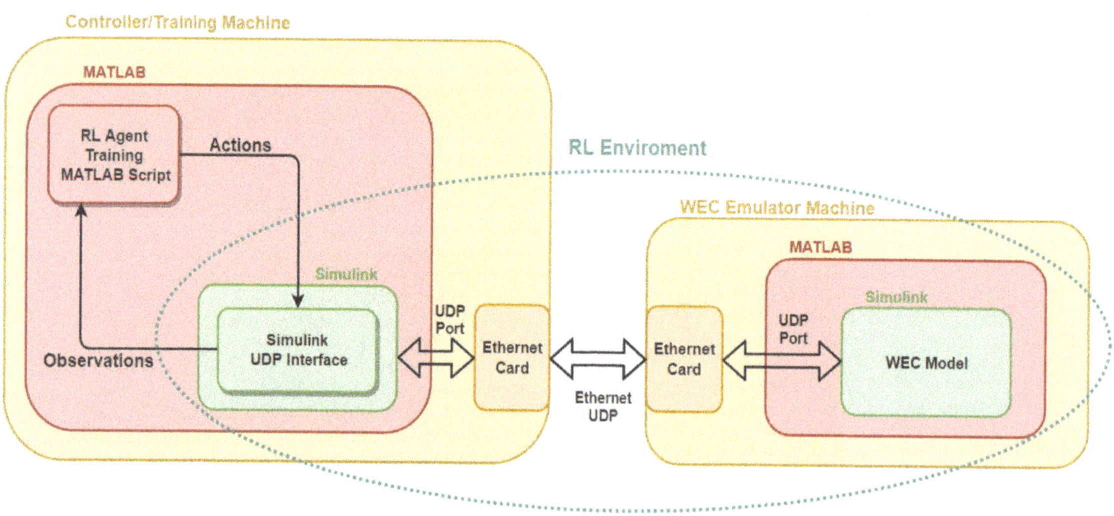

Figure 9. RL agent training through custom Simulink environment interfaced to WEC Emulator via Ethernet UDP.

The Simulink model used to implement the RL agent is shown in Figure 10. A separate RL agent is implemented for heave and pitch DoFs for each of the three pods of Centipod WEC in Figure 2. For this study, body-to-body interactions and array effects are neglected, and all pods are assumed to be identical; therefore, the DDPG RL agent is trained for the heave and pitch axes for one pod, and the trained policy is deployed for each PTO in Figure 10. The training policy is implemented in a Speedgoat Performance real-time target machine (Intel Core 3.1 GHz, 4-core, 8 GB). The velocity data (observations) of each pod are collected over the UDP link, parsed, and observed by respective policies to generate control actions, which are packed and transmitted to the WEC emulator machine over the UDP channel, as shown in Figure 10.

Two agents are trained for each DoF, one for the unconstrained PTO force and the other with a 40 kN upper bound on the PTO force magnitude. The three pods in Figure 10 are identical, so agents are trained for a single pod and the resulting agents are duplicated for the other two pods. The training stats for the heave agents for Pod 1 with constrained and unconstrained PTO forces are shown in Figures 11a and 11b, respectively. In either case, RL training converges in about 40 episodes with a 100 s simulation time for each episode. The training stats for the pitch agents for Pod 1 with constrained and unconstrained PTO forces are shown in Figures 12a and 12b, respectively. The constrained force pitch agent converges in 25 episodes, and the uncontained force pitch agent converges in 60 episodes.

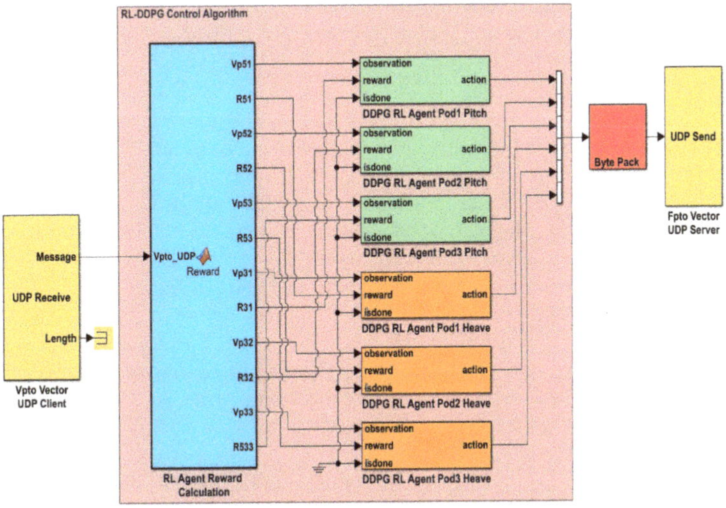

Figure 10. RL-DDPG controller for 2-DoF 3-pod CENTIPOD WEC.

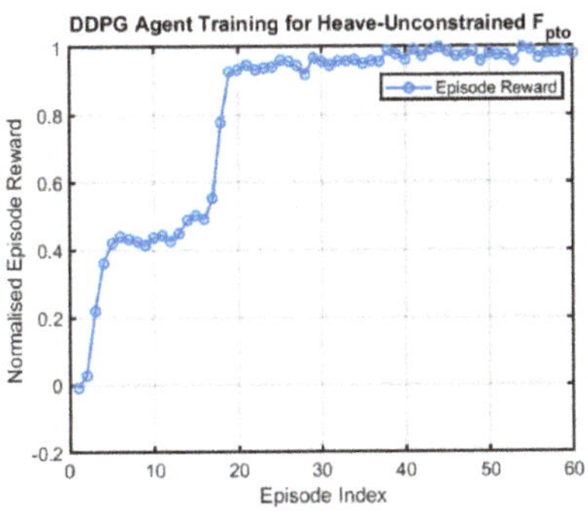

(a)

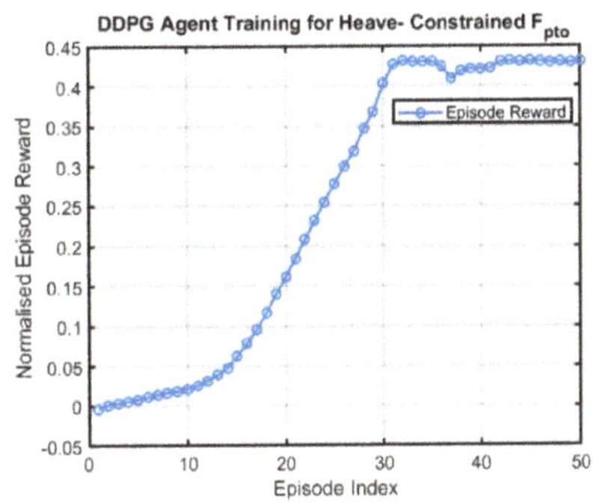

(b)

Figure 11. RL DDPG agent training for heave PTO: (**a**) training with unconstrained PTO force and (**b**) training with constrained PTO force, $|F_p| \leq 40$ kN.

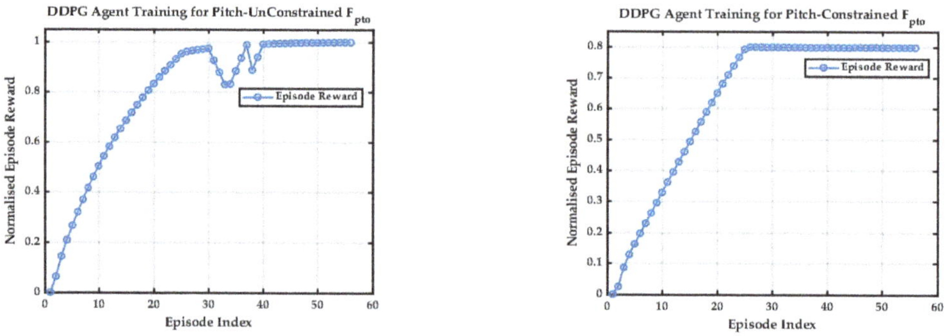

(**a**) (**b**)

Figure 12. RL DDPG agent training for pitch PTO: (**a**) training with unconstrained PTO force and (**b**) training with constrained PTO force, $|F_p| \leq 40$ kN.

5. Results

The training setup shown in Figure 9 is also used for executing system simulations with the trained agents. When executing simulations, the agents in Figure 10 are replaced with the trained agents, and the training machine in Figure 9 plays the role of a controller machine. A controller model similar to the RL model in Figure 9 is developed to simulate the NMPC by extending the scheme in [13] to the NMPC designed in Section 4, as shown in Figure 13. The RL and NMPC controllers in Figures 10 and 13, respectively, are tested with the WEC-Sim model of Figure 10 running on the emulator machine in Figure 9, with the same sea-state parameters given in Table 7. Tests are run with constrained and unconstrained PTO force conditions. Each case is simulated with linear wave conditions as well as with nonlinear buoyancy and Froude–Krylov excitations enabled in WECSim. The mechanical velocity constraints of the power take-off machines are enforced as follows: $|\text{heave velocity}| \leq 2$ m/s and $|\text{pitch velocity}| \leq 0.5$ rad/s.

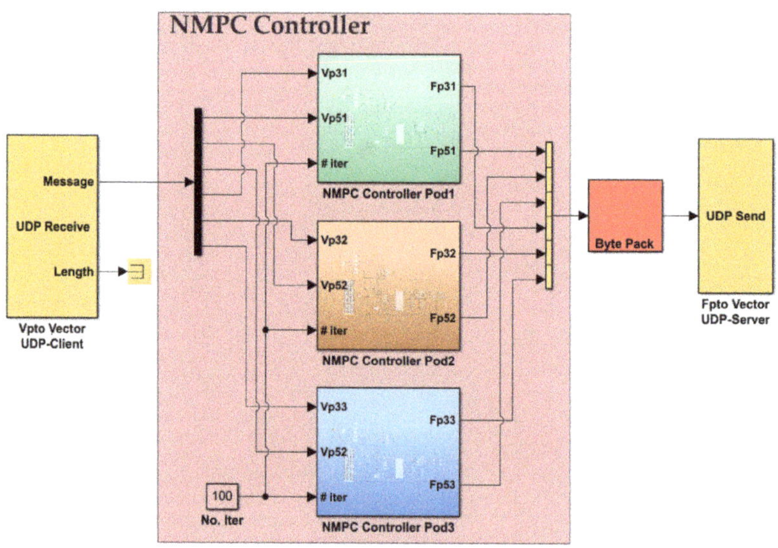

Figure 13. Nonlinear MPC controller for digital twin of three-pod Centipod device.

The moving averaged electrical power outputs with nonlinear MPC and RL are plotted in heave and pitch for Pod 1 with unconstrained and constrained PTO force cases subject to linear wave conditions, as shown in Figures 14 and 15, respectively. It is also important to evaluate the performance of the two algorithms under nonlinear buoyancy and Froude–Krylov wave excitations in WEC-Sim. The average electrical heave and pitch power outputs for Pod 1 with unconstrained and constrained PTO force cases under nonlinear hydrodynamic conditions are shown in Figures 16 and 17, respectively.

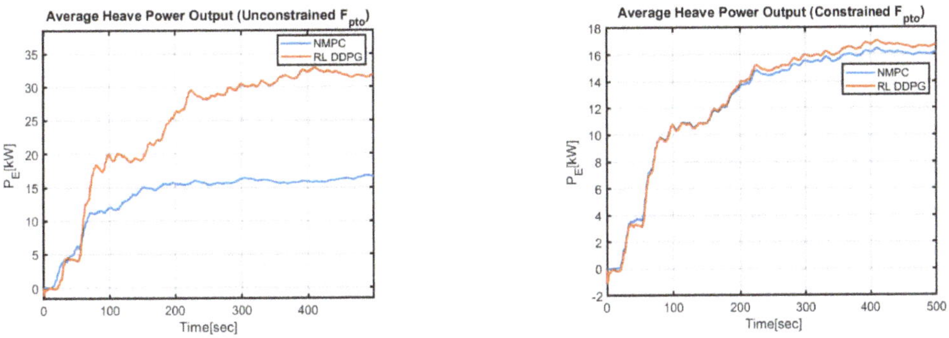

(a) (b)

Figure 14. Heave average electrical power output per pod with linear waves enabled in WECSim: (**a**) with unconstrained PTO force and (**b**) with constrained PTO force, $|F_p| \leq 40$ kN.

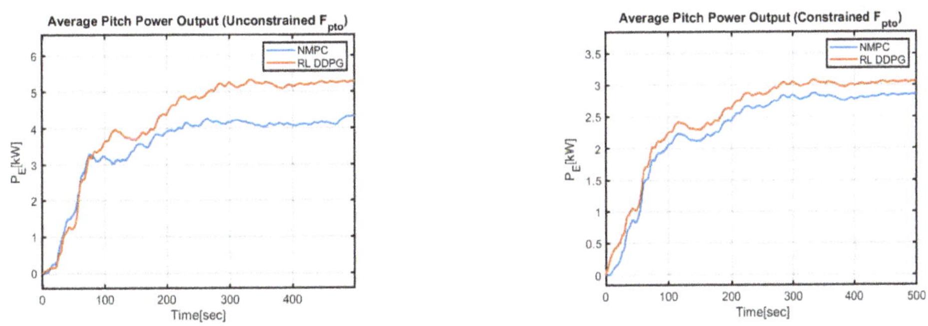

(a) (b)

Figure 15. Pitch average electrical power output per pod with linear waves enabled in WECSim: (**a**) with unconstrained PTO force and (**b**) with constrained PTO force, $|F_{pto}| \leq 40$ kN.

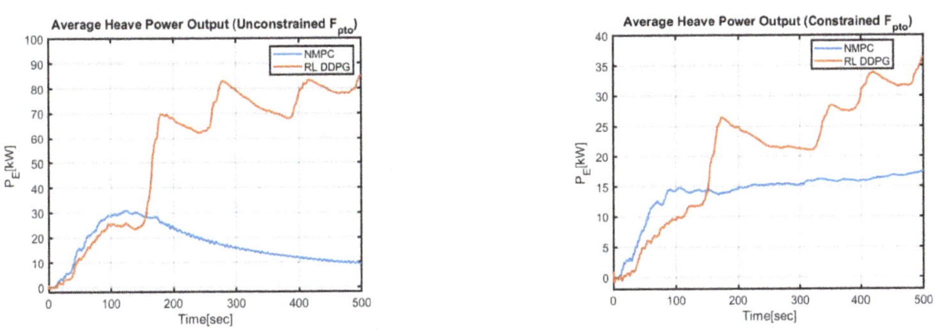

(a) (b)

Figure 16. Heave average electrical power output per pod with nonlinear buoyancy and Froude–Krylov excitations enabled in WECSim: (**a**) with unconstrained PTO force and (**b**) with constrained PTO force, $|F_{pto}| \leq 40$ kN.

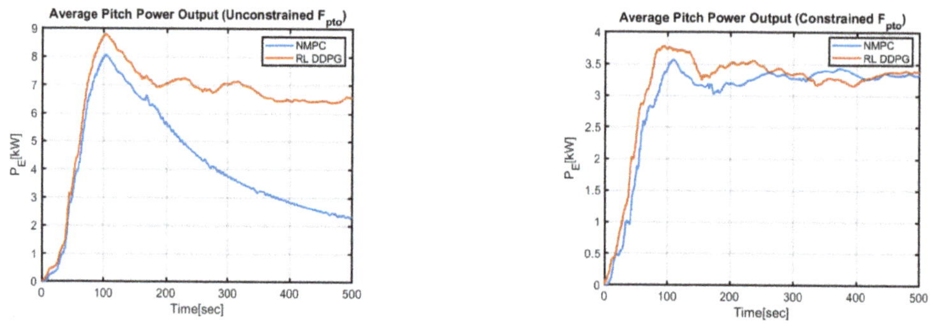

(**a**) (**b**)

Figure 17. Pitch average electrical power output per pod with nonlinear buoyancy and Froude–Krylov excitations enabled in WECSim: (**a**) with unconstrained PTO force and (**b**) with constrained PTO force, $|F_{pto}| \leq 40$ kN.

The results for the PTO power capture performance in heave and pitch from Figure 14 through Figure 17 are summarized in Table 8. The summary of the computational performance statistics for the NMPC and RL-DDGP control algorithms is given in Table 9, where the average Task Execution Time (TET) for each algorithm and RL agent training times for the PTO machines in heave and pitch axis are listed.

Table 8. Moving mean of electrical output power [kW] per PTO in WEC-Sim with irregular waves.

Controller Algorithm	Linear Wave Conditions		Nonlinear Buoyancy and Froude–Krylov Excitation	
	F_{pto} Unconstrained	$\lvert F_{pto}\rvert \leq 40$ kN	F_{pto} Unconstrained	$\lvert F_{pto}\rvert \leq 40$ kN
NMPC L-DDPG	17 32	16 17	Unstable 85	17 37
		Average electrical power [kW] for pitch		
NMPC RL-DDPG	4.50 5.25	2.80 3.10	Unstable 6.50	3.25 3.30

Table 9. Timings stats for NMPC and RL DDPG control.

	Training Time [h]	Task Execution Time (TET) [s]
NMPC heave and pitch combined	-	7.92×10^{-3}
NMPC per DoF	-	3.96×10^{-3}
RL-DDPG heave	1.12	4.32×10^{-4}
RL-DDPG pitch	1.67	7.52×10^{-4}

6. Discussion

The observations of the moving mean of electrical output power from the PTO mechanisms in Figures 14–17 reveal an improvement in the power output in the case of the RL-DPPG agent compared to the NMPC. This observation may be attributed to the fact that the operation of the NMPC is based upon the prediction model of the WEC plant, which, from the definition of the NMPC method, is an approximate representation of the actual process. On the other hand, the RL-DPPG agent was trained on the actual process in Figure 9 and observed the full process dynamics to determine how to act accordingly. In the constrained linear hydrodynamic cases in Figures 14b and 15b, the performance of the RL-DPPG very closely resembles the performance of NMPC. However, in the cases with unconstrained PTO forces (Figures 14a and 15a), nonlinear effects in the process dynamics become prominent as large PTO force magnitudes emerge and the NMPC performance degrades. This degradation can be attributed to the NMPC algorithm being susceptible to unmodelled or poorly modeled nonlinearities. The plot of the instantaneous electrical power output for the heave DoF is shown in Figure 18a, which corresponds to the average power output plot in Figure 14a. A significantly improved RL-DPPG control strategy operation is evident compared to the NMPC.

The performance degradation of the NMPC is attributed to the unmodeled process nonlinearities and becomes fully visible when observing the system's operation under nonlinear hydrodynamic wave conditions in WEC-Sim, as detailed in Figures 19 and 17a. The NMPC performs poorly in these figures, and the controller becomes unstable. The plot of instantaneous electrical power output for heave is shown in Figure 18b, corresponding to the average power output plot in Figure 19 under unconstrained PTO force conditions. It can be observed in Figure 18b that the NMPC algorithm fails to converge after around 170 s because it is unable to respond appropriately to nonlinear wave hydrodynamic conditions. On the other hand, RL-DPPG remains stable under the same nonlinear wave conditions. The heave PTO force plots corresponding to Figure 18b for NMPC and RL-DPPG are shown in Figure 19.

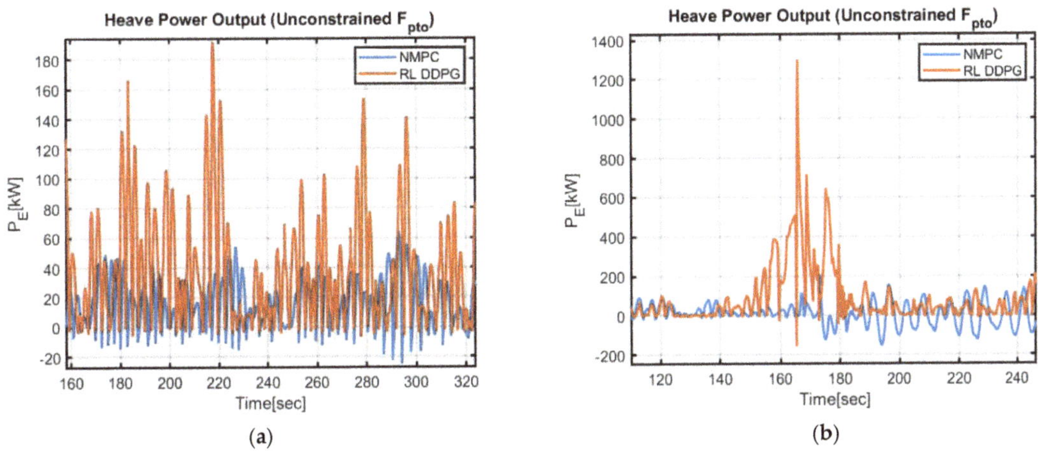

Figure 18. Heave instantaneous electrical power output per pod with unconstrained PTO force: (**a**) with linear wave conditions in WECSim and (**b**) with nonlinear buoyancy and Froude–Krylov excitations enabled in WECSim.

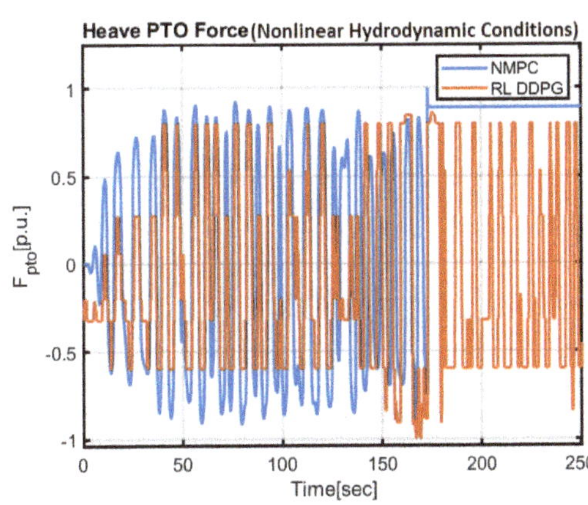

Figure 19. Heave PTO force output of NMPC and RL controllers with nonlinear buoyancy and Froude–Krylov excitations enabled in WECSim.

In Figure 19 the unstable output of the NMPC can be observed. The performance of the RL-DPPG remains stable and robust against process uncertainties in the same conditions. This might also be attributed to the fact that no underlying online optimization problem is being solved at every time step in the RL-DPPG policy deployment stage as opposed to the case of the NMPC, which also explains the significantly reduced TET in Table 9 for the RL-DDPG agent compared to the NMPC.

7. Conclusions

This article presents a comparison of two strategies, (a) NMPC and (b) RL-DPPG, for controlling the power-capture dynamics of the nonlinear WEC device by Dehlsen Associates' (the three-pod CENTIPOD WEC), with a PTO operating simultaneously in the heave axis and pitch axis. A state-space model of the WEC plant is formulated, including nonlinear quadratic viscous drag, and we consider a case study PTO model with a non-quadratic cost function. Two controllers are designed to optimize power capture from the PTO, (a) NMPC and (b) RL-DPPG, by training agents for the PTO machines in the heave and pitch axis. Both control algorithms are tested against the same simulated WEC model in WEC-Sim running on an external emulator machine. The heave and pitch PTO power output results are obtained for the linear wave conditions as well as with nonlinear buoyancy, and Froude–Krylov excitations are enabled in WECSim for cases where the PTO force is constrained or unconstrained. Results depict a significant enhancement in the performance of the proposed RL-DDPG algorithm when compared to the NMPC controller, based on various performance metrics, including a reduction in the Task Execution Time (TET), an increase in the power extraction, an improvement in the robust operation when subject to exogenous conditions, and more overall flexibility and ease of design.

Author Contributions: Conceptualization, A.S.H. and A.M.; methodology, A.S.H., K.B. and A.M.; software, A.M. and K.B.; validation, A.S.H., A.M. and K.B.; formal analysis, A.S.H. and K.B.; investigation, A.S.H., A.M. and K.B.; resources, A.M. and K.B.; data curation, A.M.; writing—original draft preparation, A.S.H. and K.B.; writing—review and editing, A.S.H., A.M. and K.B.; visualization, A.S.H., K.B. and A.M.; supervision, K.B. and A.M.; project administration, K.B. and A.M.; funding acquisition, K.B. All authors have read and agreed to the published version of the manuscript.

Funding: This research was funded by the McCain Foundation postdoctoral fellowship award in innovation.

Institutional Review Board Statement: Not applicable.

Informed Consent Statement: Not applicable.

Data Availability Statement: Not applicable.

Acknowledgments: The authors appreciate technical support from Dehlsen Associates, LLC, California, and the provision of experimental data for the PTO machines from McCleer Power, Michigan.

Conflicts of Interest: The authors declare no conflict of interest.

References

1. Muetze, A.; Vining, J.G. Ocean Wave Energy Conversion—A Survey. In Proceedings of the Conference Record of the 2006 IEEE Industry Applications Conference Forty-First IAS Annual Meeting, Tampa, FL, USA, 8–12 October 2006; Volume 3, pp. 1410–1417.
2. Richter, M.; Magana, M.E.; Sawodny, O.; Brekken, T.K.A. Nonlinear Model Predictive Control of a Point Absorber Wave Energy Converter. *IEEE Trans. Sustain. Energy* **2013**, *4*, 118–126. [CrossRef]
3. Genest, R.; Ringwood, J.V. A Critical Comparison of Model-Predictive and Pseudospectral Control for Wave Energy Devices. *J. Ocean Eng. Mar. Energy* **2016**, *2*, 485–499. [CrossRef]
4. Falcão, A.F.O.; Henriques, J.C.C. Effect of Non-Ideal Power Take-off Efficiency on Performance of Single- and Two-Body Reactively Controlled Wave Energy Converters. *J. Ocean Eng. Mar. Energy* **2015**, *1*, 273–286. [CrossRef]
5. Brekken, T.K.A. On Model Predictive Control for a Point Absorber Wave Energy Converter. In Proceedings of the 2011 IEEE Trondheim PowerTech, Trondheim, Norway, 19–23 June 2011; pp. 1–8.
6. Bubbar, K.; Buckham, B.; Wild, P. A Method for Comparing Wave Energy Converter Conceptual Designs Based on Potential Power Capture. *Renew. Energy* **2018**, *115*, 797–807. [CrossRef]

7. What Is Reinforcement Learning?—MATLAB & Simulink—MathWorks United Kingdom. Available online: https://uk.mathworks.com/help/reinforcement-learning/ug/what-is-reinforcement-learning.html (accessed on 2 October 2023).
8. Anderlini, E.; Forehand, D.I.M.; Bannon, E.; Xiao, Q.; Abusara, M. Reactive Control of a Two-Body Point Absorber Using Reinforcement Learning. *Ocean. Eng.* **2018**, *148*, 650–658. [CrossRef]
9. Control of a Realistic Wave Energy Converter Model Using Least-Squares Policy Iteration. Available online: https://ieeexplore.ieee.org/document/7911321 (accessed on 2 October 2023).
10. Zadeh, L.G.; Glennon, D.; Brekken, T.K.A. Nonlinear Control Strategy for a Two-Body Point Absorber Wave Energy Converter Using Q Actor-Critic Learning. In Proceedings of the 2020 IEEE Conference on Technologies for Sustainability (SusTech), Santa Ana, CA, USA, 23–25 April 2020; pp. 1–5.
11. Anderlini, E.; Husain, S.; Parker, G.G.; Abusara, M.; Thomas, G. Towards Real-Time Reinforcement Learning Control of a Wave Energy Converter. *J. Mar. Sci. Eng.* **2020**, *8*, 845. [CrossRef]
12. Rij, J.; Yu, Y.-H.; McCall, A.; Coe, R.G. Extreme Load Computational Fluid Dynamics Analysis and Verification for a Multi-Body Wave Energy Converter. In Proceedings of the International Conference on Offshore Mechanics and Arctic Engineering. American Society of Mechanical Engineers, Glasgow, UK, 9–14 June 2019.
13. Haider, A.S.; Brekken, T.K.A.; McCall, A. A State-of-the-Art Strategy to Implement Nonlinear Model Predictive Controller with Non-Quadratic Piecewise Discontinuous Cost Index for Ocean Wave Energy Systems. In Proceedings of the 2020 IEEE Energy Conversion Congress and Exposition (ECCE), Detroit, MI, USA, 11–15 October 2020; pp. 1873–1878.
14. Houska, B.; Ferreau, H.J.; Diehl, M. An Auto-Generated Real-Time Iteration Algorithm for Nonlinear MPC in the Microsecond Range. *Automatica* **2011**, *47*, 2279–2285. [CrossRef]
15. Lu, H.; Chang, S.; Chen, C.; Fan, T.; Chen, J. Replacement of Force-to-Motion Relationship with State–Space Model for Dynamic Response Analysis of Floating Offshore Structures. *Appl. Ocean. Res.* **2022**, *119*, 102977. [CrossRef]
16. Barradas-Berglind, J.J.; Muñoz-Arias, M.; Wei, Y.; Prins, W.A.; Vakis, A.I.; Jayawardhana, B. Towards Ocean Grazer's Modular Power Take-Off System Modeling: A Port-Hamiltonian Approach. *IFAC-PapersOnLine* **2017**, *50*, 15663–15669. [CrossRef]
17. WEC-Sim (Wave Energy Converter SIMulator)—WEC-Sim Documentation. Available online: https://wec-sim.github.io/WEC-Sim/ (accessed on 27 March 2021).
18. Wamit, Inc. The State of the Art in Wave Interaction Analysis. Available online: https://www.wamit.com/ (accessed on 28 March 2021).
19. Falnes, J. Wave-Energy Conversion through Relative Motion between Two Single-Mode Oscillating Bodies. *J. Offshore Mech. Arct. Eng.* **1999**, *121*, 32–38. [CrossRef]
20. Haider, A.S.; Brekken, T.K.A.; McCall, A. Real-Time Nonlinear Model Predictive Controller for Multiple Degrees of Freedom Wave Energy Converters with Non-Ideal Power Take-Off. *J. Mar. Sci. Eng.* **2021**, *9*, 890. [CrossRef]
21. Technology—Centipod. Available online: https://centipodwave.com/technology/ (accessed on 24 October 2023).
22. Lillicrap, T.P.; Hunt, J.J.; Pritzel, A.; Heess, N.; Erez, T.; Tassa, Y.; Silver, D.; Wierstra, D. Continuous Control with Deep Reinforcement Learning. *arXiv* **2019**. [CrossRef]
23. PacWave—TESTING WAVE ENERGY FOR THE FUTURE. Available online: https://pacwaveenergy.org/ (accessed on 24 October 2023).
24. Dunkle, G.; Zou, S.; Robertson, B. Wave Resource Assessments: Spatiotemporal Impacts of WEC Size and Wave Spectra on Power Conversion. *Energies* **2022**, *15*, 1109. [CrossRef]

Journal of
Marine Science and Engineering

MDPI

Article

Measuring the Robustness of Optimal Design Solutions for Wave Energy Converters via a Stochastic Approach

Filippo Giorcelli *, Sergej Antonello Sirigu, Giuseppe Giorgi, Nicolás Faedo, Mauro Bonfanti, Jacopo Ramello, Ermanno Giorcelli and Giuliana Mattiazzo

Marine Offshore Renewable Energy Lab, Department of Mechanical and Aerospace Engineering, Politecnico di Torino, Via Paolo Borsellino 38/16, 10138 Torino, Italy; sergej.sirigu@polito.it (S.A.S.); giuseppe.giorgi@polito.it (G.G.); nicolas.faedo@polito.it (N.F.); mauro.bonfanti@polito.it (M.B.); jacopo.ramello@polito.it (J.R.); ermanno.giorcelli@polito.it (E.G.); giuliana.mattiazzo@polito.it (G.M.)
* Correspondence: filippo.giorcelli@polito.it

Abstract: Among the challenges generated by the global climate crisis, a significant concern is the constant increase in energy demand. This leads to the need to ensure that any novel energy systems are not only renewable but also reliable in their performance. A viable solution to increase the available renewable energy mix involves tapping into the potential available in ocean waves and harvesting it via so-called wave energy converters (WECs). In this context, a relevant engineering problem relates to finding WEC design solutions that are not only optimal in terms of energy extraction but also exhibit robust behavior in spite of the harsh marine environment. Indeed, the vast majority of design optimization studies available in the state-of-the-art consider only perfect knowledge of nominal (idealized) conditions, neglecting the impact of uncertainties. This study aims to investigate the information that different robustness metrics can provide to designers regarding optimal WEC design solutions under uncertainty. The applied methodology is based on stochastic uncertainty propagation via a Monte Carlo simulation, exploiting a meta-model to reduce the computational burden. The analysis is conducted over a dataset obtained with a genetic algorithm-based optimization process for nominal WEC design. The results reveal a significant deviation in terms of robustness between the nominal Pareto set and those generated by setting different thresholds for robustness metrics, as well as between devices belonging to the same nominal Pareto frontier. This study elucidates the intrinsic need for incorporating robust optimization processes in WEC design.

Keywords: wave energy converter; robustness quantification; uncertainty; surrogate model; Gaussian process regression; robust design optimization

Citation: Giorcelli, F.; Sirigu, S.A.; Giorgi, G.; Faedo, N.; Bonfanti, M.; Ramello, J.; Giorcelli, E.; Mattiazzo, G. Measuring the Robustness of Optimal Design Solutions for Wave Energy Converters via a Stochastic Approach. *J. Mar. Sci. Eng.* **2024**, *12*, 482. https://doi.org/10.3390/jmse12030482

Academic Editor: Eugen Rusu

Received: 2 February 2024
Revised: 5 March 2024
Accepted: 11 March 2024
Published: 13 March 2024

1. Introduction

In 2020, the United Nations emphasized the urgent need for every nation to address a critical goal for the near future: achieving carbon neutrality by 2050 [1]. However, the pursuit of this objective faces a significant obstacle due to the rising energy demand, which is currently largely reliant on fossil fuels [2]. For many years, various renewable energy sources have been under the scope of bridging this gap, with ocean wave energy emerging as a large and almost untapped resource. The global wave energy resource is estimated to lie in the range of 1 to 10 TW, but the exact extractable wave power remains the subject of ongoing research. Theoretical calculations, as presented in [3], suggest that this global resource totals around 32,000 TWh/y with a mean power output of 3.65 TW. Using the same assessment methodology and data, the Oceanic Energy System and the International Renewable Energy Agency report a value of 29,500 TWh/y [4,5]. These numbers demonstrate that wave energy could play a pivotal role in the race towards achieving carbon neutrality by 2050.

From a historical perspective, the origins of technology designed to harness wave energy date back to 1799 in France.[1] Today, wave energy is gaining attention due to its potential to become a competitive new player in the renewable energy market. One of the primary reasons for this is the absence of marketable devices for energy production, essential for establishing viable wave energy, mainly due to the lack of conversion concepts' convergence. A broad overview of the topic can be found in [6].

Wave energy presents some advantages (therefore sparking interest) when its power density is compared to that of other well-known renewable energy sources. Waves are mainly generated via the interaction between wind and the sea surface, in which the constant mechanical action of the wind, acting as a tangential effort, leads to the formation of waves. On the other hand, the deployment of wave energy converter (WEC) technology has yet not reached a competitive level compared to other clean energy industries. Different key points have been mapped to proceed towards WEC's successful development and commercialization. Concerning the latter topic, various studies (see, e.g., [6,7]) underline that, for a WEC investment to be considered appealing to stakeholders and investors, it should meet the following criteria: it should offer energy production costs that are competitive on the market; it should present a secure and dependable investment opportunity; it should demonstrate reliability in supporting grid operations; it should contribute positively to society or have societal benefits; it should align with permitting and certification requirements; it should prioritize operational safety; and it should have the potential for global deployment. A general aspect that encompasses all the issues listed immediately above is the common emphasis on the techno-economic reliability of WEC technology. This, in fact, is consistent with the conclusions argued for in [6]: to reduce investor risk, it is necessary to reduce both the levelized cost of energy (LCoE) and its associated uncertainty. The highlighted challenges are classically addressed via robust optimization (RO). RO frameworks aim to find solutions to optimization problems while incorporating uncertainty in the evaluation of the fitness function. RO techniques' purpose is to achieve optimal solutions that are less prone to being affected by uncertainty.

To illustrate this, in Figure 1, it is possible to distinguish among different local optima, i.e., the minima of the objective function $f(x)_{obj}$. Each one is characterized by a different behavior. Point B is referred to as a robust optimum since it remains locally optimal even when a relatively large variation Δx occurs. Conversely, despite the fact that A is a global optimum, if the same variation Δx perturbs point A, the objective function increases sharply, losing optimality with respect to point B; i.e., it is effectively more sensitive to variations in x. A conventional formulation of the associated RO problem is provided in Section 2.

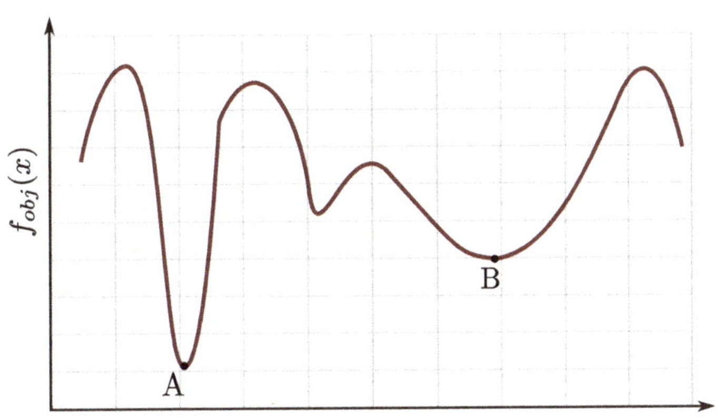

Figure 1. Notional difference between a global optimum (A) and a robust optimum (B). Image adapted from [8].

The developments presented within this study were effectively motivated by the limited literature available on the analysis of the performance robustness of WECs from a design perspective. In particular, and with respect to the classification of robustness metrics outlined in a previous paper [9], it can be asserted that there exists a lack of comprehensive and systematic research in the field of WEC design addressing all four subcategories that underlie the idea of robustness. The first subcategory is the so-called *robustness of a concept*, and it relies on robustness metrics that assess the impact of design parameters or independent noise factors on the investigated performance. Subsequently, when focusing on the size of the feasible design space, the *robustness of a design* and its indicators are defined. Another branch of robustness measures assesses the different expectations and dispersion of the system's output, i.e., the *robustness of a function*. Finally, the *robustness of a product* classification exploits metrics that assess the likelihood of satisfactory performance conformity under the influence of ongoing variations. To date, within the realm of wave energy conversion, the literature on robust optimization has predominantly focused on the development of robust control frameworks [10–17]. Furthermore, concerning the study of wave energy and WEC uncertainties, the existing literature primarily covers environmental [18–22] and full/high-scale prototyping or experimental uncertainties and subsystems' influence on performance [23–29]. The neighboring Reliability-Based Design Optimization (RBDO) field of research directs its attention to WEC structural and maintenance cost uncertainties [30] (in which the so-called Wavestar device is considered as a case study) and the power take-off (PTO) reliability relationship with hull geometry [31]. A review of RBDO studies and their application in the offshore renewable energy sector is presented in [32].

Motivated by the necessity of WEC systems' RO, the present study evaluates different robustness indices (RIs) for WEC design solutions and assesses their impact on devices' overall normalized power performance (Equation (18); i.e., it performs a stochastic Robustness Quantification (RQ) investigation for WEC systems. To achieve this objective, the present study conducted an analysis to gather information regarding the robustness of these devices generated in terms of a dataset processed during a multi-objective nominal optimization process. First, nominal Pareto set devices RI were evaluated, and information about their statistical behavior was obtained accordingly. This nominal classification is discussed in accordance with the standard Pareto criterion. Threshold criteria were then imposed on robustness indices and new Pareto research, highlighting the main differences between obtained and nominal limit sets. Furthermore, due to the computational burden associated with the uncertainty propagation, performed via a Monte Carlo simulation, a surrogate-assisted RQ method is proposed. The reason for presenting the results and devices' data for the case study examined in normalized form is the confidential and industrial nature of the analyzed technology. Despite this limitation, the results presented in this non-dimensional form are sufficient to draw conclusions regarding robustness and, thus, adequate for the purpose of this study.

The remainder of the paper is structured as follows. In Section 2, an overview of the WEC robust optimization problem is provided, together with a description of different RQ approaches and robustness indices. Sections 3.2 and 3.1, respectively, discuss and provide details on the specific framework applied in this study and the applicative case of analysis, i.e., the Inertial Sea Wave Energy Converter (ISWEC) device. Section 4 presents the main results and discussion. Finally, in Section 5, key conclusions of this study are provided, including future lines of research.

2. Robustness Quantification

In this section, a brief overview of robust optimization and its formulation is presented, together with a brief review of different robust quantification metrics.

2.1. Robust Optimization Overview

In the literature, the robustness concept has been articulated in different ways, according to different fields of study. One of the first definitions of robustness was provided by Taguchi for the quality improvement of industrial processes: *"a product whose performance is minimally sensitive to factors causing variability (at the lowest possible cost)"* [33]. The robustness concept also plays a central role in the already-mentioned robust control field, where different applications can be found. In this specific case, the aim is to design and synthesize a control a strategy that is ideally insensitive to either parametric or non-parametric uncertainty, able to guarantee both a prescribed set of performance specifications (e.g., maximizing energy harvesting) and the robust stability of the controlled system [34]. Regarding WEC control, in [10], a hierarchical robust strategy was devised to mitigate the controller susceptibility to modeling inaccuracies and potentially unmodeled nonlinear influences. Model sensitivity issues for different WEC control system architectures are addressed in [11], leveraging the inherent parametric uncertainty in standard device modeling. Moreover, a moment-based framework to design robust energy-maximizing optimal controllers for WEC is presented in [12].

The pivotal role of robustness is also clear in the context of optimization problems, in which the RO research field is relevant. As a historical note, consider that in 1973, Soyster [35] formulated a linear programming (LP) optimization model, yielding feasible solutions for all data belonging to a convex set. However, this approach leaned towards excessive conservatism, compromising optimality for enhanced robustness. In subsequent studies by Ben-Tal, Nemirovski, El-Ghaoui, and Lebert [34,36], spanning from 1997 to 2000, less conservative solutions were proposed. These strategies tackled uncertain linear problems with ellipsoidal uncertainties, requiring the solution of robust counterparts in a conic quadratic form. Despite offering reduced conservatism, this approach introduced practical drawbacks, as it led to inherently nonlinear (yet convex) formulations with increased computational demand compared to Soyster's earlier linear framework. Expanding on this historical note, Bertsimas and Sim [37] integrated the insights from Soyster and Ben-Tal et al. into a fully controllable framework. This latter approach allowed for the adjustment of conservatism levels with probabilistic guarantees. Notably, their proposed robust counterparts were linear optimization problems, enabling a seamless generalization to discrete optimization problems. A key novelty was the introduction of the so-called Γ parameter, which refers to a protective threshold that plays a relevant role in fine-tuning robustness against conservatism levels.

The second main methodology for solving RO problems is called "Robust Design Optimization" (RDO) [38]. This method, complementary to RO, employs a stochastic approach based on different robustness metrics, aiming to achieve a system design insensitive to uncertainty, external noises, perturbations, potential model inaccuracies, and design tolerances [8]. The latter is of significant importance within the realm of simulation-based optimization, particularly when addressing complex engineering challenges, such as the design of WECs. Typically, the RDO framework is based on the RQ process, which is a stochastic approach to uncertainty propagation that can be outlined in the following main steps.

1. Parameter uncertainty vectors' ($\vec{u}$) and design ($\vec{x}$) parameter vectors' definition:
    - $\vec{u} = [u_1 ... u_N]^\intercal \in \mathbb{R}^N$;
    - $\vec{x} = [x_1 ... x_D]^\intercal \in \mathbb{R}^D$,

    where N is the number of system parameters affected by uncertainty, whose magnitude is described by the relative n^{th}-element of the uncertainty vector $\vec{u}$, and D is the dimension of the design parameter vector $\vec{x}$, i.e., the number of decision elements that define each possible optimization problem solution.
2. Uncertainty propagation via sampling techniques.
3. Robustness metric quantification.

A description of the proposed RQ framework for the present study is provided in Section 3.2. Therefore, it is common that, in the RDO field, the objective function is not known exactly but is evaluated in a computationally expensive fashion, leveraging stochastic logic. Consequently, efforts have been directed towards studies concerning evolutionary algorithms [39,40], especially via multi-objective optimization, to find a robust Pareto frontier [41,42].

2.2. Problem Formulation and Robustness Indices

As discussed within Section 1, this paper is focused on the realm of RDO, and it is primarily aimed at serving as a quantitative exploration of the challenges involved when incorporating uncertainty in optimal WEC design, together with any consequent implications. In particular, a stochastic approach was adopted, and the motivation behind this choice lay in avoiding the adoption of assumptions potentially restrictive to the system and exploring computer simulation-experiments' viability. Therefore, a general RDO problem formulation can be mathematically stated as follows:

$$\min_{\vec{x} \in \mathscr{X}} \vec{f}(\vec{x}, \vec{u}), \tag{1}$$

which is subject to the following:

$$\begin{aligned} h(\vec{x}, \vec{u}) &= 0, \\ g(\vec{x}, \vec{u}) &< 0, \end{aligned} \tag{2}$$

where the following points apply:

- $\vec{u}^{min} < \vec{u} < \vec{u}^{max}$ are the upper and lower bounds of the uncertainty parameters' vector. The space resulting from limiting $\vec{u}$ can be referred to as the parameter uncertainty space $\mathscr{U}$.
- $\vec{x}^{min} < \vec{x} < \vec{x}^{max}$ are the upper and lower bounds of the design vector. The space resulting from limiting $\vec{x}$ can be referred to as the parameter design space $\mathscr{X}$.
- $h(\vec{x}, \vec{u}) \in \mathbb{R}^q$ represent q equality constraints.
- $g(\vec{x}, \vec{u}) \in \mathbb{R}^k$ represent k inequality constraints.
- $\vec{f}(\vec{x}, \vec{u}) = [f_1(\vec{x}, \vec{u}) ... f_m(\vec{x}, \vec{u})]^\top \in \mathbb{R}^m$ is the vector containing the potentially different m objective functions.

A comprehensive survey, centered on the methods for solving the RDO problem and the associated robustness quantification metrics, is presented in [38]. In particular, RDO problems are solved via single-objective robust design optimization or multi-objective robust design optimization. Single-objective RDO frameworks involve setting the robustness metric as part of the optimization objective function. On the other hand, multi-objective RDO processes incorporate the RI as an additional objective function [43]. Furthermore, another option is to rely on supplementary constraints to enhance robustness, i.e., RBDO [44]. A well-chosen robustness measure offers a valuable means to discern a system's responsiveness (or lack thereof) to factors that might result in sub-optimal performance. A glimpse of some of the robustness metrics employed in the literature [43,45], and considered within the present study, is provided in Table 1, together with their formulations. The notation used in Table 1 refers to the so-called zero-uncertainty vector $\vec{u_0}$ as the parameters uncertainty vector with all elements equal to zero, i.e., $\vec{u_0} = [0...0]^\top \in \mathbb{R}^N$. Therefore, the performance evaluated for $\vec{x_i}$ and $\vec{u_0}$ ($f_m(\vec{x_i}, \vec{u_0})$) is named the mth-performance nominal expected value.

Definition **I** in Table 1 is consistent with the idea of optimizing the average m^{th}-performance of the analyzed system. This option could direct the final choice (as well as the process of exploration of the optimization itself) towards individuals characterized by good average performance but potentially highly dispersed around the nominal expected value.

Following Definition **II** in Table 1 means considering only the samples' distribution concerning the mean value of the PDF. Due to its limited possibility to describe a system's performance in terms of an absolute value, $\hat{\sigma}_i$ is mainly adopted in multi-objective RDO frameworks as part of the objective functions, together with other parameters, e.g., $\hat{\mu}_i$, or the performance nominal value $f(\vec{x_i}, \vec{u_p})$ directly.

The "k-sigma" approach establishes a range of performance deviations from the PDF's statistical average value by merging the approaches specified in Definitions **I** and **II** within Table 1. This is accomplished by applying a multiplicative coefficient on $\hat{\sigma}_i$. The smaller this deviation, the more robust the individual examined.

Table 1. Robustness metrics and indices.

Definition	**Index**
(I) Average value of the mth -performance probability density function (PDF).	$\hat{\mu}_i = \frac{1}{N_{sample}} \sum_{p=1}^{N_{sample}} \frac{f_m(\vec{x}_i,\vec{u}_p)}{f_m(\vec{x}_i,\vec{u}_0)} = \frac{\mu_i(\vec{x}_i,\vec{u})}{f_m(\vec{x}_i,\vec{u}_0)}$, where N_{sample} is the quantity of samples used in order to evaluate the discrete PDF mean value. The same formulation results are valid for all the other **Definitions** in the present table.
(II) Standard deviation of the mth-performance PDF.	$\hat{\sigma}_i = \frac{1}{f_m(\vec{x}_i,\vec{u}_0)} \sqrt{\frac{\sum_{p=1}^{N_{sample}} [f_m(\vec{x}_i,\vec{u}_p) - \mu_i(\vec{x}_i,\vec{u})]^2}{N_{sample}-1}}$.
(III) The "k-sigma" approach.	$\hat{\mu}_i + k\hat{\sigma}_i$, where k is an integer number commonly set equal to 3 or 6^2
(IV) Symmetric robustness index.	$R_i = \frac{s_{f_m(\vec{x}_i,\vec{u}_p)} + \left\| \mu_{f_m(\vec{x}_i,\vec{u}_p)} - f_m(\vec{x}_i,\vec{u}_0) \right\|}{f_m(\vec{x}_i,\vec{u}_0)}$, with: $s_{f_m(\vec{x}_i,\vec{u}_p)} = \sqrt{\frac{\sum_{p=1}^{N_{sample}} [f(\vec{x}_i,\vec{u}_p) - f_m(\vec{x}_i,\vec{u}_0)]^2}{N_{sample}-1}}$ and $\mu_{f_m(\vec{x}_i,\vec{u}_p)} = \frac{1}{N_{sample}} \sum_{p=1}^{N_{sample}} f_m(\vec{x}_i,\vec{u}_p)$.
(V) Asymmetric robustness index.	$Q_i = \frac{Q_0}{f_m(\vec{x}_i,\vec{u}_0)}$.
(VI) The upper limit of the mth-performance PDF.	$\sup \hat{f}_i = \max_{\vec{u}} \frac{f_m(\vec{x}_i,\vec{u}_p)}{f_m(\vec{x}_i,\vec{u}_0)} \forall p \in [1 \ldots N_{sample}]$.

In Definition **IV** (Table 1), R_i is defined as "symmetric" since its formulation weights every possible deviation from the nominal expected value. In particular, a smaller value of R_i indicates a higher level of robustness in the system being considered. The framework in which R_i is usually used can involve its optimization, together with the robust performance objective. Thanks to the influence of both the distance between the average value of the distribution and its spread with respect to the expected nominal value $f_m(\vec{x}_i, \vec{u}_0)$, Definition **IV** is well suited for metrics for which the purpose of the designer is to keep the performance very close to a specific target.

On the other hand, the Q_i "asymmetric" robustness index (Definition **V** in Table 1) penalizes only the deviations from the nominal value that result in poorer performances and, consequently, worse values of the objective function in question. In the case in which the mth performance needs to be minimized, Q_0 represents the objective function value below the $q\%$ (typically greater than 90%) of observed occurrences. In this scenario, a smaller value of Q_i indicates a better risk measure. Conversely, if the mth performance needs to be maximized, Q_0 is the objective function value above the $q\%$ of observed occurrences. In this case, a larger value of Q_i indicates a better risk measure. To enable a comparison across different devices, the nominal value $f_m(\vec{x}_i, \vec{u}_0)$ is used in the index for normalization purposes. The Q_i index offers insights into device quality concerning the mth objective function. Given this capability, it can be used for systems belonging to various stages of product development and their potential for exploitation. Nonetheless, this measurement does not inform the designer about specific occurrences' values; it reports only cumulative information.

To perform a robustness quantification that considers the upper limit in a minimization problem (respectively, the lower limit in a maximization case), Definition **VI** in Table 1 can be adopted, i.e., the Worst-Case Scenario (WCS). The WCS RI presents two main drawbacks: over-conservative solutions are often reached, and an extensive exploration phase of the uncertain parameter's N-dimensional space (to evaluate the real WCS) is required, leading to computationally expensive evaluations. The interested reader can find comprehensive examinations of robustness metrics employed in RDO processes in the works of Beyer et al. [38] and Moritz Göhler et al. [9]. Moreover, in [41,42], Deb et al. define two different robust solution approaches to optimization, focusing on evolutionary algorithms.

When quantifying robustness, it is crucial to take into account the nature of uncertainties. Typically, in the realm of RDO, the primary types of uncertainties under scrutiny are epistemic and aleatory. The first type of uncertainty, i.e., epistemic uncertainty, stems from a lack of knowledge that can be mitigated by acquiring more information about the parameter. For instance, in the context of renewable energy systems, epistemic uncertainty in the typical annual electricity demand may arise from unknown occupant behavior. On the other hand, aleatory uncertainty is associated with the unpredictable variation of parameter values and is, therefore, inherent and irreducible. For example, aleatory uncertainty regarding wholesale electricity prices results from the unpredictable variations of this parameter in the future. Both types of uncertainties can be characterized using interval-valued statistical moments [47].

It is readily evident that adopting a stochastic approach to propagating uncertainties in quantifying the robustness of the previously described system presents several challenges, with the most prominent being elevated computational expenses. To address this challenge, so-called surrogate-assisted RDO has been developed. These methods rely on the use of mathematical models that approximate a time-demanding model to reduce the computational cost of the objective function (i.e., a so-called surrogate model or metamodel), making the computational expense required to solve the optimization problem tractably. Various types of mathematical models are employed in the literature, with some of the most common ones including Gaussian process regression (GPR), the Kriging Model, and polynomial chaos expansion [48,49]. An additional avenue for reducing the computational cost of RDO processes could involve a shift in the problem-solving approach, transitioning from a stochastic framework to a deterministic framework. Assuming the system exhibits behavior such that the WCS relies on the extremes of the uncertainty-affected parameter range, the designer could choose to evaluate the objective function solely at the vertices of the resulting N-dimensional polytope. For a single uncertain variable, there would be two vertex points (the uncertain space will be described by a single line); for two variables, there would be four evaluation points (forming a square); for three variables, eight points (describing a cube), and so forth. This approach would result in a reduction in computational costs by diminishing the number of expensive evaluations required. The advantages and disadvantages associated with the various approaches outlined earlier are condensed in Table 2.

Table 2. Main advantages and disadvantages of different approaches to optimization in the presence of uncertainty, inspired by [50].

Method	Advantage	Disadvantage
Stochastic RDO	Provide trade-off solutions and detailed information on the output under examination distribution.	Very computationally expensive.
Deterministic RDO	Less computationally expensive.	Need the assumption of good behavior in the N-dimensional polytope. Overly conservative.

3. Methodology

The ISWEC device and its model under analysis are described in more detail in the present section, along with a technical introduction to the technology. Furthermore, the objectives of the analyses conducted, the framework, and the methodologies used to achieve them are introduced. The main goal of this work was to compare different stochastic robustness metrics and evaluate the information that each of them provides to the designer. The robustness metrics used included the six previously introduced in Table 1, namely $\hat{\mu}_i$, $\hat{\sigma i}$, $sup(\hat{f}_i)$, $\hat{\mu}_i + k\hat{\sigma i}$, R_i and Q_i.

3.1. Model

The ISWEC device (Figure 2) is a WEC characterized by a pitch-floating hull, featuring a fully enclosed design. The device was designed to perform as well as the Pendulum Wave Energy Converter device [51] in the demanding conditions of the Mediterranean Sea, and its performance was optimized with this aim. A concise and general overview of ISWEC and its operational principles is provided below.

The heart of this system lies within the floater, housing a PTO mechanism that leverages the precession motion of a gyroscope system to capture the mechanical energy induced via wave action. This precession motion is initiated through the dynamic coupling between pitch motion and flywheel rotation. To prevent gyroscope stabilization, an eccentric mass is integrated, serving the purpose of inducing elastic recoil. For the enhancement of energy harvesting and resonance tuning, a well-tailored PTO control logic can be incorporated [52]. In the scope of this study, the PTO control logic description is provided in subsequent paragraphs. In the pursuit of investigating the role of uncertainty in the WEC design process, it is imperative to conduct a series of numerical experiments through simulations. Consequently, consideration must be given to the selection of an appropriate numerical model. A state-of-the-art perspective on the topic of numerical modeling applied to the WEC case is presented in [53].

In this section, the ISWEC modeling details are described. In Figure 3, the ISWEC functional diagram is presented. The block chart provides a concise visual representation of the device's behavior, encompassing forces and reactions between the floater and the gyroscope, as well as the control actions and signal flow of the PTO system [54]. In the image, we can recognize the following elements:

- $\vec{f}_f$ is the combined gravitational and fluid forces acting on the floater.
- T_ε and $\vec{f}_{gf}$ represent the gyroscopic effects exchanged between the floater and the gyroscope units.
- T_{PTO} is used to represent the PTO reaction acting on the gyroscope shaft.
- $\mathbf{X}_f$, $\mathbf{X}_\varepsilon$, and $\mathbf{X}_{PTO}$ describe the motion of the floater, gyroscope, and PTO, respectively.
- $\mathbf{C}_{PTO}$ stands for the control signals generated via the ISWEC's control system.

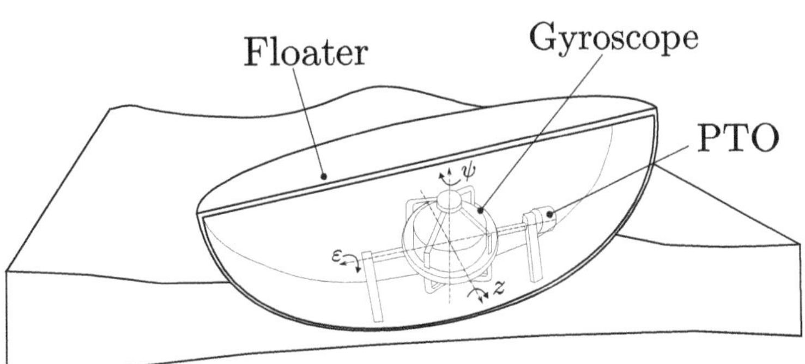

Figure 2. ISWEC device technology graphical representation.

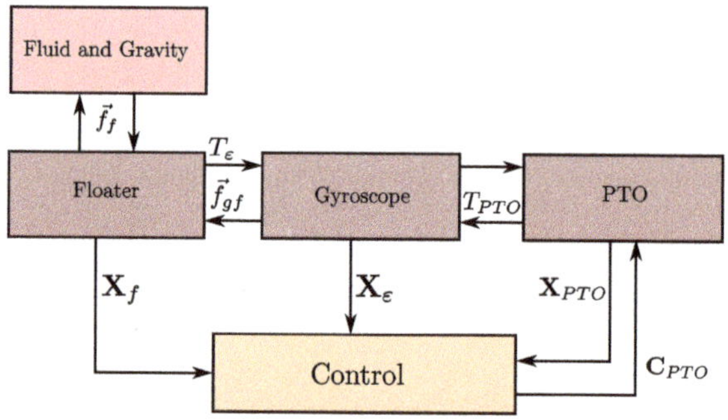

Figure 3. ISWEC block diagram, adapted from [54].

The mathematical equations that govern the ISWEC model were derived from the coupling of the WEC hydrodynamic with the gyroscope and PTO dynamics. As suggested in [54], a classic time domain representation is well suited for a more realistic and advanced simulation phase of WEC development than the frequency domain approach; that is, via the time domain, it is possible to consider non-linearities in the model [55]. Following the formulation presented in [54], the time domain system was composed with the involvement of the floater equation and the radiation forces' approximation, together with the gyro-pendulum model and the consideration of gyroscopic effects. Therefore, the non-linear state-space Time Domain Model (TDM) equations can be written in the following form:

$$\mathbf{M}_T\ddot{\mathbf{X}}_T(t) + \mathbf{B}_T\dot{\mathbf{X}}_T(t) + \mathbf{K}_T\mathbf{X}_T(t) + \mathbf{\Theta}_T(\ddot{\mathbf{X}}_T(t), \dot{\mathbf{X}}_T(t), \mathbf{X}_T(t)) = \mathbf{f}_T(t). \quad (3)$$

For a system of dimension n_T, the TDM model formulation of Equation (3) includes the kinematic variable vector $\mathbf{X}_T(t) \in \mathbb{R}^{n_T}$, the mass matrix $\mathbf{M}_T \in \mathbb{R}^{n_T \times n_T}$, the damping matrix $\mathbf{B}_T \in \mathbb{R}^{n_T \times n_T}$, the stifness matrix $\mathbf{K}_T \in \mathbb{R}^{n_T \times n_T}$, the non-linear function $\mathbf{\Theta}_T : \mathbb{R}^{n_T} \to \mathbb{R}^{n_T}$, and the external time-dependent excitation forces $\mathbf{f}_T(t) \in \mathbb{R}^{n_T}$. Several non-linearities affect the term $\mathbf{\Theta}_T$, deriving from the floater, gyro-pendulum and PTO subsystems.

Alternatively, Frequency Domain Models (FDMs) offer a computationally efficient solution via the boundary element method. To adopt an FDM means to accept some assumptions related to the realistic non-linear system behavior. The two presupposition characteristics of the ISWEC FDM regard the gyroscopic torques and pendulum elastic recall and the exclusion of saturations from the PTO model. That is, the ISWEC system is assumed to work only in the linear PTO model zone and the gyroscopic torques and pendulum elastic recall are linearized around the equilibrium point of the precession axis. After those assumptions, the model can be written in a matrix form concerning a harmonic input at the intended angular frequency ω:

$$[-\omega^2\mathbf{M}_F(\omega) + i\omega\mathbf{B}_F(\omega) + \mathbf{K}_F]\hat{\mathbf{X}}_F = \mathbf{H}_{\eta F}(\omega)\hat{\eta}, \quad (4)$$

where $\mathbf{M}_F(\omega) \in \mathbb{R}^{n_F \times n_F}$ is the frequency-dependent mass matrix (which includes the floater mass matrix, the gyroscope inertia, and the added mass contribution), $\mathbf{B}_F(\omega) \in \mathbb{R}^{n_F \times n_F}$ is the frequency-dependent damping matrix, $\mathbf{K}_F \in \mathbb{R}^{n_F \times n_F}$ is the stiffness matrix, and $\mathbf{H}_{\eta F}(\omega) \in \mathbb{C}^{n_F}$ is the complex transfer function between the wave profile η and the resultant excitation forces F, while n_F is the dimension of the frequency domain system. The FDM presentation also allows us to fully describe the statistical properties of ISWEC outputs in terms of the power spectral density (PSD) of WEC displacements $\mathbf{S}_{X_F X_F}(\omega)$ [54]. The latter statement is explained because, when a Gaussian stochastic input passes through a linear transfer function, the output also follows a Gaussian distribution [53]. Furthermore, the PDF of sea wave elevation η can be considered to be in accordance with the assumption of a Gaussian-distribution sea [56]. Therefore, when adopting the FDM representation,

Gaussian PDF properties can be assumed for both floater and gyroscope displacements and evaluated, via the transfer function $\mathbf{H}_{\eta \mathbf{X}_F}(\omega)$, using Equation (6) [54]:

$$\mathbf{S}_{X_F X_F}(\omega) = \mathbf{H}_{\eta X_F}(\omega) S_{\eta\eta}(\omega) \mathbf{H}^*_{\eta X_F}(\omega). \tag{5}$$

In Equation (5), the notation * stands for the complex conjugate operator. Moreover, $S_{\eta\eta}$ represents the wave elevation PSD.

$$\mathbf{H}_{\eta X_F}(\omega) = \frac{\mathbf{H}_{\eta F}(\omega)}{-\omega^2 \mathbf{M}_F(\omega) + i\omega \mathbf{B}_F(\omega) + \mathbf{K}_F}. \tag{6}$$

In order to incorporate non-linear factors to enhance the model's accuracy, the Spectral Domain Model (SDM) representation can be considered. SDM's purpose is to obtain an approximate solution of a generic TDM while avoiding the extensive computational overhead associated with it. SDM's practical implementation is accomplished by describing the system as probabilistic, considering its inputs as a stochastic ergodic process [53]. Hence, as observed earlier in the context of an FDM, when the wave spectrum is subjected to a suitable transfer function, it generates a probabilistic prediction of the WEC response [54]. Therefore, in terms of the PSD of WEC displacements, the following applies:

$$\mathbf{S}_{X_S X_S}(\omega) = \mathbf{H}_{\eta X_S}(\omega) S_{\eta\eta}(\omega) \mathbf{H}^*_{\eta X_S}(\omega), \tag{7}$$

where $\mathbf{H}_{\eta X_S}(\omega) \in \mathbb{C}^{n_S}$ is the SDM transfer function between WEC displacement and wave elevation, n_S is the dimension of the spectral domain system, and $\mathbf{S}_{X_S X_S}(\omega)$ is the ISWEC outputs' PSD. In SDM, the generic system equations can be written as follows:

$$\mathbf{M}_S \ddot{\mathbf{X}}_T(t) + \mathbf{B}_S \dot{\mathbf{X}}_T(t) + \mathbf{K}_S \mathbf{X}_T(t) = \mathbf{f}_T(t), \tag{8}$$

where the $\mathbf{X}_T$ vector identifies the hull's motion ($\mathbf{X}_f$), and the matrix $\mathbf{M}_S \in \mathbb{R}^{n_S \times n_S}$ is the equivalent mass matrix, $\mathbf{B}_S \in \mathbb{R}^{n_S \times n_S}$, is the equivalent damping matrix, and $\mathbf{K}_S \in \mathbb{R}^{n_S \times n_S}$ is the equivalent stiffness matrix.

Moreover, the following applies:

$$\mathbf{M}_S = \mathbf{M}_T + \mathbf{M}_{eq}, \tag{9}$$
$$\mathbf{B}_S = \mathbf{B}_T + \mathbf{B}_{eq}, \tag{10}$$
$$\mathbf{K}_S = \mathbf{K}_T + \mathbf{K}_{eq}. \tag{11}$$

These matrices are composed of a first linear part related to the linear terms of Equation (3) and a second part that represents the statistical time-varying behavior of the time domain system's non-linear terms. Therefore, this second part of each matrix is defined using the gradient and the expected value $\langle . \rangle$ operator as follows:

$$\mathbf{M}_{eq} = \langle \nabla_{\ddot{\mathbf{X}}} \mathbf{\Theta} \rangle, \tag{12}$$
$$\mathbf{B}_{eq} = \langle \nabla_{\dot{\mathbf{X}}} \mathbf{\Theta} \rangle, \tag{13}$$
$$\mathbf{K}_{eq} = \langle \nabla_{\mathbf{X}} \mathbf{\Theta} \rangle. \tag{14}$$

where the $\nabla_{\mathbf{X}}$ operator defines the gradient with respect to $\mathbf{X}$. The SDM approach involves the minimization of the mathematical expectation of the discrepancy between the non-linear system and its linearized counterpart, which is represented as e_S:

$$e_S = \Theta_T(\mathbf{X}_T(t), \dot{\mathbf{X}}_T(t), \ddot{\mathbf{X}}_T(t)) - \mathbf{M}_{eq} \ddot{\mathbf{X}}_T(t) - \mathbf{B}_{eq} \dot{\mathbf{X}}_T(t) - \mathbf{K}_{eq} \mathbf{X}_T(t). \tag{15}$$

After the statistical linearization of the system is performed, it is possible to evaluate the SDM transfer function:

$$\mathbf{H}_{\eta X_S}(\omega) = \frac{\mathbf{H}_{\eta S}(\omega)}{-\omega^2\mathbf{M}_S(\omega) + i\omega\mathbf{B}_S(\omega) + \mathbf{K}_S}. \tag{16}$$

This problem lacks an analytical solution, necessitating an iterative procedure, which is presented in [54]. The use of the spectral model is justified as long as the assumption of the system's Gaussian behavior remains valid. If the previously described assumption were to fail (e.g., in the presence of dominant non-linear forces), the system would exhibit non-Gaussian behavior, making the spectral model inaccurate and erroneous.

While both spectral and frequency domain models may sacrifice some degree of accuracy compared to time domain simulations, they are recommended for analyses seeking to compare different WEC devices and design solutions, as suggested in previous works [54] and the literature [53]. For the forthcoming analyses, we opted for the spectral domain as our mathematical model of choice. Given the importance of the process of converting wave energy into the electrical energy of the PTO, a brief description of its model is provided below. The mechanical PTO consists of a one-stage gearbox connecting the gyroscope shaft to the electrical generator. The control logic used in this work for the PTO was founded on the principle of impedance matching, modeled as a spring-damping system with a fixed saturation constraint to avoid the electro-mechanical generator nominal work area. This kind of control logic allowed us to define the PTO torque in Equation (17):

$$T_{PTO} = \begin{cases} k\varepsilon + c\dot{\varepsilon} & \text{if } -T_s \leq k\varepsilon + c\dot{\varepsilon} \leq T_s \text{ and } -P_s \leq k\varepsilon\dot{\varepsilon} + c\dot{\varepsilon}^2 \leq P_s, \\ T_s & \text{if } k\varepsilon + c\dot{\varepsilon} > T_s \text{ and } -\frac{P_s}{T_s} \leq \dot{\varepsilon} \leq \frac{P_s}{T_s}, \\ -T_s & \text{if } k\varepsilon + c\dot{\varepsilon} < -T_s \text{ and } -\frac{P_s}{T_s} \leq \dot{\varepsilon} \leq \frac{P_s}{T_s}, \\ \frac{P_s}{\dot{\varepsilon}} & \text{if } k\varepsilon\dot{\varepsilon} + c\dot{\varepsilon}^2 > P_s \text{ and } \dot{\varepsilon} \geq \frac{P_s}{T_s} \text{ or } \dot{\varepsilon} \leq -\frac{P_s}{T_s}, \\ -\frac{P_s}{\dot{\varepsilon}} & \text{if } k\varepsilon\dot{\varepsilon} + c\dot{\varepsilon}^2 < -P_s \text{ and } \dot{\varepsilon} \geq \frac{P_s}{T_s} \text{ or } \dot{\varepsilon} \leq -\frac{P_s}{T_s}. \end{cases} \tag{17}$$

Parameters k and c represent stiffness and damping coefficients, respectively. T_s corresponds to the saturation torque, and P_s the saturation power. The ISWEC system employs a "slow-tuning" approach [54], adjusting control parameters such as the flywheel rotational velocity $\dot{\psi}$, c, and k in response to the spectral characteristics of the incoming sea state to maximize the energy harvested. This method simplifies control compared to wave-by-wave strategies, optimizing power production by adapting parameters for the entire duration of a sea state determined through its PSD. The "slow-tuning" concept signifies parameter adjustments based on evolving sea-state spectral properties while disregarding instantaneous wave profiles.

3.2. Framework

A classic Monte Carlo process was adopted with the aim of propagating uncertainties and avoiding assumptions inherent in the procedure. Two variables were defined as being subject to uncertainty. The first parameter was the device's pitch inertia $M_{(5,5)}$. Previous analyses [45] highlight its impact on device performance. To propagate uncertainty concerning this variable, it was necessary to define a probability distribution for the associated uncertainties. A uniform distribution within the probability box defined by the interval $[-10\%M_{(5,5),0}, +10\%M_{(5,5),0}]$ was chosen for the uncertainty of $M_{(5,5)}$, where $M_{(5,5),0}$ is the nominal expected value for the parameter. The choice was made after considering the effects on the pitch inertia of the device due to variations in the distribution of masses of the stern and bow ballasts. Consistent with the considerations for the first parameter described above, the second parameter subjected to real-world uncertainties was the position along the z-axis of the device's center of gravity CoG_z. For this parameter, the uncertainty was modeled as an additive variation from the nominal value $CoG_{z,0}$: $[CoG_{z,0} - \Delta_{CoG_z}, CoG_{z,0} + \Delta_{CoG_z}]$, describing again a uniform distributed probability-box. Different values of Δ_{CoG_z} additive noise were set, and their relative impact in the robustness quantification was observed. The output of the robustness analyses was the normalized power ($P_{out,adim}$) generated through the device via harvesting wave energy, denoted as P_{out}, normalized with respect to the maximum generated power in the whole dataset of the analyzed device $P^{max}_{out,adim}$:

$$P_{out,adim} = \frac{P_{out}}{P^{max}_{out}}. \tag{18}$$

While the literature includes studies on optimizing sampling in terms of quantity, distribution, and parameter space geometry [49], the focus for this study was solely calibrating the number of samples. To determine the minimum number of samples necessary to achieve a good system PDF representation, the focus was the convergence of standard deviation, as illustrated in Figure 4; that is, a good convergence ratio was reached with 1000 samples. The notation symbol ^ represents the normalization of the nominal expected value of the output under examination, $f_m(\vec{x}_i, \vec{u}_0)$.

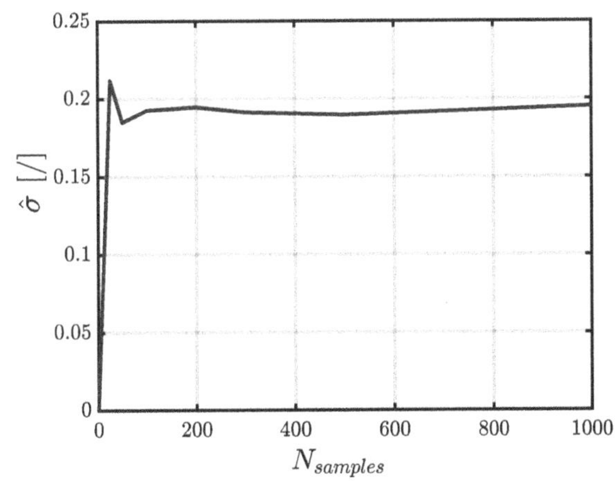

Figure 4. $\hat{\sigma}$ convergence concerning the number of samples for the Monte Carlo stochastic simulation approach.

Simultaneously, when observing that the trend of the standard deviation converged with an increasing number of samples, a significant rise in computational costs could also be noticed. This is visually reported in Figure 5: the computational burden follows a likely linear trend, also highlighting the impracticality of investing such a substantial amount of time (8 h) to conduct a complete robustness quantification process for a single device. This drawback becomes even more relevant when considering the necessity of repeating a high number of simulations, e.g., for an RDO framework, for which the evaluation of the objective function might need to be performed thousands of times for the RQ phase. To mitigate this challenge and maintain a stochastic RQ approach, the next step was to develop a suitable surrogate model that was also able to reduce the computational costs for the uncertainty propagation process. A surrogate model, also known as a metamodel, can take the form of a mathematical relationship or an algorithm that captures the relationships between input and output variables; i.e., the procedure typically involves studying the output and input relationships and then fitting the right metamodels to represent that behavior. That is, a surrogate model $\tilde{\mathcal{M}}(\vec{v})$ approximates the computationally expensive model $\mathcal{M}(\vec{v})$ [57].

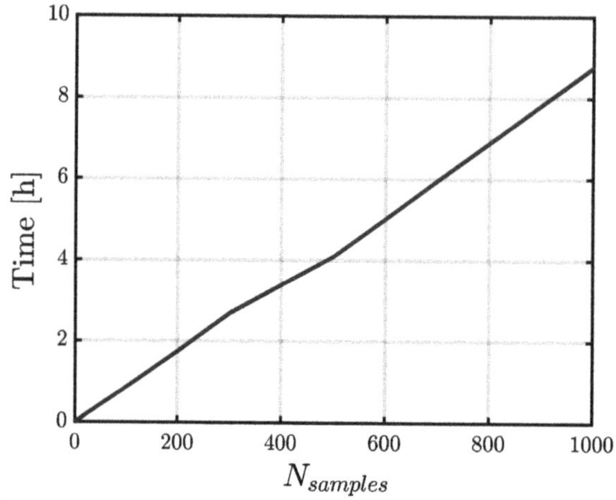

Figure 5. Computational time concerning the number of samples used for the Monte Carlo stochastic simulation approach. The computational equipment used for the present study was a 12thGenIntel(R)Core(TM)i7-12700H-2.30 GHz with 32.0 GB of installed RAM.

Highlighting the crucial aspects of achieving an efficient metamodel, it becomes possible to list the characteristics of what we define as an ideal surrogate model (a gold model, as referred to in [57]). The surrogate model we obtained is superior not only if it is characterized by good accuracy but also if it has the ability to be trained with a limited number of computationally expensive model samples, i.e., saving computational time.

The technique chosen was Gaussian process regression (GPR), which can be considered an extension of the Kriging method [58] and was already used in [59], where a data-driven and GPR-based approach was developed in order to address the optimization problem for the PTO control strategy of the Pendulum Wave Energy Converter. Considering a dataset containing N_{sample} joint observations of the extracted normalized power y_p and uncertain parameters $\vec{v}_p \in \mathbb{R}^b$, the following equation was used:

$$\mathscr{D} = \{(\vec{v}_p, y_p) : p = 1 \dots N_{samples}\}. \tag{19}$$

In the present study, $b = 2$ and each $\vec{v}_p$ represents a pair of $M_{(5,5)}$ and CoG_z. Our goal was to develop a metamodel $\tilde{\mathscr{M}}$, assuming that the observed extracted normalized power values followed a joint Gaussian distribution perturbed by an additional Gaussian noise ζ:

$$\mathscr{M}(\vec{v}) = \tilde{\mathscr{M}}(\vec{v}) + \zeta, \qquad \zeta \sim \mathscr{N}(0, \lambda). \tag{20}$$

Here, $\mathscr{M}(\vec{v})$ is the computational expensive model underlying the observations y_i, and $\tilde{\mathscr{M}}(\vec{v})$ is a Gaussian process defined as follows:

$$\tilde{\mathscr{M}}(\vec{v}) \sim GP(E(\vec{v}), \Omega(\vec{v}, \vec{v'})). \tag{21}$$

This Gaussian process is characterized by the mean function $E(\vec{v})$ and covariance function $\Omega(\vec{v}, \vec{v'})$. The fundamental concept behind GPR is that any set of observations has a joint normal distribution with expected values provided via $E(\vec{v_p})$ and a covariance matrix with entries represented by $\Omega(\vec{v_p}, \vec{v_j'})$ for $p, j = 1 \dots N_{sample}$. In practical terms, it is possible to designate $f_p(\vec{v}_p) = \tilde{\mathscr{M}}(\vec{v}_p)$ as the observed values of the Gaussian process (distinct from y_p due to noise subtraction) and $\tilde{\mathscr{M}}_*(\vec{v}_*)$ as a random variable associated with the value of interest $f_*(\vec{v}_*)$, signifying the prediction for $\vec{v}_*$. Therefore, it is possible to compute the model prediction as the mean value of $\tilde{\mathscr{M}}_*(\vec{v}_*)$. In [60], a formal demonstration of the GPR mathematical aspects can be found.

To test the GPR approach's accuracy and efficiency, a comparison with an expensive Monte Carlo simulation with 1000 samples was performed. Two different metamodels were tested with different expensive observations (50 and 100), and Latin Hypercube Sampling (LHS) was chosen as a sampling method. LHS enables the creation of an N-dimensional grid, with its density varying based on the desired number of samples. In the present case of study, N remained constant at 2. This grid was comprised of a set number of cells, corresponding to the total number of desired samples, ensuring that each cell was evaluated with a sample. This approach prevents the potential concentration of samples in specific regions of the parameter space (oversampling) or the under-sampling of certain areas, resulting in a uniform and well-distributed sampling pattern.

In Figure 6, the two obtained surfaces are represented, and in Table 3 the results are summarized. For the procedure, the surrogate model training and test datasets were different and without common samples. All the results were normalized with respect to the nominal expected value due to a nondisclosure agreement about the technology.

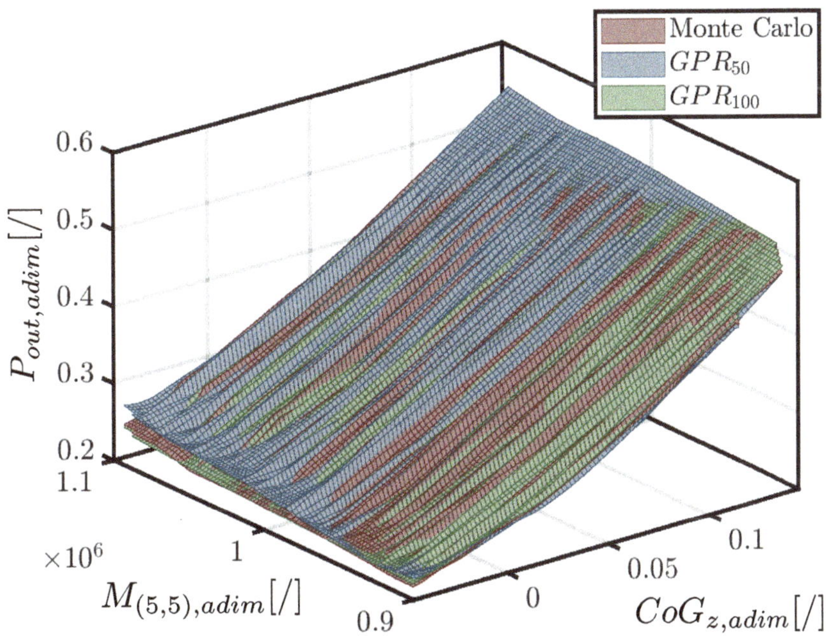

Figure 6. Expensive model $\mathcal{M}(\vec{v})$ surface (in red) compared with the model surfaces obtained via metamodels $\tilde{\mathcal{M}}_{50}(\vec{v})$ (in blue) and $\tilde{\mathcal{M}}_{100}(\vec{v})$ (in green). The Figure reports a third-model accuracy comparison visualization after the one described in Figure 7. The surfaces were obtained by testing the three different models for the same device when affected by the same uncertainty dataset.

The outcomes show a clear overlap among the three surfaces. This means that the GPR potential is saturated; that is, increasing the number of observations resulted in a very similar surface. On the other hand, the accuracy of the metamodel can be considered good enough for statistical purposes. When looking at Figure 7, it is possible to reach the same conclusion. The distributions on the right (Figure 7b) differ for some localized values, but the general statistical behavior is similar among the two surrogates' PDF and the expensive models. Moreover, this is consistent with the left image (Figure 7a) of the same Figure 7: the majority of data points align closely with the bisector of the plane defined by the real model's normalized power output ($P_{out,adim}$—True on the y-axis) and the normalized power values predicted via surrogate models ($P_{out,adim}$—Pred on the x-axis). In the case of a perfectly accurate model, all data points would lie precisely on the bisector. Furthermore, there is evident minimal dispersion among the data points. The most significant deviations from the bisector (which describes the ideal prediction) are confined to a specific, limited region within the domain populated by only a few data points (<1%). The lack of samples made it difficult for the surrogate model to predict the output variable in those areas. When looking at Figure 7b, it can be seen that the green (GPR_{100}) and blue (GPR_{50}) curves do not follow the course of the red curve ($MonteCarlo_{1000}$) with the same accuracy as with the other intervals.

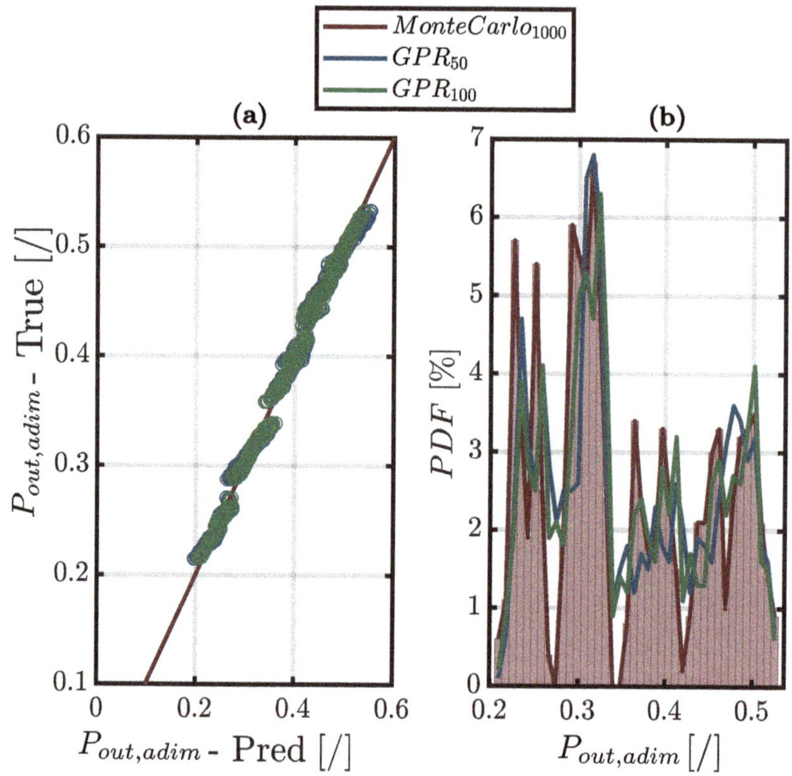

Figure 7. Accuracy tests on $\tilde{\mathcal{M}}_{50}(\vec{v})$, $\tilde{\mathcal{M}}_{100}(\vec{v})$, and $\mathcal{M}(\vec{v})$ PDF and prediction. Subfigure (**a**) describes the prediction accuracy by comparing the GPR prediction ($P_{out,adim}$—Pred) with the computationally expensive mathematical model, both affected by the same uncertainties. Subfigure (**b**) reports the statistical behavior in terms of PDF for the three different models.

Table 3. Comparison of an expensive Monte Carlo model $\mathcal{M}(\vec{v})$ simulation and a GPR surrogate model $\tilde{\mathcal{M}}(\vec{v})$ simulation built with 50 and 100 samples, with an associated relative error $\varepsilon_{rr,50}$ $\varepsilon_{rr,100}$.

Index	$\mathcal{M}(\vec{v})$	$\tilde{\mathcal{M}}_{50}(\vec{v})$	$\tilde{\mathcal{M}}_{100}(\vec{v})$	$\varepsilon_{rr,50}$	$\varepsilon_{rr,100}$
NRMSE [/]	—	0.03	0.02	—	—
Time [min]	527.66	27.00	52.80	—	—
$\hat{\mu}$ [/]	1.0811	1.0784	1.0821	0.25 [%]	0.09 [%]
$\hat{\sigma}$ [/]	0.2760	0.2798	0.2743	1.38 [%]	0.62 [%]
$\sup \hat{f}$ [/]	0.6456	0.6216	0.6180	3.72 [%]	4.28 [%]
$\hat{\mu} + k\hat{\sigma}$ [/]	2.7369	2.7570	2.7279	0.73 [%]	0.33 [%]
R [/]	0.3687	0.3690	0.3684	0.08 [%]	0.08 [%]
Q [/]	0.7219	0.7230	0.7293	0.15 [%]	0.96 [%]

Another indicator of the good statistical approximation of the obtained surrogate model is provided via the value of the normalized root mean squared error (NRMSE).

As suggested in [49], an NRMSE < 0.1 is characterized by a reasonable predictive capacity. Thus, the surrogate model $\mathscr{M}(\vec{v})$ is suitable for analysis, which would capture the overall system behavior, but accuracy in the prediction of specific points cannot be guaranteed. The trade-off between the speed and accuracy of the model is significantly directed towards the choice of $\tilde{\mathscr{M}}_{50}(\vec{v})$ due to the metamodel accuracy, which is very similar to the one obtained via $\tilde{\mathscr{M}}_{100}(\vec{v})$ with savings in computational costs equal to 25.8 min. The conservation of time resources is equal to 96% of the total amount of time requested for the RQ of a single device exploiting the $\tilde{\mathscr{M}}_{50}(\vec{v})$ model: using 50 samples, it was possible perform the RQ for about twice the number of devices as would have been processed in the same time frame with $\tilde{\mathscr{M}}_{100}(\vec{v})$ and with comparable and acceptable accuracy.

4. Results

The previously delineated framework was applied to the results of a nominal optimization process performed via a genetic algorithm; the selected site is near Pantelleria Island. In Figure 8, the scatter diagram for both the normalized annual sea energy density and occurrences is presented. Moreover, in the same Figure 8, the red dots represent the sea state set of representative waves (modeled as irregular waves according to a Jonswap spectrum) that were used in order to simulate the performances for each individual of the optimization algorithm. The waves were chosen to cover the relevant regions in terms of the energy and frequency of the scatter diagrams. The same representative sea states were employed in the RQ framework.

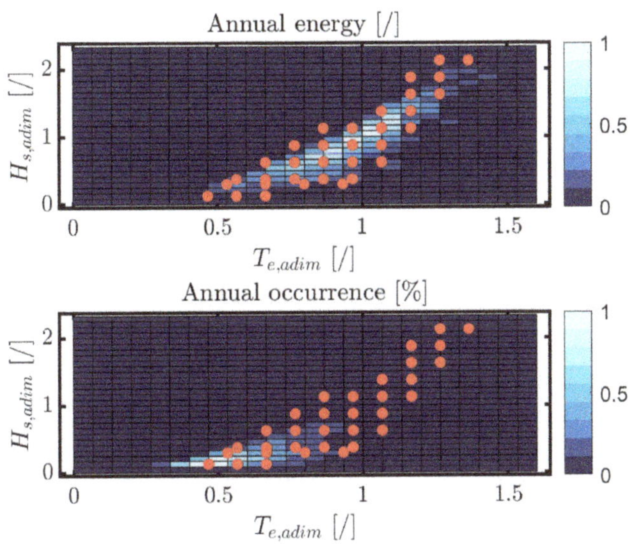

Figure 8. Pantelleria Island scatter plot of occurrences and annual marine energy per meter. The selection of representative waves (red dots) was performed to comprehensively cover significative regions in terms of the energy and occurrence of the site scatter diagrams. Both the annual energy and the occurrences were normalized with respect to their maximum values.

In Figure 9, the optimization results are reported in terms of $P_{out,adim}$ and $CapEx_{adim}$ (Equation (22)) as objective functions:

$$CapEx_{adim} = \frac{CapEx}{CapEx^{max}}. \tag{22}$$

The initial phase of this analysis centered on the examination of Pareto frontier devices with a direct comparison of outcomes following the assessment of previously expounded robustness indices for each WEC. Three different values of Δ_{CoG_z} were used to compare the responsiveness of the system robustness.

When examining Figure 10c, related to index $\hat{\mu}$ in Figure 10, the results show that the majority of devices were located in the neighborhood of the normalized value of the mean performance ($\hat{\mu}$) around 1, exhibiting good average behavior when subjected to the disturbances imposed on uncertain parameters.

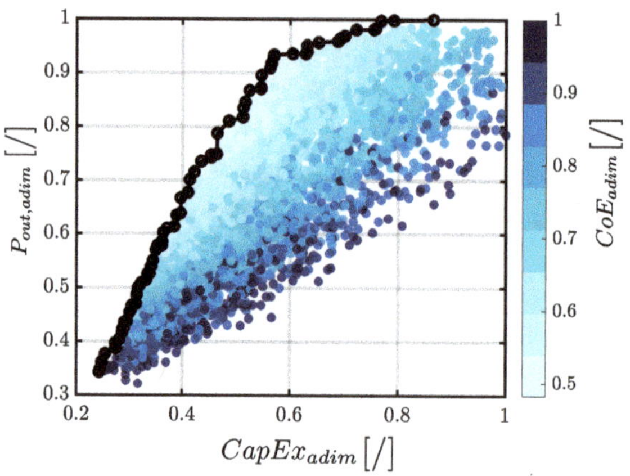

Figure 9. Devices obtained during the nominal optimization via GA. In black, the Pareto frontier devices are depicted.

As anticipated in the preceding sections, it was crucial to consider additional robustness metrics to obtain comprehensive information regarding the statistical trends of the examined performance.

When analyzing the Q_i measurements in Figure 10b, it becomes evident that devices belonging to the nominal Pareto-set present PDF, in 90% of the occurrences, exceed values even lower than 65% of their relative $f_m(\vec{x}_i, \vec{u}_0)$, highlighting a difference between the claimed nominal performance and the performance achieved when taking into account uncertainties.

The achieved outcomes could be helpful during the decision-making phase. When a choice needs to be made in order to identify a singular device among the one belonging to the nominal Pareto set, the designer can rely on the robustness classification performed via the studied RIs.

The analysis now moves towards a broader pool of devices, which was selected from those involved in the whole nominal optimization process. The selection criteria were based on the normalized cost of energy (CoE_{adim}), defined in Equations (23) and (24) as the ratio between the device's capital cost ($CapEx$ in $[MEuro]$) and the associated device's annual energy production (AEP in $[\frac{MWh}{y}]$), considering an average device life time ($life_{25}$) of 25 years:

$$CoE = \frac{CapEx}{life_{25}AEP}, \tag{23}$$

and normalized regarding the maximum dataset CoE:

$$CoE_{adim} = \frac{CoE}{CoE^{max}}. \tag{24}$$

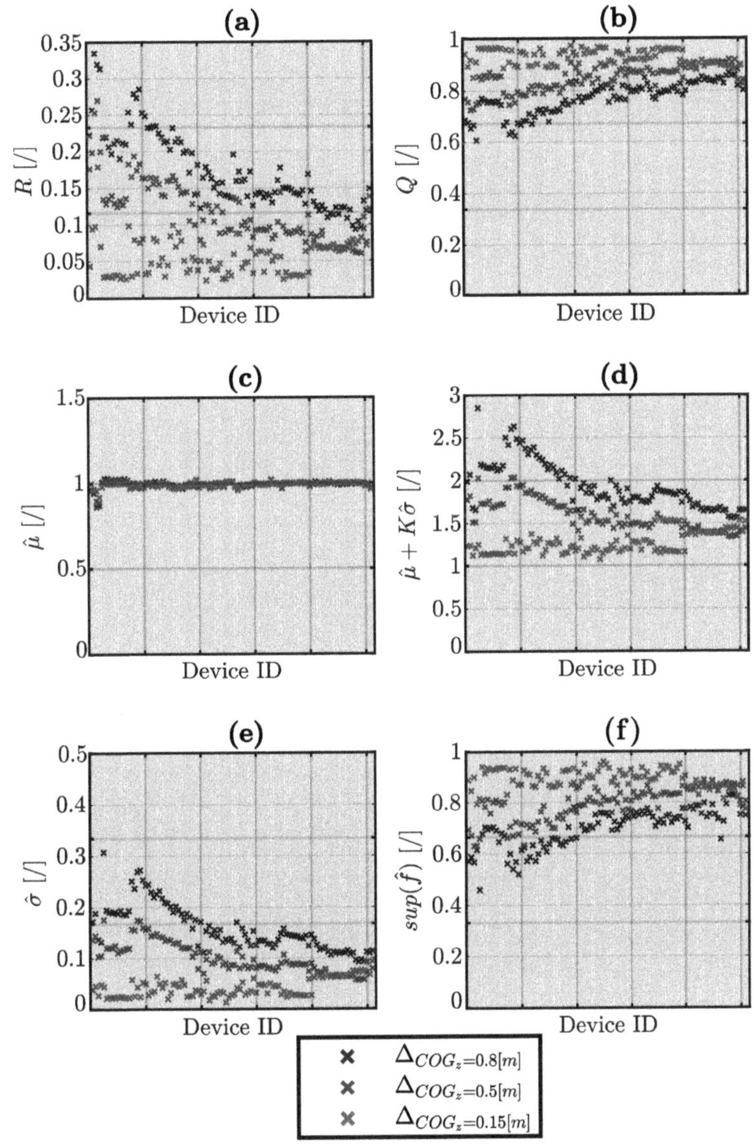

Figure 10. Pareto frontier devices' RI comparison for three different Δ_{CoG_z} values. Each cross corresponds to one singular device, and the different colors refer to different Δ_{CoG_z} parameters. Each subfigure (**a**–**f**) shows the Pareto set devices' comparison for each of the robustness metrics described in Table 1.

The capital expenditures for the device $CapEx$ are composed of three primary factors: hull materials and construction, unit materials and construction, and PTO costs. Mooring system and operational and maintenance expenditures were excluded from the cost evaluation in this study due to their device-specific nature. The robustness of five thousand devices was quantified.

Also, in this case, the same three different Δ_{CoG_z} perturbations used above were applied. In the context of uncertainty propagation and result acquisition, the delineation of new Pareto frontiers occurred by establishing threshold values for each index. The limit values established above (for indices for which maximization indicated superior performance,

Figure 11b,c,f) or below (for indices for which minimization indicated optimal performance, Figure 11a,d,e), for which robustness values were referred to, an individual WEC system's performance can be considered acceptable. Therefore, these frontiers are not contingent upon the normalized nominal power output ($P_{out,adim}$) but are associated with the intrinsic robustness index pertinent to this parameter. From the results in Figure 11, it is evident that the imposed uncertainties play a significant role in shaping the Pareto set, considering the threshold set for the studied metrics. The most notable deviations were observed among the Pareto sets related to indices Q, $\hat{\sigma}$, and sup $\hat{f}$, respectively, in Figure 11b,e,f).

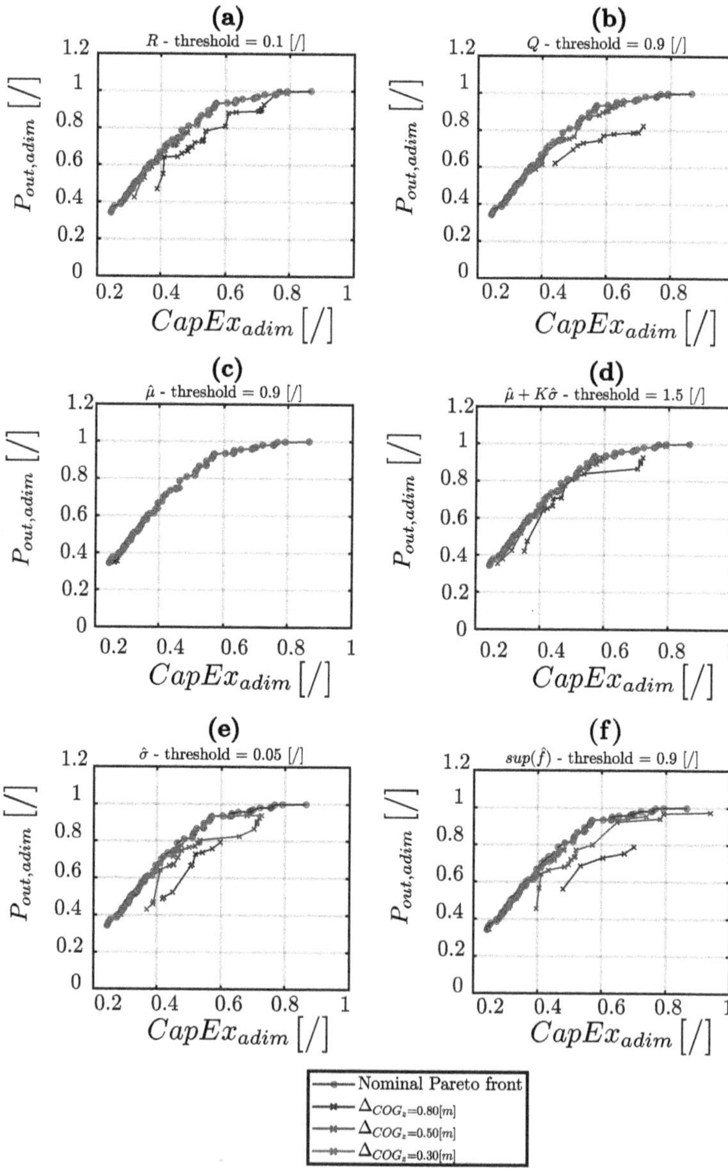

Figure 11. New Pareto frontiers for fixed RI thresholds. Each subfigure (**a**–**f**) illustrates the different Pareto sets for each of the robustness metrics described in Table 1, with a fixed RI's threshold and three Δ_{CoG_z} values.

The standard deviation threshold was set to constrain its value equal to 5% of the nominal expected value of the ith device. Moreover, the WCS for the singular system was bound to be equal to or greater than the 90% of $f_m(\vec{x}_i, \vec{u}_0)$. The Q threshold was imposed to ensure that each perturbed WEC guaranteed a statistical trend with the 90% of the occurrences defined by a $P_{out,adim}$ greater than the 90% of the nominal expected values. In these three cases, the new robust-optimal devices exhibited significantly lower power levels compared to those belonging to the nominal Pareto set, revealing a trade-off between robustness and optimal performances. In Figure 11e, nearly 21% of the performance reduction could be achieved, and the $P_{out,adim}$ could decrease from 0.94–0.78 to 0.74–0.56 for Q and sup $\hat{f}$, respectively, in Figure 11b,f. Moreover, imposing a threshold concerning device robustness led not only to a degradation in performance but also to a reduction in the Pareto set dimension itself, compelling the designer to make choices within a more limited range of objective functions ($P_{out,adim}$ and $CapEx_{adim}$). The mentioned Pareto size reduction is pronounced in Figure 11b,e,f.

Then, the study's focus was directed toward emphasizing how the thresholds' stringency magnitude impacts the robustness metric Q. Three distinct levels were investigated: 0.85 (the least stringent), 0.9 (medium-stringent), and 0.95 (the most rigorous). The outcomes are represented in Figure 12. Also, in this case, a trade-off between performance and robustness threshold severity was present.

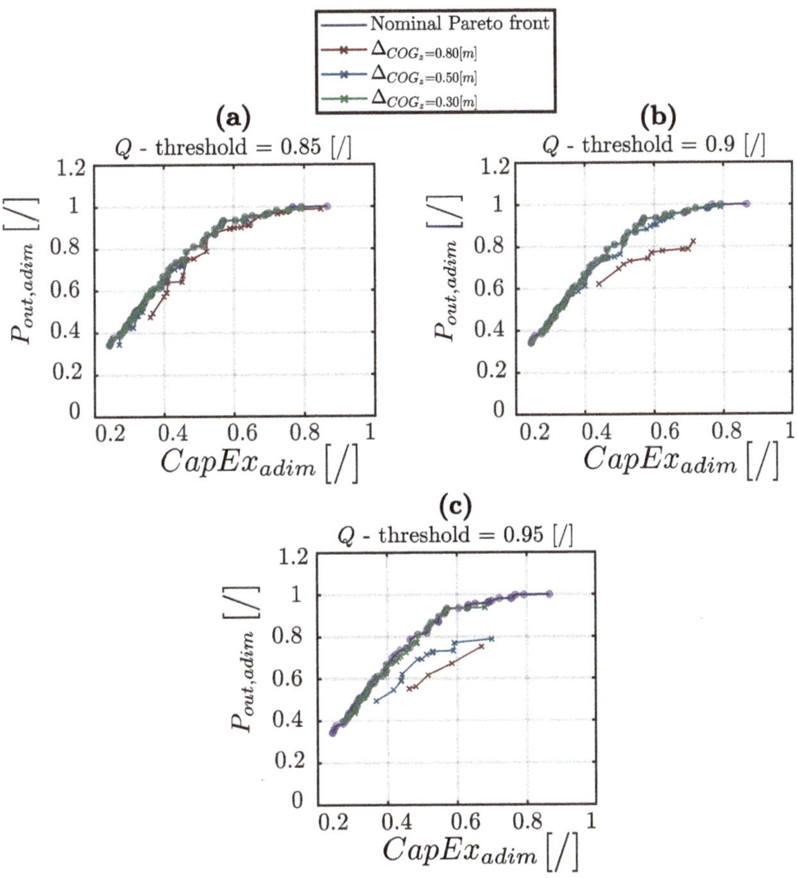

Figure 12. New Pareto frontiers for different Q thresholds' stringency. Each subfigure (**a–c**) illustrates the different Pareto sets for each of the three fixed Q threshold and three Δ_{CoG_z} values.

At last, two different devices were selected from the whole pool of systems, namely D1 and D2, and a robustness performance assessment for the technology with respect to the site under analysis was carried out, i.e., Pantelleria Island (Figure 8). The case study was investigated in order to compare the performance robustness of the two different systems (D1 and D2), which both belong to the Pareto frontier and, therefore, are not dominant with respect to each other; i.e., neither of the two is univocally optimal compared to the other in relation to the performance examined ($CapEx_{adim}$, $P_{out,adim}$). The devices' data are reported in Table 4.

Table 4. D1 and D2 data.

Device	$CapEx_{adim}$ [/]	$P_{out,adim}$ [/]	$\hat{\sigma}$ [/]
D1	0.45	0.74	0.13
D2	0.27	0.39	0.31

The D1–D2 set of devices is characterized by a good trade-off between $CapEx_{adim}$ and performance for both systems: a medium-high normalized power for medium $CapEx$ for D1 and medium-low $P_{out,adim}$ but a low cost for D2. Although the two WECs are characterized by a large gap in terms of robustness, for those two devices, this value was quantified via the $\hat{\sigma}$ index, i.e., $\Delta\sigma = 0.18[/]$.

In Figure 13, the annual energy distribution over the normalized wave period and height is compared between D1 and D2. Here, light colors illustrate the variability in distribution when uncertainties were taken into account. On the x-axis, the wave period and significant height were normalized relative to the site's most energetic period and wave height, denoted as the ratio between the relative sea state data and the site's most energetic period ($T_{e,site}^{maxEn}$) and significant wave height ($H_{s,site}^{maxEn}$):

$$T_{e,adim} = \frac{T_e}{T_{e,site}^{maxEn}}, \tag{25}$$

$$H_{s,adim} = \frac{H_s}{H_{s,site}^{maxEn}}. \tag{26}$$

The devices' AEP was normalized relative to its respective distribution peak value (AEP^{max}):

$$AEP_{adim} = \frac{AEP}{AEP^{max}}. \tag{27}$$

The plot clearly shows the difference in terms of the standard deviation $\hat{\sigma}$ for the two systems. The shading representative of the distribution fluctuation was wider for D2 than for D1. On the contrary, D1 showed a smaller range of variability. In this scenario, the decision-making process may favor the utilization of the D1 device due to its reduced susceptibility to uncertainties and higher productivity despite an increase in $CapEx_{adim}$ amounting to 18% of the maximum device cost in the whole dataset $CapEx^{max}$.

5. Conclusions and Future Works

The exploration of uncertain parameters' impacts in real-world scenarios is in its early stages within the context of WEC design optimization processes. In the current literature addressing uncertainty treatment in WECs, the majority of attention is directed toward modeling control logics that are insensitive to uncertainties (i.e., robust control) and addressing uncertainties in the experimental phases of technology development.

When broadening the perspective to other research domains, it is observed that the most commonly employed methodologies in optimization processes taking into account perturbation are RO and RDO. Within the latter, particular emphasis is placed on robustness metrics aimed at quantifying the impact of parameter perturbations on the system's performance.

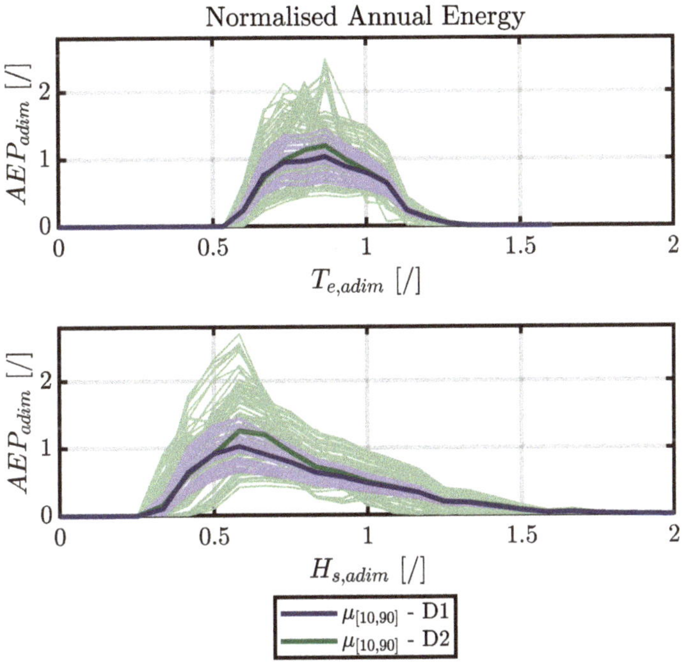

Figure 13. D1 (green) and D2 (violet) normalized annual energy distribution. The bold lines represent the distributions' mean value, excluding the 10th and 90th percentiles. Light colors describe the distributions' variability when uncertainties were taken into account.

The objective of this study was to investigate the information that various robustness indices can provide regarding optimal devices obtained following a nominal optimization process in the design of a WEC. For this purpose, a methodology based on stochastic uncertainty propagation via the Monte Carlo method was employed, exploiting a surrogate model (GPR) to reduce the computational cost of the RQ process.

The results highlight the necessity for attention to robustness evaluation in optimal WEC designs' optimization process. To support this thesis, the results of the present study indicate that devices belonging to the nominal Pareto set are characterized by a $P_{out,adim}$ statistical trend that, in 90% of occurrences, was found to be higher than just 65% of the expected nominal value. Robustness indices were then evaluated for all individuals in a dataset obtained during the various generations of a design optimization process conducted using GA. These additional results reveal a significant deviation between the nominal Pareto set and the new one sought by setting differently stringent thresholds for robustness metrics. Around a 20% robustness performance reduction was achieved in the analyzed case when considering the standard deviation $\hat{\sigma}$ index. Moreover, imposing a threshold on the device's robustness can decrease the $P_{out,adim}$ from 0.94–0.78 to 0.74–0.56 for Q and $\sup \hat{f}$, respectively. These findings represent novel information that was not obtainable previously from the nominal optimization process, and it can constitute valuable data in decision-making. Furthermore, outcomes justified the intention to proceed toward the implementation of RO processes for WEC design.

Ultimately, a performance robustness assesment analysis was conducted for two selected devices chosen from the nominal Pareto set, namely D1 and D2. The aim of the case study was to delineate the annual productivity distribution over the sea $H_{s,adim}$ and $T_{e,adim}$. D1 and D2 were distinguished by a favorable balance between performance metrics and different degrees of robustness; i.e., the distribution fluctuation was wider for D2 than for D1. Therefore, the outcomes show that D1 WEC might be the better choice during a decision-making phase due to its reduced sensitivity to uncertainties and improved AEP despite a cost increase of around 18% of the maximum device $CapEx$ in the whole dataset.

The main limitations of this study were the high computational costs resulting from the necessary expensive numerical simulations to build the employed GPR model and the restriction of uncertain parameters to only two. The confidentiality constraints present in this study imposed restrictions concerning both the quantification of uncertainties' impact on technology performance in absolute terms and the disclosure of devices' design data. Nonetheless, the results obtained from analyzing different metrics effectively convey information to the designer about the robustness of different devices with respect to their expected performance ratings, and they allow for comparisons between different devices, providing insights into their PDF performance and robustness.

Moreover, the present research has revealed a certain trade-off between performance and robustness, which is not a new topic in the WEC literature. This sort of balance has been widely investigated in the robust control (e.g., WCS) and robust energy maximizing approaches [12,17]. Similarly, the performance compromise was investigated in WECs' array field, e.g., with lower energy production costs, against the aggregate power performance degradation and the system sensitivity with respect to the wave interaction effects [61] or optimal array layout and efficiency [62].

Combining the obtained results and considering the limitations and trade-off highlighted in this study, future work, in addition to developing a WEC design RO framework, could involve a study of the impact that different types of control logic might have on the robustness of the conceptual design of a technology. Finally, the research should consider evaluating uncertainties in environmental and economic parameters, thus shifting attention towards the optimization of techno-economic indices such as the LCoE.

Author Contributions: Conceptualization, F.G.; methodology, F.G.; software, F.G., S.A.S. and M.B.; validation, F.G. formal analysis, F.G.; investigation, F.G. and S.A.S.; writing—original draft preparation, F.G. and G.G.; writing—review and editing, F.G., S.A.S., G.G., N.F., M.B., J.R., E.G. and G.M.; supervision, S.A.S., G.G., N.F. and G.M.; funding acquisition, G.M. All authors have read and agreed to the published version of the manuscript.

Funding: This project received funding from the Ministry of University and Research and PNRR's PhD scholarship program defined in DM-352 in collaboration with Eni S.p.A. This publication is part of the PNRR-NGEU project, which has received funding from the MUR—DM 352/2022. Moreover, this study was also carried out under Ministerial Decree no. 1062/2021, and it received funding from the FSE REACT-EU-PON Ricerca e Innovazione 2014–2020. This manuscript reflects only the authors' views and opinions; neither the European Union nor the European Commission can be considered responsible for them. Funding information: Ministero dell'Università e della Ricerca (FSE REACT-EU—PON Ricerca e Innovazione 2014–2020).

Institutional Review Board Statement: Not applicable.

Informed Consent Statement: Not applicable.

Data Availability Statement: The data obtained in this study are not eligible for sharing due to the confidential nature of the examined technology. Consequently, all research results were presented in a normalized format, maintaining alignment with the confidentiality requirements governing the study.

Acknowledgments: The authors thank Dario Basile from Eni S.p.a. for support during the present study.

Conflicts of Interest: The authors declare no conflicts of interest. The funders had no role in the design of the study, in the collection, analyses, or interpretation of data, in the writing of the manuscript, or in the decision to publish the results.

Abbreviations

The following abbreviations are used in this manuscript:

WEC	Wave energy converter
LCoE	Levelized cost of energy
RO	Robust optimization
RI	Robustness index
RQ	Robustness quantification
RDO	Robust design optimization
RBDO	Reliability-based design optimization
PDF	Probability density function
GPR	Gaussian process regression
LHS	Latin hypercube sampling
WCS	Worst-case scenario
PTO	Power take-off
FDM	Frequency domain model
TDM	Time domain model
SDM	Spectral domain model
PSD	Power spectral density
CoG	Center of gravity

Notes

1 A comprehensive overview of the historical and commercial efforts to develop these technologies can be found in [6].

2 For a normal distribution, approximately 68% of the data is within one standard deviation σ from the mean distribution value, roughly 95% within two σ, around 99.7% within three, and so on, almost reaching 100% with a six σ distance. This method is commonly used in the industrial product quality field [46].

References

1. Guterres, A. Carbon Neutrality by 2050: The World's Mosturgent Mission. In *United Nation Treaty Collection*; United Nations Secretary General: New York, NY, USA, 2020. Available online: https://www.un.org/sg/en/content/sg/articles/2020-12-11/carbon-neutrality-2050-the-world%E2%80%99s-most-urgent-mission(accessed on 2 February 2024).
2. IEA. *World Energy Outlook 2019*; International Energy Agency: Paris, France, 2019. Available online: https://www.iea.org/reports/world-energy-outlook-2019(accessed on 2 February 2024).
3. Reguero, B.; Losada, I.; Méndez, F. A global wave power resource and its seasonal, interannual and long-term variability. *Appl. Energy* **2015**, *148*, 366–380. [CrossRef]
4. IRENA. *Innovation Outlook: Ocean Energy Technologies*; IRENA: Abu Dhabi, United Arab Emirates, 2020. Available online: https://www.irena.org/-/media/Files/IRENA/Agency/Publication/2020/Dec/IRENA_Innovation_Outlook_Ocean_Energy_2020.pdf(accessed on 2 February 2024).
5. OES. *An Internation Vision for OCcean Energy*; Oceanic Energy System: Lisbon, Portugal, 2017. Available online: https://www.ocean-energy-systems.org/documents/24845-oes-vision-2017.pdf/(accessed on 2 February 2024).
6. Guo, B.; Ringwood, J.V. A review of wave energy technology from a research and commercial perspective. *IET Renew. Power Gener.* **2021**, *15*, 3065–3090. [CrossRef]
7. Babarit, A.; Bull, D.; Dykes, K.; Malins, R.; Nielsen, K.; Costello, R.; Roberts, J.; Ferreira, C.B.; Kennedy, B.; Weber, J. Stakeholder requirements for commercially successful wave energy converter farms. *Renew. Energy* **2017**, *113*, 742–755. [CrossRef]
8. Park, G.J.; Lee, T.H.; Lee, K.H.; Hwang, K.H. Robust design: An overview. *AIAA J.* **2006**, *44*, 181–191. [CrossRef]
9. Moritz Göhler, S.; Eifler, T.; Howard, T.J. Robustness metrics: Consolidating the multiple approaches to quantify robustness. *J. Mech. Des.* **2016**, *138*, 111407. [CrossRef]
10. Fusco, F.; Ringwood, J.V. Hierarchical robust control of oscillating wave energy converters with uncertain dynamics. *IEEE Trans. Sustain. Energy* **2014**, *5*, 958–966. [CrossRef]
11. Ringwood, J.V.; Mérigaud, A.; Faedo, N.; Fusco, F. An analytical and numerical sensitivity and robustness analysis of wave energy control systems. *IEEE Trans. Control Syst. Technol.* **2019**, *28*, 1337–1348. [CrossRef]
12. Faedo, N.; García-Violini, D.; Scarciotti, G.; Astolfi, A.; Ringwood, J.V. Robust moment-based energy-maximising optimal control of wave energy converters. In Proceedings of the 2019 IEEE 58th Conference on Decision and Control (CDC), Nice, France, 11–13 December 2019; IEEE: Piscataway, NJ, USA, 2019; pp. 4286–4291.
13. Schoen, M.P.; Hals, J.; Moan, T. Wave prediction and robust control of heaving wave energy devices for irregular waves. *IEEE Trans. Energy Convers.* **2011**, *26*, 627–638. [CrossRef]
14. Wahyudie, A.; Jama, M.; Saeed, O.; Noura, H.; Assi, A.; Harib, K. Robust and low computational cost controller for improving captured power in heaving wave energy converters. *Renew. Energy* **2015**, *82*, 114–124. [CrossRef]

15. Na, J.; Li, G.; Wang, B.; Herrmann, G.; Zhan, S. Robust optimal control of wave energy converters based on adaptive dynamic programming. *IEEE Trans. Sustain. Energy* **2018**, *10*, 961–970. [CrossRef]
16. Garcia-Violini, D.; Ringwood, J.V. Energy maximising robust control for spectral and pseudospectral methods with application to wave energy systems. *Int. J. Control* **2021**, *94*, 1102–1113. [CrossRef]
17. Farajvand, M.; Grazioso, V.; García-Violini, D.; Ringwood, J.V. Uncertainty estimation in wave energy systems with applications in robust energy maximising control. *Renew. Energy* **2023**, *203*, 194–204. [CrossRef]
18. Hiles, C.E.; Beatty, S.J.; De Andres, A. Wave energy converter annual energy production uncertainty using simulations. *J. Mar. Sci. Eng.* **2016**, *4*, 53. [CrossRef]
19. Mackay, E.B.; Bahaj, A.S.; Challenor, P.G. Uncertainty in wave energy resource assessment. Part 1: Historic data. *Renew. Energy* **2010**, *35*, 1792–1808. [CrossRef]
20. Mackay, E.B.; Bahaj, A.S.; Challenor, P.G. Uncertainty in wave energy resource assessment. Part 2: Variability and predictability. *Renew. Energy* **2010**, *35*, 1809–1819. [CrossRef]
21. López-Ruiz, A.; Bergillos, R.J.; Lira-Loarca, A.; Ortega-Sánchez, M. A methodology for the long-term simulation and uncertainty analysis of the operational lifetime performance of wave energy converter arrays. *Energy* **2018**, *153*, 126–135. [CrossRef]
22. Ambühl, S.; Kofoed, J.P.; Sørensen, J.D. Stochastic modeling of long-term and extreme value estimation of wind and sea conditions for probabilistic reliability assessments of wave energy devices. *Ocean Eng.* **2014**, *89*, 243–255. [CrossRef]
23. Berque, J.; Minguela, P.R.; Crooks, D.; Thiebaut, F. Uncertainty in Wave Energy Converter Power Performance Assessment. *Opera2020* **2019**. Available online: https://ec.europa.eu/research/participants/documents/downloadPublic?documentIds=080166e5c480ec53&appId=PPGMS (accessed on 2 February 2024).
24. Orphin, J.; Penesis, I.; Nader, J.R. Uncertainty analysis for a wave energy converter: The Monte Carlo method. In Proceedings of the 4th Asian Wave and Tidal Energy Conference AWTEC 2018, Taipei, Taiwan, 9–13 September 2018; p. 444.
25. ITTC. Recommended Procedures General Guideline—Uncertainty Analysis in Resistance Tests. In Proceedings of the International Towing Tank Conference, Online, 13–18 June 2021.
26. ITTC. Recommended Procedures and Guidelines—Guideline to Practical Implementation of Uncertainty Analysis. In Proceedings of the International Towing Tank Conference, Online, 13–18 June 2021.
27. ITTC. Recommended Procedures and Guidelines—Uncertainty Analysis for a Wave Energy Converter. In Proceedings of the International Towing Tank Conference, Wuxi, China, 17–23 September 2017.
28. Fenu, B.; Bonfanti, M.; Bardazzi, A.; Pilloton, C.; Lucarelli, A.; Mattiazzo, G. Experimental investigation of a Multi-OWC wind turbine floating platform. *Ocean Eng.* **2023**, *281*, 114619. [CrossRef]
29. Sirigu, S.A.; Bonfanti, M.; Passione, B.; Begovic, E.; Bertorello, C.; Dafnakis, P.; Bracco, G.; Giorcelli, E.; Mattiazzo, G. Experimental investigation of the hydrodynamic performance of the ISWEC 1:20 scaled device. In *Technology and Science for the Ships of the Future*; IOS Press: Amsterdam, The Netherlands, 2018; pp. 551–560.
30. Ambühl, S.; Kramer, M.; Sørensen, J.D. Reliability-based structural optimization of wave energy converters. *Energies* **2014**, *7*, 8178–8200. [CrossRef]
31. Clark, C.E.; Garcia-Teruel, A.; DuPont, B.; Forehand, D. Towards reliability-based geometry optimization of a point-absorber with PTO reliability objectives. In Proceedings of the European Wave and Tidal Energy Conference, Naples, Italy, 1–6 September 2019; pp. 1–6.
32. Clark, C.E.; DuPont, B. Reliability-based design optimization in offshore renewable energy systems. *Renew. Sustain. Energy Rev.* **2018**, *97*, 390–400. [CrossRef]
33. Taguchi, G. *Introduction to Quality Engineering: Designing Quality into Products and Processes*; Asian Productivity Organization: Tokyo, Japan, 1986.
34. Ben-Tal, A.; El Ghaoui, L.; Nemirovski, A. *Robust Optimization*; Princeton University Press: Princeton, NJ, USA, 2009; Volume 28.
35. Soyster, A.L. Convex programming with set-inclusive constraints and applications to inexact linear programming. *Oper. Res.* **1973**, *21*, 1154–1157. [CrossRef]
36. Ben-Tal, A.; Nemirovski, A. Robust optimization–methodology and applications. *Math. Program.* **2002**, *92*, 453–480. [CrossRef]
37. Bertsimas, D.; Sim, M. The price of robustness. *Oper. Res.* **2004**, *52*, 35–53. [CrossRef]
38. Beyer, H.G.; Sendhoff, B. Robust optimization–a comprehensive survey. *Comput. Methods Appl. Mech. Eng.* **2007**, *196*, 3190–3218. [CrossRef]
39. Jin, Y.; Sendhoff, B. Trade-off between performance and robustness: An evolutionary multiobjective approach. In *Evolutionary Multi-Criterion Optimization, International Conference on Evolutionary Multi-Criterion Optimization, Faro, Portugal, 8–11 April 2003*; Springer: Berlin/Heidelberg, Germany, 2003; pp. 237–251.
40. Jin, Y.; Branke, J. Evolutionary optimization in uncertain environments-a survey. *IEEE Trans. Evol. Comput.* **2005**, *9*, 303–317. [CrossRef]
41. Deb, K.; Gupta, H. Introducing Robustness in Multi-Objective Optimization. *Evol. Comput.* **2006**, *14*, 463–494. [CrossRef]
42. Deb, K.; Gupta, H. Searching for Robust Pareto-Optimal Solutions in Multi-objective Optimization. In *Evolutionary Multi-Criterion Optimization, International Conference on Evolutionary Multi-Criterion Optimization, Guanajuato, Mexico, 9–11 March 2005*; Coello Coello, C.A., Hernández Aguirre, A., Zitzler, E., Eds.; Springer: Berlin/Heidelberg, Germany, 2005; pp. 150–164.

43. Korolev, Y.; Toropov, V.; Shahpar, S. Design optimization under uncertainty using the multipoint approximation method. In Proceedings of the 58th AIAA/ASCE/AHS/ASC Structures, Structural Dynamics, and Materials Conference, Grapevine, TX, USA, 9–13 January 2017; p. 1934.
44. Paiva, R.M.; Crawford, C.; Suleman, A. Robust and reliability-based design optimization framework for wing design. *AIAA J.* **2014**, *52*, 711–724. [CrossRef]
45. Giorcelli, F.; Sirigu, S.A.; Basile, D. Relevance of Robustness and Uncertainties Analysis in the Optimal Design of Wave Energy Converters. In Proceedings of the European Wave and Tidal Energy Conference, Bilbao, Spain, 3–7 September 2023; Volume 15.
46. Kwak, Y.H.; Anbari, F.T. Benefits, obstacles, and future of six sigma approach. *Technovation* **2006**, *26*, 708–715. [CrossRef]
47. Coppitters, D.; De Paepe, W.; Contino, F. Robust design optimization of a photovoltaic-battery-heat pump system with thermal storage under aleatory and epistemic uncertainty. *Energy* **2021**, *229*, 120692. [CrossRef]
48. Coppitters, D.; De Paepe, W.; Contino, F. Surrogate-assisted robust design optimization and global sensitivity analysis of a directly coupled photovoltaic-electrolyzer system under techno-economic uncertainty. *Appl. Energy* **2019**, *248*, 310–320. [CrossRef]
49. Forrester, A.; Sobester, A.; Keane, A. *Engineering Design via Surrogate Modelling: A Practical Guide*; John Wiley & Sons: Hoboken, NJ, USA, 2008.
50. Coppitters, D.; Contino, F. Optimizing upside variability and antifragility in renewable energy system design. *Sci. Rep.* **2023**, *13*, 9138. [CrossRef]
51. Pozzi, N.; Bonetto, A.; Bonfanti, M.; Bracco, G.; Dafnakis, P.; Giorcelli, E.; Passione, B.; Sirigu, S.A.; Mattiazzo, G. PeWEC: Preliminary design of a full-scale plant for the mediterranean sea. In *Technology and Science for the Ships of the Future*; IOS Press: Amsterdam, The Netherlands, 2018; pp. 504–514.
52. Bonfanti, M.; Bracco, G.; Dafnakis, P.; Giorcelli, E.; Passione, B.; Pozzi, N.; Sirigu, S.A.; Mattiazzo, G. Application of a passive control technique to the ISWEC: Experimental tests on a 1:8 test rig. In *Technology and Science for the Ships of the Future, Proceedings of NAV 2018: 19th International Conference on Ship & Maritime Research, Trieste, Italy, 20–22 June 2018*; IOS Press: Amsterdam, The Netherlands, 2018; p. 60.
53. Folley, M. *Numerical Modelling of Wave Energy Converters*; Elsevier: Amsterdam, The Netherlands, 2016; Volume 2. [CrossRef]
54. Bonfanti, M.; Sirigu, S.A. Spectral-domain modelling of a non-linear wave energy converter: Analytical derivation and computational experiments. *Mech. Syst. Signal Process.* **2023**, *198*, 110398. [CrossRef]
55. Faedo, N.; Giorgi, G.; Ringwood, J.; Mattiazzo, G. Optimal control of wave energy systems considering nonlinear Froude–Krylov effects: Control-oriented modelling and moment-based control. *Nonlinear Dyn.* **2022**, *109*, 1777–1804. [CrossRef]
56. Mérigaud, A.; Ringwood, J.V. Free-surface time-series generation for wave energy applications. *IEEE J. Ocean. Eng.* **2017**, *43*, 19–35. [CrossRef]
57. Fang, K.T.; Li, R.; Sudjianto, A. *Design and Modeling for Computer Experiments*; CRC Press: Boca Raton, FL, USA, 2005.
58. Krige, D.G. A statistical approach to some basic mine valuation problems on the Witwatersrand. *J. South. Afr. Inst. Min. Metall.* **1951**, *52*, 119–139.
59. Gioia, D.G.; Pasta, E.; Brandimarte, P.; Mattiazzo, G. Data-driven control of a pendulum wave energy converter: A gaussian process regression approach. *Ocean. Eng.* **2022**, *253*, 111191. [CrossRef]
60. Williams, C.K.; Rasmussen, C.E. *Gaussian Processes for Machine Learning*; MIT Press: Cambridge, MA, USA, 2006; Volume 2.
61. Zhong, Q.; Yeung, R.W. On optimal energy-extraction performance of arrays of wave-energy converters, with full consideration of wave and multi-body interactions. *Ocean. Eng.* **2022**, *250*, 110863. [CrossRef]
62. López-Ruiz, A.; Bergillos, R.J.; Raffo-Caballero, J.M.; Ortega-Sánchez, M. Towards an optimum design of wave energy converter arrays through an integrated approach of life cycle performance and operational capacity. *Appl. Energy* **2018**, *209*, 20–32. [CrossRef]

MDPI AG
Grosspeteranlage 5
4052 Basel
Switzerland
Tel.: +41 61 683 77 34

Journal of Marine Science and Engineering Editorial Office
E-mail: jmse@mdpi.com
www.mdpi.com/journal/jmse

MDPI

www.ingramcontent.com/pod-product-compliance
Lightning Source LLC
Chambersburg PA
CBHW040827250926
51850CB00007BA/1275
* 9 7 8 3 7 2 5 8 1 8 9 9 0 *